Analytics in Meteorology

Data Transformation for Weather Intelligence

NIPA® GENX ELECTRONIC RESOURCES & SOLUTIONS P. LTD.
New Delhi-110 034

About the Authors

BV Ramana Rao taught Agricultural Meteorology in the University of Agricultural Sciences, Bangalore during the years 1966 to 1979 and later joined the Indian Council of Agricultural Research and superannuated from service during April, 2000. He served as Head, Division of Climatology, Wind Power and Solar Energy Utilization at ICAR-Central Arid Zone Research Institute, Jodhpur (Rajasthan); founder Project Coordinator, All India Coordinated Research Project on Agrometeorology at ICAR Central Research Institute for Dryland Agriculture, Hyderabad (Telangana) during December 1984 to March 1994. His major contributions in the field of Agrometeorology include Agroclimatic Characterization, Delineation of Efficient Cropping Zones, Crop Weather Modeling, Classification and Management of Agricultural Droughts, Operational Agrometeorology, etc. He published about 150 research papers in peer reviewed journals with 50 of his papers appearing in high impact International Journals. He authored 15 books and monographs and a number of scientific and technical bulletins. He served as Editor in Chief, Journal of Agrometeorology during the years 2009 to 2023.

He is the recipient of British Council Fellowship during the year 1986-87 for advanced training at University of Reading, United Kingdom. He was conferred as Honorary Fellow (2011) and Life Time Achievement Award (2025) by the Association of Agrometeorologists for his valuable contribution to the growth and development of Agricultural Meteorology in India.

Surender Singh has about thirty five years of experience with PhD in the field of Agrometeorology. Worked in Agrometeorology encompassing teaching, research, extension and administration at various capacities *viz*., Principal Scientist, Professor & Head, Senior Scientist, Assistant Scientist, Advisor (Recruitments) and Associate Director (Students' Counseling and Placement) at CCS Haryana Agricultural University, Hisar, India. Recipient of Two Gold Medals for Best Research Work during PhD Program and for Best PhD Dissertation (2001-02) awarded by then President of India Dr APJ Abdul Kalam. Offered UK Universities Vice Chancellors' Scholarship for Higher Studies twice under Commonwealth Fellowship Program. Post Doctoral Fellow at Abdus Salam International Centre for Theoretical Physics, Trieste, Italy; IAC – Instituto Agronômico, Campinas, Brazil and BOKU (Met) - Institute of Meteorology and Climatology, Vienna, Austria.

He is life member of Association of Agrometeorologists, Indian Meteorological Society, Indian Society of Remote Sensing, International Society for Agrometeorology, Society for Sustainable Agric& Resource Management, Environment Science Published for Everybody Round the Earth (ESPERE), European Drought Centre, South Asian Forum on Agriculture Meteorology (SAFOAM) and The Geographical Society of India. Organized quite a few trainings/symposia in Agrometeorology and associated domains at National level. Published more than 250 research papers, popular articles, manuals/teaching aids and research bulletins/book chapters etc. Performed responsibilities as Managing Editor - Journal of Agrometeorology; Research Collaborator, Monsoon Asia Agro-Environmental Research Consortium (MARCO), Japan; SPARC Research Collaborator, ETH, Zurich, Switzerland; Member, Expert Team - CAgM-WMO, Geneva, Switzerland (2018-20). Core Committee Member on Agriculture – Drought at WMO, Geneva, Switzerland (2021-24 & 2024-27). External Expert for evaluation of research projects, COST-Association, European Union, Brussels, Belgium. Many students (UG, PG and PhD) have been trained under him and placed across the country and well settled in various positions in the field of agrometeorology. He is widely travelled researcher/expert across the globe on many academic pursuits. He has been conferred upon Fellow of Association of Agrometeorologists (FAAM) for the year 2024 in recognition of his exemplary achievements and enduring impact on agrometeorological science.

V Uma Maheswara Rao has forty years of experience with PhD in the broader field of Agrometeorology. Worked in Agrometeorology encompassing teaching, research, extension and administration at various capacities viz., Project Coordinator, Principal Scientist, Professor & Head, Agrometeorologist, and Assistant Scientist at CCS Haryana Agricultural University and Indian Council of Agriculture Research in India.

He is life member of Indian Meteorological Society, The Arid Zone Research Association of India, Indian Society of Remote Sensing, Association of Agrometeorologists, International Society for Agrometeorology, and The Indian Society of Dryland Agriculture. Organized several trainings/workshops in Agrometeorology at the International and National level. Published more than 345 research papers, popular articles, manuals/teaching aids and research bulletins/ book chapters. Member of the CCI-ET-IIC, WMO, Geneva (2014-2017). Member of XI Plan High Powered Review Committee of Ministry of Earth Sciences, Government of India (2012). In 2014, the prestigious Chaudhary Devi Lal Award, ICAR's Best AICRP on Agrometeorology was received under his dynamic leadership as Project Coordinator. Many students (UG, PG and PhD) have been trained under his guidance are placed across the country and settled in various positions in the field of Agrometeorology. He has visited numerous countries in official capacity.

Analytics in Meteorology
Data Transformation for Weather Intelligence

BV Ramana Rao
MSc; Tech; FAAM
Founder Project Coordinator - AICRP on Agrometeorology
ICAR-CRIDA, Hyderabad, Telangana, India

Surender Singh
MSc; PhD; FAAM
Professor/Principal Scientist (Agrometeorology)
CCS HAU, Hisar, Haryana, India

V Uma Maheswara Rao
MSc; PGDIAS; PhD; FAAM
Former Project Coordinator - AICRP on Agrometeorology
ICAR-CRIDA, Hyderabad, Telangana, India

NIPA® GENX ELECTRONIC RESOURCES & SOLUTIONS P. LTD.
New Delhi-110 034

NIPA® GENX ELECTRONIC RESOURCES & SOLUTIONS P. LTD.

101,103, Vikas Surya Plaza, CU Block
L.S.C. Market, Pitam Pura, New Delhi-110 034
Ph : +91-11-43860225, Mob.: +91 9717133558, 9540816132
E-mail: newindiapublishingagency@gmail.com
Website: www.nipaersources.com

Print ISBN: 978-93-72195-89-7

ebook ISBN: 978-93-72198-03-4

Composed and Designed by NIPA®.

The Book

Analytics in Meteorology

Data Transformation for Weather Intelligence

is dedicated to

Ms Anna M Mani

(23 August, 1918 – 16 August, 2001)

Ms Anna Modayil Mani, the trailblazing physicist and meteorologist whose pioneering contributions laid the scientific foundations for modern weather instrumentation and atmospheric data analytics in India.

A woman far ahead of her time, **Anna Mani** devoted her life to the pursuit of precision in meteorological observations. Her landmark work at the India Meteorological Department led to the **indigenization and standardization of over 100 weather instruments**, enabling India to develop its own robust weather monitoring systems. Her meticulous commitment to data quality and measurement accuracy echoes through her enduring words: "Wrong measurements are worse than no measurements."

As a scientist, she championed **solar radiation studies, ozone monitoring, and wind energy assessments**—long before these became global priorities. Her work enabled the collection of high-quality datasets, which now power modern weather models, climate analytics, and renewable energy planning. She bridged the gap between instrumentation and actionable meteorological intelligence with clarity, courage, and scientific discipline.

By dedicating this book to **Anna Mani**, we pay tribute to a remarkable woman whose unwavering passion for measurement science and weather research continues to illuminate the path for data-driven meteorology, climate resilience, and environmental stewardship.

Dr KPR Vittal

Founder Director

ICAR- National Institute of Abiotic Stress Management

Former Director

ICAR- Central Arid Zone Research Institute

Former Project Coordinator

ICAR-CRIDA-AICRP on Dryland Agriculture

भाकृअनुप
ICAR

Foreword

In the annals of scientific endeavor, few fields have witnessed as profound a transformation as meteorology. From the earliest human attempts to predict the skies based on folklore and observation, to the complex numerical models run on today's supercomputers, our quest to understand and anticipate the atmosphere has been relentless. Yet, even with these remarkable advancements, we are now entering an entirely new frontier, one where the sheer volume and complexity of atmospheric data demand a paradigm shift in how we approach weather and climate science. This shift is precisely what '**Analytics in Meteorology: Data Transformation for Weather Intelligence**' so eloquently addresses.

As someone who has witnessed the evolution of atmospheric science over decades, I can attest to the critical need for a resource like this. The traditional pillars of meteorology—physics, dynamics, and observation—remain indispensable. However, the modern meteorologist, climate scientist, and environmental decision-maker must now also be fluent in the language of data: its collection, curation, analysis, and, crucially, its transformation into actionable intelligence. The advent of satellite constellations, dense ground-based sensor networks, high-resolution radar, and the burgeoning Internet of Things (IoT) has unleashed a torrent of information. Without sophisticated analytical frameworks, this data remains largely untapped potential.

This book serves as an exceptional guide to unlocking that potential. It masterfully bridges the gap between the foundational principles of atmospheric science and the cutting-edge methodologies of data analytics. Readers will discover how seemingly disparate concepts—from the intricacies of atmospheric physics to the power of machine learning algorithms—converge to yield unprecedented insights. The comprehensive nature of this volume is a testament to the valuable efforts of its esteemed authors, **Professors BV**

Ramana Rao, Surender Singh, and V Uma Maheswara Rao. Their collective expertise and dedication have culminated in a text that not only illuminates how data science is a critical engine driving advancements in areas such as precision forecasting and climate change understanding, but also demonstrates its vital role in informing environmental resilience, optimizing renewable energy systems, and designing smarter, more livable urban environments.

What truly sets this book apart is its comprehensive yet accessible approach. It recognizes that the future of meteorology lies in interdisciplinary collaboration, fostering a common understanding between atmospheric scientists, data scientists, and decision-makers. The discussions on data privacy, ethical AI, and the imperative for open science are particularly timely, underscoring the responsibility that accompanies such powerful analytical capabilities. The authors have meticulously crafted a narrative that is both scientifically rigorous and practically relevant, ensuring that readers gain not just theoretical knowledge but also a profound appreciation for the real-world implications of their work.

For anyone seeking to navigate the complexities of modern atmospheric data, to master the analytical tools that transform raw numbers into strategic foresight, or to contribute meaningfully to the grand challenges of climate and environment, this book is an indispensable companion. It is a testament to the transformative power of data in the hands of thoughtful and skilled practitioners. I am confident that '**Analytics in Meteorology: Data Transformation for Weather Intelligence**' will become a cornerstone text, shaping the next generation of leaders and innovators in atmospheric science, a truly remarkable achievement and a profound contribution for many more times to come.

KPR Vittal

(**KPR Vittal**)
Fellow - National Academy of Agricultural Sciences (FNAAS)
Honorary Fellow - Society of Applied Biotechnology (FSAB)
Freelance Dryland Agriculture Consultant, Hyderabad, India
ICAR-ISDA Lifetime Achievement Award-2025

Preface

In an era increasingly defined by rapid environmental change and the pervasive influence of data, the discipline of meteorology stands at a pivotal juncture. Our world is more connected, more vulnerable, and more reliant on accurate environmental information than ever before. From the daily rhythms of local weather to the long-term shifts of global climate, the atmosphere profoundly impacts every facet of human endeavor—our economies, our health, our infrastructure, and our very way of life.

Historically, meteorology has been a science of meticulous observation, rigorous physical modeling, and expert human interpretation. While these core tenets remain absolutely foundational, the sheer volume, velocity, and variety of meteorological data generated today have created an unprecedented opportunity. This deluge of information—streaming from satellite constellations, dense ground sensor networks, sophisticated numerical models, cutting-edge drone deployments, and even burgeoning citizen science initiatives—is transforming our understanding of the atmosphere. However, this raw data, in its unrefined state, is not inherently intelligent. Its true power, its capacity to unlock deeper insights and drive smarter decisions, is realized only through the rigorous and strategic application of data science and analytics.

This book, "Analytics in Meteorology: Data Transformation for Weather Intelligence," is conceived as a comprehensive and timely guide to navigating this exciting new frontier. It is born from the conviction that the future of atmospheric science lies not just in collecting more data, but in our ability to effectively transform that data into actionable "Weather Intelligence." This intelligence is a holistic understanding that empowers more precise forecasts, more robust climate projections, and more informed decision-making across all scales of human activity.

The need for a book of this type has never been more pressing. Traditional meteorological textbooks often provide deep dives into atmospheric physics but may not fully integrate the latest advancements in data science methodologies. Conversely, general data science texts, while excellent for core techniques, frequently lack the specific domain context and unique challenges presented by meteorological data—its spatial and temporal complexities, its inherent uncertainties, and its critical societal implications. This book aims to bridge that crucial gap, offering a unified narrative that seamlessly connects the foundational principles of atmospheric science with the cutting-edge tools and techniques of modern data analytics.

Our journey within these pages will be comprehensive yet accessible. We begin by exploring the diverse sources and fundamental characteristics of meteorological data, emphasizing the paramount importance of data quality, consistency, and homogeneity for reliable analysis. We then delve into the core analytical methodologies, from statistical techniques and machine learning algorithms that uncover hidden patterns and relationships, to the intricate processes of numerical weather prediction and long-term climate modeling. The book emphasizes not just how these analyses are performed, but why they are essential, showcasing how data transformation underpins everything from real-time severe weather warnings to long-term climate change attribution. Furthermore, we explore the vital role of effective visualization in communicating complex atmospheric phenomena and examine how these analytical insights are applied in specialized domains such as air quality management, water resource planning, renewable energy integration, and urban climatology. Finally, we confront the inherent challenges, critical ethical considerations, and exciting future directions that will shape the next generation of meteorological data science.

This book is intended for a diverse and growing audience: seasoned meteorologists seeking to enhance their analytical toolkit with modern data science techniques; aspiring data scientists eager to apply their skills to a dynamic, impactful, and data-rich domain; environmental researchers grappling with complex, multi-dimensional datasets; and students aspiring to contribute to the future of atmospheric and climate science. Whether your interest lies in improving forecast accuracy, understanding the nuances of climate change, or leveraging weather intelligence for critical societal benefit, this volume aims to provide the foundational knowledge, practical context, and forward-looking perspective necessary to embark on this transformative journey.

The path ahead for meteorological data science is filled with both immense challenges and boundless opportunities. By embracing the power of data transformation, we can not only deepen our scientific understanding of our planet but also forge a more resilient, sustainable, and intelligent relationship with its ever-changing atmosphere.

The authors are very much grateful to Dr. KPR Vittal, Founder Director, ICAR-National Institute of Abiotic Stress Management, Baramati (MS), India for having kindly spared time for writing the foreword for the book which will undoubtedly inspire the readers to go through the text for updating themselves on the importance of data science to generate weather wisdom.

We are equally grateful to Sh Sumit Jain, Chief Executive Officer, NIPA who inspired us to write this book recognizing the need for a text of this type to cater to the needs of students, researchers, data scientists and policy makers particularly during an era when the prediction of weather is much more challenging than before.

Welcome to the era of Weather Intelligence.

BV Ramana Rao
Surender Singh
V Uma Maheswara Rao

Acknowledgements

This book, ***Analytics in Meteorology: Data Transformation for Weather Intelligence,*** is the culmination of years of research, collaboration, and exploration at the intersection of atmospheric science and data analytics. We are deeply grateful to the many individuals and institutions who have supported us throughout this journey.

First and foremost, We extend my sincere thanks to our academic colleagues, whose cpllaboration and encouragement have been instrumental in shaping both the direction and the rigor of this work. Their insights into meteorology, climatology, and statistical modeling were invaluable.

We would like to express our appreciation to the data scientists, meteorologists, and researchers whose groundbreaking work laid the foundation for this book. Their innovation continues to inspire the integration of analytics and meteorological science.

To our colleagues and collaborators, thank you for your support, feedback, and camaraderie. Your diverse perspectives enriched this project beyond measure.

Finally, We wish to thank our family and friends for their unwavering patience, encouragement, and belief in the value of this endeavor. Their support carried us through the many long hours of writing and revision.

This book is dedicated to all those who strive to turn data into understanding—and understanding into action—in the service of better forecasting, planning, and climate resilience.

BV Ramana Rao
Surender Singh
V Uma Maheswara Rao

Contents

List of Figures

1

Introduction to Meteorology and Data Science

1.1 Introduction

Meteorology, at its heart, is the scientific endeavor dedicated to understanding the Earth's atmosphere and the myriad phenomena that occur within it. More than just the art of predicting whether it will rain tomorrow, meteorology is a robust interdisciplinary science. It intricately weaves together principles from physics, chemistry, and mathematics to unravel the complexities of atmospheric processes. From the gentle caress of a breeze to the formidable power of a hurricane, meteorology seeks to explain the underlying mechanisms that drive our ever-changing weather and long-term climate patterns.

The study of meteorology encompasses a wide array of atmospheric characteristics. Meteorologists delve into the composition of the atmosphere, examining the various gases and tiny particles known as aerosols that reside within it. They also investigate its vertical structure, distinguishing between layers like the troposphere, where most weather occurs, and the stratosphere, home to the ozone layer. The motion of the atmosphere, manifested as wind systems ranging from local gusts to global circulations, is a central focus. Furthermore, meteorology explores the thermodynamics of the atmosphere, including how heat is transferred and how temperature changes, as well as the role of water in the atmosphere, from humidity to the formation of clouds and precipitation – the realm of atmospheric hydrology. Even the captivating optical phenomena we sometimes witness, such as rainbows and halos, fall under the purview of meteorological inquiry.

The pursuit of meteorological knowledge has a rich history. While early humans undoubtedly observed the weather for survival, often intertwining these observations with mythology and astrological beliefs, the advent of scientific meteorology began to take shape with the invention of key instruments. The 17th century witnessed pivotal developments like the refinement of the thermometer, attributed to Galileo Galilei and later scientists, allowing for the quantitative measurement of temperature, and the invention of the barometer by Evangelista Torricelli, enabling the measurement of atmospheric pressure. These tools provided the foundation for a more empirical understanding of the atmosphere.

The 19th century marked a significant turning point with the rise of telegraphy. For the first time, it became possible to rapidly collect weather observations from geographically dispersed locations. This capability led to the creation of the first weather maps, allowing scientists to visualize large-scale weather patterns and, consequently, to make rudimentary forecasts. The 20th century ushered in an era of unprecedented progress. The development of radiosondes, balloon-borne instruments that transmit upper-air measurements of temperature, humidity, and wind, provided a crucial three-dimensional view of the atmosphere. Radar technology allowed for the detection and tracking of precipitation, while the latter half of the century saw the birth and rapid advancement of numerical weather prediction models. These sophisticated computer programs utilize the fundamental laws of physics to simulate the evolution of the atmosphere. Today, Earth-orbiting satellites provide a continuous, global perspective of clouds, radiation, and other atmospheric variables, and these data feed into increasingly complex computer models that form the bedrock of modern weather forecasting and long-term climate projections.

Meteorology is not an island; it is deeply intertwined with other scientific disciplines. Physics provides the foundational principles that govern atmospheric motion, the transfer of heat, and the interaction of radiation with the atmosphere. Chemistry is indispensable for understanding the composition of the air, the complexities of air pollution, and the delicate balance of the ozone layer. Mathematics serves as the essential language, allowing meteorologists to formulate and solve the equations that describe atmospheric processes and to perform rigorous statistical analyses of meteorological data.

Fig. 1.1: Meteorology – as a science

The oceans play a critical role in the Earth's climate system, making oceanography a closely allied field. Furthermore, geography provides the crucial spatial context for understanding the distribution of weather and climate patterns across the globe (Fig. 1.1).

The field of meteorology itself is quite broad, encompassing numerous areas of specialization. **Synoptic meteorology** focuses on the large-scale weather systems that dominate our mid-latitudes, such as high and low-pressure systems and fronts. **Mesoscale meteorology** delves into regional weather phenomena, like individual thunderstorms, sea and land breezes, and mountain-valley wind circulations. **Climatology** takes a longer-term perspective, studying the average weather conditions over extended periods and investigating the phenomenon of climate change. **Boundary-layer meteorology** concentrates on the lowest part of the atmosphere, the layer directly influenced by the Earth's surface. **Physical meteorology** examines the fundamental physical processes occurring within the atmosphere, such as the formation of clouds and the transfer of radiation. Finally, **operational meteorology** is the practical application of all this knowledge to the daily task of weather forecasting.

1.1.1 Key Meteorological Variables and their Significance

Let's explore some of the key meteorological variables and discuss the significance of each:

- **Temperature:** This measures the degree of hotness or coldness of the air. Significance: Temperature profoundly influences all sorts of atmospheric processes, including the formation of clouds and precipitation, the stability of the atmosphere, and the rate of evaporation. It's a primary factor in determining what kind of weather to expect and is crucial for human comfort, agriculture, and many other aspects of life.
- **Pressure:** Atmospheric pressure is the force exerted by the weight of the air above a given point. Significance: Pressure differences in the atmosphere drive the wind. High-pressure systems are often associated with stable, fair weather, while low-pressure systems can bring clouds, precipitation, and storms. Monitoring pressure changes is vital for forecasting weather systems.
- **Wind (Speed and Direction):** Wind is the movement of air. Wind speed is how fast the air is moving, and wind direction is the direction from which the wind is blowing. Significance: Wind affects the transport of heat, moisture, and pollutants in the atmosphere. It plays a crucial role in the development and movement of weather systems. Strong winds can cause damage, while gentle breezes can influence local temperatures and humidity.

- **Humidity:** This refers to the amount of water vapor in the air. Significance: Humidity is critical for cloud formation and precipitation. High humidity can make warm temperatures feel more oppressive (heat index), while low humidity can lead to dry conditions. It also affects evaporation rates and atmospheric stability.
- **Precipitation:** Any form of water that falls from clouds to the Earth's surface, such as rain, snow, sleet, and hail. Significance: Precipitation is essential for water resources and agriculture. The type and amount of precipitation are key elements of our daily weather and can also lead to hazardous conditions like flooding or snowstorms.
- **Solar Radiation:** This is the energy emitted by the sun. Significance: Solar radiation is the primary driver of Earth's weather and climate. It heats the Earth's surface and atmosphere, leading to temperature gradients that drive atmospheric circulation. The amount of solar radiation reaching the surface influences temperature, evaporation, and photosynthesis.
- **Clouds:** Visible masses of water droplets or ice crystals suspended in the atmosphere. Significance: Clouds play a crucial role in the Earth's energy balance by reflecting incoming solar radiation and trapping outgoing terrestrial radiation. They are also the source of precipitation and can indicate atmospheric stability and the potential for storms.

These key variables interact in complex ways to produce the weather we experience. Meteorologists observe and measure these variables using a variety of instruments (as discussed earlier) to understand the current state of the atmosphere and to forecast future conditions.

Do any of these variables pique your interest more than others? We can delve deeper into their measurement or significance if you like.

1.2 Meteorological Data Base: Components and Structure

A meteorological database serves as a crucial repository for the vast amounts of data generated by various observational platforms and numerical models. These databases are meticulously designed to efficiently store, manage, and retrieve meteorological information for a wide range of applications, from real-time weather forecasting to long-term climate analysis. Understanding their components and structure is fundamental to effectively utilizing this wealth of atmospheric data.

At its core, a meteorological database comprises several key components. Firstly, there's the **raw data**, which originates from diverse sources. This includes surface observations from weather stations detailing temperature, pressure, wind, and precipitation; upper-air soundings from radiosondes providing vertical profiles of atmospheric conditions; satellite imagery capturing cloud cover, radiation, and atmospheric composition; radar data depicting precipitation intensity and motion; and the output from numerical weather prediction models, which are gridded datasets of numerous atmospheric variables at various time steps and levels. The sheer volume and heterogeneity of this raw data necessitate a structured approach for storage.

Secondly, the database incorporates **metadata**, which is essentially "data about data." This crucial component provides context and information about the raw data, such as the source of the observation (e.g., a specific weather station or satellite), the units of measurement (e.g., degrees Celsius, meters per second), the time and location of the observation, and quality control flags indicating the reliability of the data. Without comprehensive metadata, the raw data would be significantly less useful, as it would lack the necessary context for interpretation and analysis.

The **structure** of a meteorological database is carefully designed to optimize data storage and retrieval. Commonly, these databases employ relational database management systems (RDBMS) or increasingly, NoSQL databases for handling the diverse and often unstructured nature of some meteorological data (like satellite imagery). In a relational model, data is organized into tables with rows representing individual records (e.g., a single weather observation at a specific time and location) and columns representing attributes (e.g., temperature, pressure, wind speed). Relationships between different types of data (e.g., linking surface observations to the station metadata) are defined to facilitate complex queries.

For instance, a simple structure for surface observations might involve a table with columns for station ID, timestamp, latitude, longitude, temperature, pressure, wind speed, and wind direction. Another table would hold station metadata, linking the station ID to the station name, elevation, and other static characteristics.

More advanced structures are needed to handle the multi-dimensional nature of model output or satellite data. These might involve the use of array-based storage or specialized formats like NetCDF or HDF, which are designed to efficiently store gridded scientific data with associated metadata. These These formats allow for the storage of variables (like temperature) that vary across space (latitude, longitude, altitude) and time.

Furthermore, meteorological databases often include **quality control information**. Automated and sometimes manual processes are applied to the raw data to identify and flag or correct errors, ensuring the integrity and reliability of the data stored in the database.This quality control layer is vital for any subsequent analysis or application of the data.

In summary, a meteorological database is a complex system designed to manage the influx of diverse atmospheric data. DIts components include the raw observations and model outputs, essential metadata providing context, a well-defined structure (often relational or array-based) for efficient organization, and quality control information to ensure data accuracy. This structured approach enables meteorologists and data scientists to effectively access, analyze, and derive meaningful insights from the vast amounts of meteorological information available.

1.3 Data Source

Meteorological data, fundamental to understanding and predicting atmospheric phenomena, is collected with varying frequencies from a diverse array of sources. These collection frequencies are meticulously designed to capture the dynamic nature of the atmosphere across different spatial and temporal scales.

Sources and Their Collection Frequencies

- **Surface Weather Stations:** These ground-based observatories are the backbone of local weather monitoring. They continuously measure parameters such as air temperature, atmospheric pressure, wind speed and direction, humidity, and precipitation. Automated stations typically report data every **5 to 10 minutes**, providing near real-time updates. Manually operated stations might report less frequently, often **hourly** or every few hours. For climatological purposes, data is also aggregated into **daily, monthly, and annual** summaries.
- **Upper-Air Observations (Radiosondes):** To understand the vertical structure of the atmosphere, weather balloons carrying instruments called radiosondes are launched. These instruments transmit data on temperature, humidity, and pressure as they ascend, and their position is tracked to determine wind speed and direction at various altitudes. Radiosonde launches are typically conducted **twice daily** from hundreds of locations worldwide, often at standard synoptic times (e.g., 0000 UTC and 1200 UTC). During severe weather events or specialized research campaigns, launches may occur more frequently.
- **Weather Satellites:** Satellites provide a crucial global perspective.
    - **Geostationary satellites**, which orbit at a fixed position relative to the Earth's surface, offer continuous monitoring of large regions. They capture imagery and derive products (like cloud top temperatures or

moisture fields) with very high frequency, often every **5 to 15 minutes** for imagery, and hourly or sub-hourly for derived data products.

- **Polar-orbiting satellites** circle the Earth from pole to pole, providing global coverage but less frequent revisits for any single location (typically **one or two passes per day**). They are vital for collecting detailed atmospheric soundings and high-resolution imagery.

- **Weather Radar:** Ground-based weather radar systems emit pulses that detect precipitation, its intensity, and motion. These systems are invaluable for monitoring severe weather. Radar data is typically collected and updated very frequently, often every **5 to 10 minutes**, allowing meteorologists to track the rapid evolution of storms.
- **Numerical Weather Prediction (NWP) Models:** These are sophisticated computer simulations of the atmosphere. Global NWP models are typically run **every 6 or 12 hours**, while regional or high-resolution models may be initialized more frequently, sometimes every **3 hours**. The output from these models consists of forecasts for various atmospheric variables at regular time intervals, often every **1, 3, or 6 hours** for forecast lead times extending days into the future.
- **Aircraft Observations:** Commercial and research aircraft are equipped with sensors that collect meteorological data during their flights. This data, including temperature, wind speed, and turbulence, is collected **irregularly** as aircraft ascend, descend, and fly en route, providing valuable real-time information, particularly in data-sparse regions.
- **Buoys (Oceanic):** Moored and drifting buoys deployed in oceans and large lakes collect surface meteorological and oceanographic data. These typically report measurements such as air temperature, sea surface temperature, wind, and wave height at frequencies ranging from **hourly to every few hours**.
- **Reanalysis Data:** Unlike real-time observations, reanalysis datasets are created retrospectively by combining historical observations with advanced NWP models through a process called data assimilation. This process generates a consistent, long-term record of atmospheric conditions. Reanalysis data is typically available at frequencies like **hourly or 3-hourly**, spanning decades or even over a century, providing a valuable resource for climate studies and research.

Here is a summary of the data types, sources, and typical collection frequencies

Data Type	Primary Source(s)	Typical Collection Frequency
Point Data	Surface Weather Stations	5-10 minutes (automated), Hourly (manual)
	Radiosondes (Upper-Air)	Twice daily (e.g., 0000 UTC, 1200 UTC)
	Aircraft Observations	Irregular (during ascent, descent, en route)
	Buoys (Oceanic)	Hourly to every few hours
Gridded Data	NWP Models	Model runs every 3, 6, or 12 hours; Output every 1, 3, or 6 hours
	Reanalysis Data	Hourly or 3-hourly (retrospective, long-term)
	Satellite (derived products)	Hourly to sub-hourly
Imagery Data	Geostationary Satellites	Every 5-15 minutes
	Polar-Orbiting Satellites	One or two passes per day (for a given location)
	Weather Radar	Every 5-10 minutes

This comprehensive and frequent collection of diverse data types from multiple sources is what enables modern meteorology to produce increasingly accurate forecasts and to deepen our understanding of Earth's atmosphere.

1.4 Data Science

To effectively leverage the vast amounts of meteorological data, a solid understanding of data science principles is essential. One of the initial and crucial stages is **data preprocessing**. Raw meteorological data, coming from various instruments and models, is rarely in a perfect state for direct analysis. It often suffers from issues such as missing observations, values that seem unusually high or low (outliers), inconsistencies in the units of measurement across different datasets, or simply being structured in a way that's not conducive to the intended analysis. Data preprocessing involves a suite of techniques aimed at cleaning, transforming, and organizing this data. For instance, if a weather station temporarily malfunctions and doesn't record temperature for a period, we have missing values that need to be addressed, perhaps by estimating them based on nearby stations or historical trends. Similarly, a faulty sensor might produce an outlier reading that needs to be identified and potentially corrected or removed to prevent it from skewing our analysis. Bringing all temperature readings to the same unit (e.g., Celsius) or scaling wind speed values to a consistent range are other examples of preprocessing steps. Integrating data from different sources, like combining surface temperature readings with satellite-derived cloud information based on their geographical location and the time of observation, is also a key aspect of preparing the data for more advanced analysis.

Before diving into building predictive models, a data scientist typically engages in **exploratory data analysis (EDA)**. This phase is about getting to know the data intimately. Through a combination of statistical summaries and visual representations, EDA helps reveal the underlying structure, patterns, and potential anomalies within the dataset. In a meteorological context, this could involve plotting how temperature or rainfall has changed over time at a specific location, creating maps to visualize spatial patterns of wind or pressure, examining the statistical distribution of variables like wind speed or humidity to understand their typical ranges and variability, or investigating whether certain meteorological variables tend to change together. EDA is crucial for formulating hypotheses and guiding subsequent modeling efforts.

Often, the raw meteorological variables might not be in the most informative format for a model. This is where **feature engineering** comes in. It involves creating new, potentially more useful features from the existing ones. For example, instead of just using the day and month, we might create a 'day of the year' feature, which could be more directly related to seasonal weather patterns. Or, considering that human comfort is affected by both temperature and humidity, we might engineer a 'heat index' feature by combining these two. In wind analysis, it's sometimes more mathematically convenient to work with the eastward (zonal) and northward (meridional) components of the wind rather than just speed and direction. Thoughtful feature engineering can significantly improve the performance and interpretability of models (Fig. 1.2).

The core of much of data science lies in **statistical modeling**. These techniques allow us to formally analyze relationships within the data, make inferences, and quantify the uncertainty in our findings. In meteorology, statistical models can be used to predict continuous variables like tomorrow's high temperature (regression) or to forecast categorical outcomes, such as whether it will rain or not (classification). Given the sequential nature of weather data, **time series analysis** techniques are particularly relevant, allowing us to model and forecast variables that evolve over time.

Fig. 1.2: Data science in meteorology

A powerful subset of artificial intelligence used extensively in data science is **machine learning (ML)**. ML algorithms enable computers to learn from data without explicit programming. In **supervised learning**, the model learns from labeled data – for instance, historical weather observations paired with what actually happened (e.g., did it rain the next day?). This allows the model to then predict future outcomes. Examples of supervised learning algorithms include linear regression for predicting continuous values, support vector machines and decision trees for classification, and more complex models like random forests and neural networks that can capture intricate relationships. In contrast, **unsupervised learning** deals with unlabeled data, aiming to discover hidden patterns. For example, clustering algorithms can be used to identify distinct weather patterns or regimes based on similarities in meteorological variables.

Once a statistical or machine learning model has been developed, it's critical to evaluate how well it performs using appropriate **model evaluation** metrics. The choice of metric depends on the type of problem. For a temperature forecasting model (regression), we might use metrics like the Mean Squared Error, which quantifies the average squared difference between the predicted and actual temperatures. For a precipitation prediction model (classification), we might look at accuracy (the overall percentage of correct predictions), precision (what proportion of predicted rain events actually occurred), recall (what proportion of actual rain events were correctly predicted), and the F1-score, which balances precision and recall. To ensure that a model can generalize to new, unseen data, it's standard practice to train the model on one part of the data (the training set) and then evaluate its performance on a separate, held-out portion (the testing set).

Finally, **data visualization** is paramount for both understanding the data and communicating findings effectively. Visual tools, such as line graphs to show trends over time, scatter plots to explore relationships between variables, bar charts to compare categories, and maps to display spatial distributions, are indispensable in meteorology for tasks ranging from exploring the characteristics of a dataset to presenting weather forecasts or illustrating climate change trends.

1.5 Applications of Data Science in Meteorology

Data science has become an indispensable tool in modern meteorology, enhancing our ability to understand, predict, and respond to atmospheric phenomena. Here are some key applications:

- **Improved Weather Forecasting:** Machine learning algorithms can learn complex patterns from vast datasets of historical weather observations and numerical model outputs. This allows for the development of more accurate and higher-resolution forecasts, including predictions of temperature, precipitation, wind, and severe weather events. Techniques like neural networks, random forests, and gradient boosting are being used to refine traditional forecasting methods.
- **Enhanced Prediction of Extreme Weather:** Data science plays a crucial role in forecasting high-impact weather such as hurricanes, tornadoes, heatwaves, and floods. By analyzing historical data and real-time observations, machine learning models can improve the prediction of their intensity, track, and timing, providing more lead time for warnings and preparedness.
- **Climate Change Analysis and Projections:** Data science techniques are vital for analyzing the massive datasets generated by climate models and historical climate records. This helps in identifying trends, understanding the impacts of climate change, and generating more reliable future climate projections. Statistical methods and machine learning are used to downscale global climate model outputs to regional levels and to assess climate risks.
- **Nowcasting:** This involves very short-term weather forecasting (minutes to a few hours). Data science, particularly machine learning applied to real-time radar and satellite data, can improve the accuracy and detail of nowcasts, which are critical for applications like aviation, public safety, and event management.
- **Data Assimilation:** Data assimilation is the process of combining observations with numerical model forecasts to produce the best possible estimate of the current state of the atmosphere (the initial conditions for the next forecast). Machine learning techniques are being explored to enhance the efficiency and accuracy of data assimilation processes.

- **Atmospheric Research:** Data science tools enable researchers to analyze large atmospheric datasets to discover new insights into atmospheric processes. Techniques like clustering can identify patterns in atmospheric circulation, while association rule mining might reveal relationships between different meteorological variables.
- **Quality Control of Meteorological Data:** Machine learning algorithms can be trained to automatically identify and flag erroneous or suspicious data points from various observation networks, improving the overall quality and reliability of meteorological datasets.

Now, let's summarize some databases employed in meteorology and their likely outputs in a table:

Database Focus	Likely Data Sources	Likely Outputs/Applications
Real-time Observational Databases	Surface stations, radiosondes, satellites (direct readings), radar	Current weather conditions, input for nowcasting and data assimilation, initial conditions for short-term forecasts, severe weather monitoring.
Numerical Weather Prediction (NWP) Archives	Output from various global and regional NWP models	Weather forecasts (temperature, precipitation, wind, etc.) at various lead times, model diagnostics, input for statistical post-processing.
Climate Archives/ Reanalysis Databases	Historical surface and upper-air observations, satellite data (reprocessed), reanalysis products	Climate trend analysis, long-term variability studies, climate model evaluation, historical weather reconstructions.
Satellite Data Repositories	Raw and processed data from geostationary and polar-orbiting satellites	Imagery (visible, infrared, water vapor), derived products (sea surface temperature, cloud properties, atmospheric composition), input for various models and analyses.
Radar Data Archives	Raw and processed data from ground-based weather radar systems	Precipitation maps, storm tracking, nowcasting of rainfall, wind information (Doppler radar), input for hydrological models.

This overview highlights the significant role data science plays in advancing the field of meteorology across various domains, from improving daily forecasts to understanding long-term climate change. The diverse databases containing meteorological information are crucial for these applications.

1.6 Dissemination of Information Generated and Time Frames

The culmination of meteorological observation, analysis, and prediction efforts lies in the effective dissemination of the generated information to a wide array of users, ranging from the general public to specialized industries and emergency services. The methods employed to convey this information are diverse, tailored to the urgency and specificity of the data, as well as the needs of the recipient. Moreover, the timeframe within which this information is delivered is critical, often directly impacting its utility and potential to inform decision-making.

Traditional avenues for disseminating weather information to the public, such as **television and radio broadcasts**, continue to play a significant role. These outlets typically provide summarized forecasts covering broader geographical areas, often on a **daily basis**, with more frequent updates during periods of significant or rapidly changing weather. For critical situations like severe thunderstorm watches or flash flood warnings, these broadcasts can provide **immediate alerts**.

The advent of the internet and mobile technology has revolutionized how meteorological information is accessed. **Websites and mobile applications** provided by national meteorological services and commercial weather providers offer a wealth of detailed, often location-specific forecasts, interactive weather maps displaying real-time observations, and a variety of analytical products. These platforms are frequently updated, sometimes on an **hourly basis** or even more often for rapidly evolving conditions. Furthermore, many offer the capability to receive **push notifications** for severe weather warnings relevant to a user's location, ensuring timely awareness of potential hazards.

Certain sectors have highly specialized meteorological information requirements. For **aviation**, precise forecasts for airports (Terminal Aerodrome Forecasts - TAFs) are crucial for safe operations and are typically updated every **three to six hours**. The **agricultural** community may rely on seasonal climate outlooks for planting decisions or daily agrometeorological bulletins providing insights into how weather conditions are affecting crops, often with **daily or seasonal** timeframes. **Emergency management agencies** require immediate, detailed, and localized information during extreme weather events to coordinate response efforts, often necessitating dissemination in **near real-time**. These specialized needs are often met through dedicated data feeds, APIs (Application Programming Interfaces) allowing direct integration into their systems, or bespoke reporting mechanisms.

For the rapid exchange of critical meteorological data and warnings between meteorological centers globally, dedicated **telecommunication systems**, such as the World Meteorological Organization's Global Telecommunication System (GTS), are employed. These networks facilitate the near **real-time** sharing of observations and model outputs, underpinning international collaboration in weather forecasting.

Finally, **social media platforms** have emerged as a significant channel for disseminating quick updates, issuing warnings, and engaging directly with the public. Meteorological agencies and weather experts often use platforms like Twitter to provide **near real-time** information on developing weather situations and to clarify forecasts.

Now, let's summarize the types of meteorological information generated and their typical lead times in the following table:

Information Generated	Typical Lead Time	Primary Users
Real-time Observations	Current conditions (disseminated within minutes)	Nowcasters, data assimilation systems, general public
Nowcasts	0-6 hours (updated frequently, e.g., every 15-60 min)	Public, aviation, emergency responders, event organizers
Short-range Forecasts	Up to 72 hours (updated every 3-6 hours)	General public, many industries (agriculture, transportation, energy)
Medium-range Forecasts	3-10 days (updated once or twice daily)	Planning sectors, agriculture, energy
Long-range Outlooks	Weeks to seasons (updated weekly, monthly, seasonally)	Agriculture, water management, energy, long-term planning
Severe Weather Warnings (e.g., tornado, flood)	Imminent or occurring (disseminated immediately)	General public, emergency responders, local authorities
Climate Summaries & Analyses	Historical (updated periodically, e.g., annually)	Researchers, policymakers, educators

This table provides a clearer overview of the temporal aspects of different types of meteorological information and who typically relies on them.

1.7 Conclusion

In this introductory chapter, we've laid the groundwork for understanding both meteorology and data science as distinct yet increasingly intertwined fields. Meteorology, the scientific study of our atmosphere, provides the context and the rich, complex data generated by natural processes and sophisticated observation systems. We've touched upon its historical development, the key atmospheric variables it examines, and the various scales at which atmospheric phenomena occur. Simultaneously, we've introduced data science as the multidisciplinary field focused on extracting knowledge and insights from data. We've highlighted core data science concepts and the typical workflow involved in turning raw data into actionable intelligence.

The convergence of these two domains marks an exciting and transformative era for atmospheric science. The vast datasets produced by meteorological instruments and models present both opportunities and challenges that data science is uniquely equipped to address. From refining the precision of weather forecasts and enhancing our ability to predict and prepare for extreme weather events, to deepening our understanding of long-term climate trends, the application of data science techniques is proving invaluable. This synergy allows us to move beyond traditional methods, leveraging the power of computation and statistical learning to unlock deeper insights from the atmospheric data that surrounds us.

As we move forward, subsequent chapters will delve into specific ways data science is applied to meteorological problems, exploring techniques in data preprocessing, statistical modeling, machine learning etc.

2

Atmospheric Physics and Measurement Principles

2.1 Introduction

Building upon the foundational understanding of meteorology and data science established in Chapter 1, we now turn our focus to the fundamental physical processes that govern our atmosphere and the essential principles behind how we measure them. The Earth's atmosphere is a dynamic fluid system, and its behavior is dictated by a complex interplay of physical laws. To truly comprehend weather phenomena, from the smallest updraft to the largest global circulation patterns, one must grasp the underlying atmospheric physics. This includes understanding how energy is transferred, how gases and water vapor behave, and how forces shape atmospheric motion.

This chapter will delve into these core physical concepts, providing the theoretical bedrock necessary for a deeper appreciation of meteorological processes. Furthermore, accurate and reliable data is the lifeblood of modern meteorology and, by extension, data science applications within the field. Therefore, we will also explore the fundamental principles that underpin the measurement of key atmospheric variables.

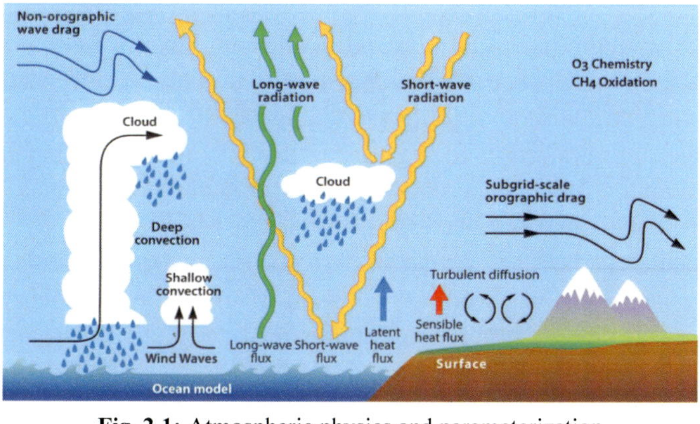

Fig. 2.1: Atmospheric physics and parameterization

Understanding how instruments work, their limitations, and the physics behind their readings is crucial for appreciating data quality and for effectively applying data science techniques. Without precise measurements of atmospheric conditions, even the most sophisticated models and algorithms would lack the necessary input to produce meaningful insights or accurate forecasts (Fig. 2.1).

Thus, Chapter 2 will bridge the gap between the theoretical understanding of the atmosphere and the practical acquisition of the data that fuels meteorological science and its data-driven advancements.

2.2 Basic Atmospheric Thermodynamics

Atmospheric thermodynamics is the branch of meteorology that applies the fundamental laws of physics, specifically those governing heat and energy, to understand the behavior of the Earth's atmosphere. It is crucial for explaining phenomena such as cloud formation, precipitation, and the vertical movement of air, which are central to weather processes. At its core, atmospheric thermodynamics deals with the transformation of energy within air parcels and the atmosphere as a whole, focusing on how heat, work, and internal energy interact.

The **First Law of Thermodynamics** is a cornerstone of this study, essentially stating the principle of energy conservation in the atmosphere. It posits that energy cannot be created or destroyed, only transformed from one form to another. In the context of an air parcel, any heat added to or removed from it will either change its internal energy (manifesting as a change in temperature) or result in work being done by or on the parcel (such as expansion or compression). This law helps us understand the energy budget of the atmosphere, where solar radiation provides the primary energy input, and various processes redistribute and transform this energy.

A particularly vital concept in atmospheric thermodynamics is the **adiabatic process**. An adiabatic process is one in which an air parcel undergoes a change in temperature due without any heat exchange with its surroundings. This is a reasonable approximation for rapidly rising or sinking air parcels, as the process occurs too quickly for significant heat transfer to take place. As an air parcel rises, it encounters lower atmospheric pressure, causing it to expand. This expansion requires the parcel to do work on its environment, and since no heat is exchanged, this work comes from its own internal energy, leading to a decrease in the parcel's temperature. Conversely, a sinking air parcel is compressed by increasing pressure, and the work done *on* the parcel increases its internal energy, causing it to warm. The rate at which a dry (unsaturated) air parcel cools as it rises, or warms as it sinks, is known as the **dry adiabatic lapse rate**, a constant value of approximately 9.8^{0}C per kilometer.

However, the atmosphere is rarely completely dry. The presence of water vapor introduces the critical role of **phase changes of water**. Water can exist in the atmosphere as a gas (water vapor), a liquid (cloud droplets, rain), or a solid (ice crystals, snow). Each transition between these phases involves the absorption or release of **latent heat**. For instance, when water vapor condenses into liquid droplets to form clouds, a significant amount of latent heat is released into the surrounding air. This release of heat warms the air parcel, counteracting the adiabatic cooling that occurs as it rises. Consequently, a rising air parcel that becomes saturated and undergoes condensation will cool at a slower rate than a dry parcel, a rate known as the **moist adiabatic lapse rate**. This rate is not constant but varies with temperature and moisture content, typically being less than the dry adiabatic lapse rate. The absorption of latent heat during evaporation (e.g., from the Earth's surface) and its release during condensation (e.g., in cloud formation) are fundamental energy transfer mechanisms that drive many atmospheric phenomena, including the development of thunderstorms and hurricanes.

Understanding these thermodynamic principles is crucial for determining **atmospheric stability**. The stability of the atmosphere refers to its tendency to resist or enhance vertical motion. If a rising air parcel becomes colder and therefore denser than its surroundings, it will sink back down, indicating a stable atmosphere. Conversely, if it remains warmer and less dense, it will continue to rise, signifying an unstable atmosphere. The comparison of an air parcel's adiabatic lapse rate (dry or moist, depending on saturation) with the actual environmental lapse rate (the rate at which the temperature of the surrounding atmosphere decreases with height) allows meteorologists to assess atmospheric stability. This assessment is vital for forecasting cloud development, the potential for convection, and the likelihood of severe weather.

Beyond the conservation of energy, the **Second Law of Thermodynamics** provides profound insights into the direction and spontaneity of atmospheric processes. This law, often stated in various forms, fundamentally describes the tendency of isolated systems to evolve towards a state of greater disorder or randomness, a concept quantified by **entropy**. In essence, it states that the total entropy of an isolated system can only increase over time, or remain constant in ideal reversible processes, but it can never decrease. Heat, for example, naturally flows from warmer bodies to colder bodies, never spontaneously in the reverse direction, because this process increases the overall entropy of the system.

In the atmosphere, the Second Law explains why certain processes occur naturally and why others do not. For instance, the mixing of different air masses, or the diffusion of gases, leads to an increase in entropy as the system moves towards a more uniform and disordered state. While the First Law tells us that energy is conserved during these processes, the Second Law tells us *which way* the energy will flow and *which processes* are thermodynamically favored. It underscores the irreversibility of many atmospheric phenomena; for example, once a cloud forms and precipitates, the energy released during condensation cannot be spontaneously re-collected to lift the water back into vapor and recreate the exact initial state without external work.

The concept of entropy also helps explain the efficiency limits of atmospheric "heat engines," such as those driving large-scale circulation patterns or hurricanes. These systems convert thermal energy differences into kinetic energy (wind). The Second Law dictates that no such engine can be perfectly efficient; some energy will always be lost to the environment as increased entropy, meaning not all available thermal energy can be converted into useful work. This inherent inefficiency is a fundamental aspect of how the atmosphere operates, constantly striving towards a state of equilibrium, albeit one that is continuously disturbed by solar energy input and the Earth's rotation. Thus, the Second Law of Thermodynamics provides a crucial framework for understanding the directionality and limitations of energy transformations and mixing processes within the Earth's complex atmospheric system.

2.3 Ideal Gas Equation

The Ideal Gas Law, also known as the Ideal Gas Equation, describes the relationship of an ideal gas to the state of its physical properties of temperature, pressure, volume, and amount of gas. The Ideal Gas Law is based on the relationships between pairs of these properties.

An ideal gas is a theoretical gas in which the volume of gas molecules does not affect its behavior, the gas particles are the same size and have no molecular interactions. Additionally, ideal gas particles move in straight lines, often running into their container and one another. These collisions do not result in any energy loss.

Ideal gases do not actually exist, but making assumptions allows for a much simpler assessment of the physical properties of a gas. The Ideal Gas Law ignores some real-world conditions but gives some general explanation of the ways that gases behave. A gas is a state of matter with low density, no specific shape or volume, and is made up of molecules that move freely and quickly spread throughout a vessel.

The Ideal Gas Law is expressed as: $PV = nRT$

In this equation, the pressure (P) multiplied by the volume (V) is equal to the moles of gas (n) multiplied by the gas constant (R) and temperature (T). The units of the variables P, V, n, R, and T are important to consider. P can be given in atm, Torr, Pa, and mmHg.

Volume can be expressed in units of L and m^3. N is given as moles, and T is in Kelvin. The gas constant (R) is given different values based on the units. R is equal to 0.082057 L atm / mol K.

Additionally, R is equal to 62.364 L Torr / mol K. R is also equal to 8.3145 m^3 Pa / mol K. When solving Ideal Gas Law problems, it is critical to choose the appropriate R-value that matches the units used for the other variables.

The ideal gas law, expressed as PV=nRT (where P is pressure, V is volume, n is the number of moles, R is the ideal gas constant, and T is temperature),holds significant importance in meteorology for several reasons:

- **Relating Atmospheric Variables:** It provides a fundamental relationship between the key state variables of the atmosphere: pressure, temperature, and volume (or density). This allows meteorologists to understand how changes in one variable can affect the others. For instance, if air is heated (increase in T) at a constant pressure (P), its volume (V) will increase, leading to a decrease in density.
- **Calculating Air Density:** By rearranging the ideal gas law to ρ=RTPM (where ρ is density and M is the molar mass of air), meteorologists can calculate air density given the pressure and temperature. Density differences in the atmosphere are crucial for understanding buoyancy and vertical motion of air parcels, which drive weather phenomena.
- **Understanding Atmospheric Processes:** The ideal gas law is fundamental to understanding various atmospheric processes:
    - **Adiabatic Processes:** When air parcels risc or sink, they expand or compress, respectively. If this occurs without heat exchange with the surroundings (adiabatically), the ideal gas law helps determine the temperature changes associated with these volume changes. This is critical for understanding cloud formation and stability.
    - **Buoyancy:** Warmer, less dense air rises, while cooler, denser air sinks. The ideal gas law helps quantify these density differences based on temperature variations.
    - **Pressure Variations with Altitude:** Combined with the hydrostatic equation, the ideal gas law helps explain how pressure decreases with height in the atmosphere.

- **Weather Forecasting:** Meteorologists use the ideal gas law, often in more complex thermodynamic equations derived from it, to model the behavior of the atmosphere. These models are essential tools for weather forecasting.By knowing the current state of the atmosphere (pressure, temperature, humidity), they can use these laws to predict how these variables will evolve over time. In summary, the ideal gas law is a cornerstone of meteorological thermodynamics, providing a crucial link because pressure, temperature, and density, which are fundamental to understanding and predicting weather and climate.

2.4 Understanding Hydrostatic Equilibrium

Hydrostatic equilibrium is a state where the forces acting on a fluid are balanced, resulting in no net acceleration. In simpler terms, it's when a fluid is at rest, or its motion is such that there are no unbalanced forces.

In meteorology, hydrostatic equilibrium specifically refers to the balance between two primary vertical forces acting on an air parcel:

- **The downward force of gravity (Fg)**: This pulls the air parcel towards the Earth.
- **The upward-directed pressure gradient force (Fp)**: This force arises because pressure in the atmosphere generally decreases with height. Thus, the pressure at the bottom of an air parcel is usually slightly higher than the pressure at the top, resulting in a net upward force.

When these two forces are equal and opposite (Fp+Fg=0), the air is said to be in hydrostatic equilibrium.

Mathematically, this balance is expressed by the **hydrostatic equation**:

$$dP/dZ = -\rho g$$

Where: dP/dZ is the rate of change of pressure (P) with respect to height (z).

- ρ is the density of the air.
- g is the acceleration due to gravity.
- **Significance in Meteorology**
  - **A Fundamental Approximation:** The atmosphere, on a large scale and away from vigorous vertical motions (like in thunderstorms or strong fronts), is very close to hydrostatic equilibrium. This makes the hydrostatic equation an excellent approximation for many meteorological applications.

- **Relating Pressure and Height:** The hydrostatic equation allows us to relate the pressure at a certain height to the height itself, given the density profile. This is crucial for:
- Determining the height of pressure levels on weather maps.
- Estimating the thickness of atmospheric layers based on pressure differences (using the hypsometric equation, which is derived from the hydrostatic equation and the ideal gas law).
- **Understanding Vertical Structure:** It helps in understanding the vertical structure of the atmosphere, showing why pressure decreases with height.
- **Input for Numerical Weather Prediction:** While modern numerical weather prediction models also account for non-hydrostatic conditions, the hydrostatic equation is often a good starting point or a component in some model formulations, especially for larger-scale atmospheric motions.

- **Limitations**

  It's important to remember that hydrostatic equilibrium is an approximation. It doesn't hold well *in situ*ations with significant vertical accelerations, such as:

  - Thunderstorms with strong updrafts and downdrafts.
  - Small-scale turbulent motions.
  - In rapidly changing weather systems.

In these cases, the vertical acceleration term cannot be neglected, and the full vertical momentum equation must be considered.

In summary, hydrostatic equilibrium and the hydrostatic equation are fundamental concepts in meteorology, providing a crucial understanding of the vertical balance of forces in the atmosphere and a vital tool for relating pressure and height.

2.5 Atmospheric Stability

Atmospheric stability refers to the atmosphere's tendency to either encourage or resist vertical motion of air parcels. It's a crucial concept in meteorology because it directly influences cloud development, precipitation, turbulence, and air pollution dispersion.

Think of it this way

- **Stable Atmosphere:** If an air parcel is displaced vertically, it will tend to return to its original level. It resists vertical motion.

- **Unstable Atmosphere:** If an air parcel is displaced vertically, it will continue to move away from its original level (upward if initially lifted, downward if initially pushed down). It encourages vertical motion.
- **Neutral Atmosphere:** If an air parcel is displaced vertically, it will remain at its new level.

2.5.1. Determining Atmospheric Stability

Atmospheric stability is primarily determined by comparing the **environmental lapse rate (ELR)**, which is the actual rate at which temperature decreases with altitude in the atmosphere, with two theoretical lapse rates:

- **Dry Adiabatic Lapse Rate (DALR):** The rate at which a dry (unsaturated) air parcel cools as it rises or warms as it sinks due to expansion or compression (approximately 9.8°C per kilometer).
- **Moist Adiabatic Lapse Rate (MALR) or Saturated Adiabatic Lapse Rate (SALR):** The rate at which a saturated air parcel cools as it rises. This rate is less than the DALR (typically around 6°C per kilometer in the lower troposphere) because the condensation of water vapor releases latent heat.

Based on the comparison of these lapse rates, we can classify atmospheric stability into a few main categories:

- **Absolutely Stable:** Occurs when the ELR is less than both the MALR and the DALR (ELR<MALR<DALR). Any lifted air parcel (whether saturated or unsaturated) will be colder and denser than its surroundings and will sink back down. This leads to suppressed vertical motion, smooth airflow, and often clear skies or stable cloud types like stratus.
- **Absolutely Unstable:** Occurs when the ELR is greater than both the MALR and the DALR (ELR>DALR>MALR). Any lifted air parcel (whether saturated or unsaturated) will be warmer and less dense than its surroundings and will continue to rise. This favors strong vertical motions, leading to the development of convective clouds (like cumulus and cumulonimbus) and potentially thunderstorms.
- **Conditionally Unstable:** Occurs when the ELR is between the DALR and the MALR (MALR<ELR<DALR). In this case, an unsaturated air parcel that is lifted will initially be colder than its surroundings and tend to sink back (stable for unsaturated air).However, if this parcel is lifted to a level where it becomes saturated, it will then cool at the MALR. If at this point, the saturated parcel becomes warmer than its environment, it will continue to rise (unstable for saturated air). This condition is crucial for the development of thunderstorms.

2.5.2. Neutral Stability

- **Dry Neutral:** ELR equals the DALR (ELR=DALR). An unsaturated air parcel will have the same temperature as its surroundings if lifted or sunk.
- **Moist Neutral:** ELR equals the MALR (ELR=MALR). A saturated air parcel will have the same temperature as its surroundings if lifted or sunk.

Significance in Meteorology

- **Cloud Formation:** Stability dictates the type and vertical extent of clouds. Unstable conditions lead to vertically developed clouds, while stable conditions result in flat, layered clouds.
- **Precipitation:** Unstable atmospheres are more likely to produce showers and thunderstorms.
- **Air Pollution:** Stable atmospheres can trap pollutants near the surface, leading to poor air quality.
- **Turbulence:** Unstable conditions often result in turbulent airflow, while stable conditions lead to smoother flow.
- **Severe Weather:** Conditional and absolute instability are key ingredients for severe weather development.

Understanding atmospheric stability is fundamental to weather analysis and forecasting. Meteorologists use various tools, including atmospheric soundings (vertical profiles of temperature, humidity, and wind), to assess the stability of the atmosphere.

2.6 Radiation and Remote Sensing Principles

2.6.1 Electromagnetic spectrum

The *electromagnetic spectrum* is the distribution of electromagnetic radiation according to energy (or equivalently, by virtue of the relations in the previous section, according to frequency or wavelength).

Regions of the Electromagnetic Spectrum

The following table gives approximate wavelengths, frequencies, and energies for selected regions of the electromagnetic spectrum.

Region	Wavelength (Angstroms)	Wavelength (centimeters)	Frequency (Hz)	Energy (eV)
Radio	> 109	> 10	< 3 x 109	$< 10^{-5}$
Microwave	109 - 106	10 - 0.01	$3 \times 109^{-3} \times 1012$	10^{-5} - 0.01
Infrared	106 - 7000	$0.01^{-7} \times 10^{-5}$	$3 \times 1012^{-4.3} \times 1014$	0.01 - 2
Visible	7000 - 4000	$7 \times 10^{-5} - 4 \times 10^{-5}$	$4.3 \times 1014^{-7.5} \times 1014$	2 - 3
Ultraviolet	4000 - 10	$4 \times 10^{-5} - 10^{-7}$	$7.5 \times 1014^{-3} \times 1017$	3 - 103
X-Rays	10 - 0.1	$10^{-7} - 10^{-9}$	$3 \times 1017^{-3} \times 1019$	103 - 105
Gamma Rays	< 0.1	$< 10^{-9}$	> 3 x 1019	> 105

The notation "eV" stands for electron-volts, a common unit of energy measure in atomic physics. A graphical representation of the electromagnetic spectrum is shown in the figure 2.2 below.

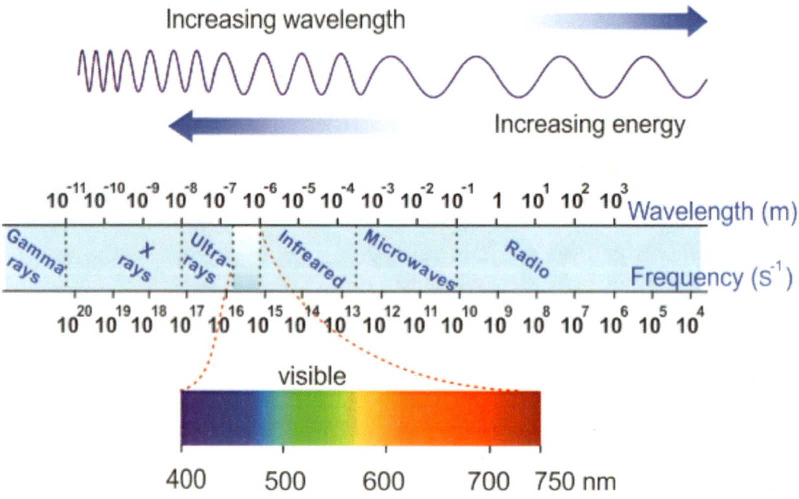

Fig. 2.2: Electromagnetic spectrum

Thus we see that visible light and gamma rays and microwaves are really the same things. They are all electromagnetic radiation; they just differ in their wavelengths.

2.6.2 The Spectrum of Visible Light

In the same way that we sense frequency of sound as pitch, we sense the frequency of light as color. Notice how small the visible spectrum is over the full range of the electromagnetic spectrum.

The visible part of the spectrum may be subdivided according to color, with red at the long wavelength end and violet at the short wavelength end, as illustrated (schematically) in the above figure.

2.7 Absorption, Emission and Scattering

When light interacts with matter, several things can happen. Three of the most important are **absorption**, **reflectance**, and **scattering**. These processes determine how we perceive the world around us and are crucial in fields like meteorology and remote sensing.

2.7.1 Absorption

- **What it is:** Absorption is the process by which the energy of a photon of light is taken up by matter. This energy is often converted into other forms, such as heat, or can cause electrons in atoms to jump to higher energy levels.
- **How it works:** When the frequency of light matches the energy difference between electron orbitals in an atom or molecule, the light can be absorbed, causing an electron to move to a higher energy state. If the frequency doesn't match, the light is usually transmitted or reflected.
- **Effects**
  - **Heating:** Absorbed light energy often increases the temperature of the material.
  - **Color:** The color of an object is determined by the wavelengths of light it *doesn't* absorb. For example, a red apple absorbs most colors of light but reflects red light.
  - **Opacity:** Materials that absorb all visible light appear opaque.

2.7.2 Reflectance

- **What it is:** Reflectance is the process where light bounces off a surface.
- **How it works:** When light hits an interface between two media (e.g., air and a mirror), some portion of the light is reflected back into the original medium. The amount of reflection depends on the properties of the two media (like their refractive indices) and the angle of incidence.
- **Types**
  - **Specular Reflection:** Occurs from smooth surfaces (like mirrors or a calm lake) where the angle of incidence equals the angle of reflection, resulting in a clear, mirror-like reflection.
  - **Diffuse Reflection:** Occurs from rough surfaces (like paper or a matte wall) where light is reflected in many different directions. This is how we see most objects.

- **Effects:** Reflectance allows us to see objects that don't produce their own light. The intensity and spectral composition of reflected light provide information about the surface properties of the object.

2.7.3 Scattering

- **What it is:** Scattering occurs when light is redirected in many different directions as it interacts with particles or irregularities in a medium.
- **How it works:** When light encounters particles (like molecules in the atmosphere, dust, or water droplets), it can be absorbed and then re-emitted in different directions, or it can be deflected by the particle.
- **Types (depending on particle size relative to the wavelength of light):**
  - **Rayleigh Scattering:** Occurs when particles are much smaller than the wavelength of light (e.g., air molecules). Shorter wavelengths (like blue) are scattered more strongly than longer wavelengths (like red). This is why the sky is blue.
  - **Mie Scattering:** Occurs when particles are comparable in size to the wavelength of light (e.g., water droplets in clouds, aerosols). Mie scattering is less wavelength-dependent than Rayleigh scattering, which is why clouds appear white.
  - **Non-selective Scattering:** Occurs when particles are much larger than the wavelength of light (e.g., large raindrops). All wavelengths of visible light are scattered equally.
- **Effects**
  - **Sky Color:** Scattering is responsible for the blue color of the sky and red sunsets.
  - **Visibility:** Scattering by particles like fog or haze reduces visibility.
  - **Remote Sensing:** The way different surfaces and atmospheric components scatter light at various wavelengths is used to gather information about them from a distance.

In summary, when light encounters matter, it can be absorbed (its energy is taken in), reflected (it bounces off), or scattered (it's redirected in multiple directions).The proportion of light that undergoes each of these processes depends on the properties of the light (e.g., wavelength) and the properties of the matter it interacts with. These three processes are fundamental to how we perceive the world and how we study it using various scientific techniques.

2.8 Remote Sensing

2.8.1 Remote Sensing Principles

Remote sensing is the science of obtaining information about an object, area, or phenomenon without being in direct physical contact. It relies on detecting and measuring the radiation emitted or reflected by the object.

- **Electromagnetic Radiation as the Information Carrier:** Most remote sensing techniques utilize EMR as the means of gathering information. The characteristics of the radiation detected by a sensor provide clues about the properties of the target.
- **The Remote Sensing Process:** A typical remote sensing process involves:
    - **Energy Source (A):** Providing electromagnetic energy (e.g., the Sun for passive sensing, or the sensor itself for active sensing).
    - **Radiation and the Atmosphere (B):** Interaction of the energy with the atmosphere.
    - **Interaction with the Target (C):** Interaction of the energy with the Earth's surface or object of interest (reflection, emission, absorption).
    - **Recording of Energy by the Sensor (D):** Detection of the radiation reflected or emitted from the target by a remote sensor.
    - **Transmission, Reception, and Processing (E):** Transmission of the data to a ground station and processing it into an image or other usable format.
    - **Interpretation and Analysis (F):** Examining the processed data to extract information about the target.
    - **Application (G):** Using the information to solve a particular problem or gain knowledge.
- **Types of Remote Sensing**
    - **Passive Remote Sensing:** Detects naturally emitted or reflected radiation (e.g., sunlight reflected from the Earth, thermal radiation emitted by the Earth). Examples include visible and infrared imagery.
    - **Active Remote Sensing:** Emits its own energy and then detects the radiation reflected or backscattered from the target. Examples include radar and lidar.
- **Key Characteristics of Remote Sensing Data**
    - **Spatial Resolution:** The size of the smallest feature that can be distinguished.

- **Spectral Resolution:** The number and width of the wavelength bands that the sensor can measure.
- **Radiometric Resolution:** The sensitivity of the sensor to differences in signal intensity.
- **Temporal Resolution:** The frequency with which data is collected for the same area.

In summary, remote sensing uses the principles of radiation to observe and measure the Earth's surface and atmosphere from a distance. By analyzing the electromagnetic radiation that interacts with our planet, we can gain valuable insights into its physical, chemical, and biological properties. These principles are fundamental to various applications in meteorology, environmental monitoring, agriculture, urban planning, and many other fields.

2.9 Fluid Dynamics and Atmospheric Motion

The movement of air in the atmosphere, which we experience as wind, is governed by a balance of several forces. These forces determine the speed and direction of the atmospheric flow at different scales. Here are the primary forces affecting atmospheric flow:

2.9.1 Forces Affecting Atmospheric Flow

- **Pressure Gradient Force (PGF)**
  - This force arises from differences in air pressure over a distance. Air naturally moves from areas of high pressure to areas of low pressure.
  - The magnitude of the PGF is proportional to the pressure gradient (the rate of change of pressure with distance). Closely spaced isobars (lines of constant pressure on a weather map) indicate a strong pressure gradient and thus a strong PGF, leading to stronger winds.
  - The direction of the PGF is always from high to low pressure and perpendicular to the isobars.
  - The PGF is the initial force that sets air in motion.
- **Coriolis Force**
  - This is an apparent force that arises due to the Earth's rotation. From the perspective of an observer on the rotating Earth, moving objects (including air parcels) appear to be deflected from a straight path.
  - In the Northern Hemisphere, the deflection is to the right of the direction of motion.
  - In the Southern Hemisphere, the deflection is to the left of the direction of motion.

  - The6 magnitude of the Coriolis force depends on the speed of the moving object and the latitude (it's zero at the Equator and maximum at the poles).
  - The Coriolis force acts perpendicular to the direction of motion and is crucial for large-scale atmospheric circulation patterns.
- **Gravitational Force**
  - Gravity pulls air molecules towards the Earth's center.
  - In the vertical direction, gravity is primarily balanced by the vertical pressure gradient force in a state of hydrostatic equilibrium, which we discussed earlier.
  - Gravity also plays a role in phenomena like cold air drainage down slopes.
- **Friction**
  - Friction acts to oppose the motion of air.
  - **Surface friction** is significant in the lower atmosphere (the atmospheric boundary layer) where air interacts with the Earth's surface (land, water, vegetation, buildings). It slows down the wind speed and alters its direction, causing air to flow slightly towards lower pressure.
  - **Internal friction** (viscosity) within the air itself is generally much smaller than surface friction at larger scales in the atmosphere. However, it becomes important at very small scales.
  - Friction is more significant near the surface and decreases with altitude.

2.9.2 Balance of Forces

The actual flow of the atmosphere is often a result of a balance (or imbalance) between these forces. For example:

- **Geostrophic Wind:** Above the friction layer, a balance between the pressure gradient force and the Coriolis force results in wind that flows parallel to the isobars.
- **Gradient Wind:** When the flow is curved (around highs and lows), the centrifugal force also becomes important, leading to a balance between the pressure gradient force, Coriolis force, and centrifugal force.
- **Surface Winds:** Near the surface, friction plays a significant role, and the balance includes the pressure gradient force, Coriolis force, and friction, resulting in winds that cross isobars towards lower pressure.

Understanding these forces and how they interact is fundamental to explaining and predicting atmospheric motion and weather patterns.

2.10 Units and Conversions

Let us break down the standard meteorological units and some common conversions. Meteorology largely uses the **International System of Units (SI)**, but some traditional units are still common, especially in certain regions.

2.10.1 Standard Meteorological Units (SI Preferred)

Here are some key meteorological variables and their commonly used units:

- **Temperature**
  - **SI:** Kelvin (K)
  - Commonly used: Degrees Celsius (°C)
  - Sometimes used: Degrees Fahrenheit (°F)
- **Pressure**
  - **SI:** Pascal (Pa)
  - Commonly used: Hectopascal (hPa) - Note: 1hPa=100Pa=1mb (millibar)
  - Sometimes used: Millibars (mb), inches of mercury (in Hg), millimeters of mercury (mmHg), kilopascals (kPa)
- **Wind Speed**
  - **SI:** Meters per second (m/s)
  - Commonly used: Knots (kt), kilometers per hour (km/h), miles per hour (mph)
- **Wind Direction**
  - Degrees clockwise from North (°) or cardinal directions (N, S, E, W, etc.)
- **Precipitation**
  - **SI:** Millimeters (mm) (for liquid equivalent)
  - Sometimes used: Inches (in)
- **Humidity**
  - Percent (%) for relative humidity
  - g/kg for specific humidity
  - kg/m3 for absolute humidity
- **Visibility**
  - Meters (m) or kilometers (km)
  - Sometimes used: Miles (mi), feet (ft)

2.10.2 Common Meteorological Conversions

Here are some useful conversion formulas:

Temperature

- Celsius to Fahrenheit: F=(C×59)+32
- Fahrenheit to Celsius: C=(F−32)×95
- Celsius to Kelvin: K=C+273.15
- Kelvin to Celsius: C=K−273.15

Pressure

- 1 hPa=1 mb
- 1 atm=1013.25 hPa
- 1 atm≈29.92 inHg
- 1 inHg≈33.86 hPa
- 1 kPa=10 hPa

Wind Speed

- 1 knot≈0.514 m/s
- 1 m/s≈1.94 knots
- 1 knot≈1.852 km/h
- 1 mph≈1.609 km/h
- 1 knot≈1.151 mph

Length/Distance (relevant for visibility, height)

- 1 inch=25.4 mm
- 1 foot=0.3048 m
- 1 mile≈1.609 km

2.11 Conclusion

Atmospheric measurements form the bedrock of our understanding of the Earth's dynamic atmosphere, providing the essential data needed to comprehend, predict, and ultimately respond to atmospheric phenomena. This chapter has explored the diverse array of parameters that define the state of the atmosphere, from fundamental variables like temperature, pressure, and humidity to complex interactions involving radiation, atmospheric stability, and the forces driving air flow. We've seen how a variety of instruments, both in-situ and remote sensing technologies, are employed to capture these crucial observations, adhering to standardized units and conversion protocols to ensure global consistency and comparability. Accurate atmospheric data allows scientists to validate theoretical models, such as the ideal gas law and hydrostatic equilibrium, and to unravel the intricate interplay of forces that govern atmospheric circulation. Beyond scientific inquiry, these measurements have tangible societal impacts, informing decisions in aviation, agriculture, disaster preparedness, and air quality management.

As technology continues to advance, the precision, spatial coverage, and temporal resolution of atmospheric measurements are constantly improving. This ongoing evolution promises even deeper insights into the complex workings of our atmosphere, further enhancing our ability to anticipate changes, mitigate risks, and adapt to a changing climate. The continuous pursuit of comprehensive and accurate atmospheric data remains indispensable for safeguarding lives, livelihoods, and the planet itself.

3

In situ Observations

3.1 Introduction

Understanding the Earth's atmosphere—its intricate dynamics, ever-changing weather patterns, and long-term climate trends—hinges critically on the quality and detail of the **observations** we collect. While remote sensing technologies offer broad perspectives from a distance, a fundamental and indispensable category of atmospheric data comes from ***in situ*** **observations**. The term "*in situ*," meaning "in place" or "on-site," precisely describes the nature of these measurements: they are direct, physical readings taken within the atmospheric volume itself.

These direct **observations** are the bedrock of meteorological science. They provide the ground truth for atmospheric conditions, offering unparalleled precision in measuring parameters such as temperature, humidity, atmospheric pressure, wind speed and direction, and the characteristics of precipitation. Unlike indirect sensing, *in situ* methods involve instruments physically interacting with the air, yielding highly accurate and localized data points. This granular detail is crucial for a multitude of applications: it allows for the precise calibration of remote sensing instruments, provides essential validation for complex numerical weather prediction models, and enables scientists to dissect and understand fine-scale atmospheric processes that influence local weather.

This chapter will focus on the diverse array of **observations** gathered through *in situ* methods. We will explore how these direct measurements are obtained across various atmospheric layers, from comprehensive surface weather station networks that provide continuous ground-level data, to radiosondes carried by weather balloons that ascend through the troposphere and lower stratosphere, and specialized instruments onboard research aircraft that probe specific atmospheric phenomena. We will delve into the types of **observations** collected by each platform, the inherent advantages and limitations of direct measurement, and the ongoing advancements in observational technologies that continue to refine our understanding of the atmosphere. By examining these foundational **observations**, we underscore their vital role in advancing meteorological research, improving forecast accuracy, and monitoring the pulse of our planet's climate.

3.2 Surface Weather Stations

Surface weather stations are the most fundamental and widespread form of *in situ* atmospheric observation. They are ground-based facilities equipped with a variety of instruments designed to measure and record atmospheric conditions at or near the Earth's surface. These stations form dense networks across continents, providing a continuous stream of vital data that serves as the backbone for local weather forecasting, climate monitoring, and atmospheric research.

3.2.1 Key Observations and Measured Parameters

Surface weather stations are typically designed to collect a comprehensive set of meteorological parameters, providing a detailed snapshot of the atmospheric state at their specific location. The primary observations include:

- **Air Temperature:** Measured by thermometers (e.g., resistance temperature detectors, thermocouples) housed in a Stevenson screen or similar radiation shield to protect them from direct sunlight and precipitation, ensuring an accurate reading of ambient air temperature.
- **Humidity:** The amount of water vapor in the air, typically measured by hygrometers. This can be expressed as relative humidity, dew point temperature, or absolute humidity.
- **Atmospheric Pressure:** Measured by barometers, indicating the weight of the air column above the station. Pressure readings are crucial for identifying high and low-pressure systems, which are key drivers of weather patterns.
- **Wind Speed and Direction:** Measured by anemometers (for speed) and wind vanes (for direction). These instruments are typically mounted on a mast at a standard height (e.g., 10 meters) to minimize local obstructions.
- **Precipitation:** Measured by rain gauges (for liquid precipitation) and snow gauges (for solid precipitation). These instruments quantify the amount of rainfall or snowfall over a specific period.
- **Solar Radiation:** Pyranometers measure incoming shortwave radiation from the sun, which is important for understanding energy budgets and agricultural applications.
- **Ground Temperature:** Thermometers placed at various depths in the soil measure soil temperature, relevant for agriculture, hydrology, and climate studies.

- **Visibility:** Measured by transmissometers or human observation, indicating how far one can see horizontally through the atmosphere. This is critical for aviation and transportation.
- **Cloud Cover and Type:** Often observed visually by trained personnel, though automated ceilometers can measure cloud base height and coverage.
- **Present Weather:** Manual or automated reporting of current weather phenomena such as fog, mist, drizzle, rain, snow, hail, thunderstorms, etc.

3.2.2 Types of Surface Weather Stations

Surface weather stations can vary in complexity and automation:

- **Manned Stations:** Historically, many stations were staffed by human observers who manually read instruments and made visual observations. While less common now, some critical stations still retain human observers for specific tasks.
- **Automated Weather Stations (AWS):** These are the most prevalent type today. They use electronic sensors to automatically measure, record, and transmit data at regular intervals (e.g., every 10 minutes, hourly). AWS can operate in remote locations and provide continuous data without human intervention.
- **Specialized Stations:** These include agricultural weather stations (measuring soil moisture, leaf wetness), hydrological stations (measuring river levels, snow depth), air quality stations (measuring pollutants), and climatological stations (focused on long-term data collection).

3.2.3 Sensors Used in Different Types of Weather Stations

The specific array of sensors deployed at a weather station is largely determined by its purpose, level of automation, and the detailed meteorological parameters it aims to observe. While many fundamental sensors are common across various station types, their sophistication, density, and integration can differ significantly.

Here's how sensor usage varies across different categories of weather stations:

3.2.4 Manned Weather Stations (Traditional/Observational)

Historically, and in some specialized or remote locations today, manned weather stations rely on human observers for certain measurements and instrument readings. While they might use some automated sensors, the emphasis is often on robust, easily readable, and sometimes manually operated instruments.

- **Temperature:** Often use **liquid-in-glass thermometers** (e.g., mercury or alcohol) housed in a Stevenson screen for dry-bulb and wet-bulb readings (for psychrometric humidity calculations). **Bimetallic thermometers** might be found in older recording devices (barographs, thermographs).
- **Humidity:Psychrometers** (wet-bulb/dry-bulb thermometers) are common, requiring manual reading and calculation. **Hair hygrometers** might also be present but are less precise.
- **Pressure:Mercury barometers** were standard for high accuracy, though **aneroid barometers** are also used. Readings are often taken manually.
- **Wind:Cup anemometers** for speed and **wind vanes** for direction are typical, with readings often taken from a dial or recorded manually.
- **Precipitation:Standard rain gauges** (e.g., 8-inch diameter) are manually emptied and read. **Snow boards** are used for measuring snow depth.
- **Visibility:** Primarily determined by **human observation** of known landmarks.
- **Cloud Cover/Type: Human observation** is paramount for identifying cloud types, coverage, and base height.
- **Present Weather: Human observation** and coding of current weather phenomena (e.g., fog, drizzle, thunderstorms).

3.2.5 Automated Weather Stations (AWS)

AWS are the most common type of weather station globally, designed for continuous, unattended operation. They rely entirely on electronic sensors connected to a data logger and communication system. The emphasis is on reliability, accuracy, and the ability to transmit data automatically.

- **Temperature:** Predominantly use highly accurate and responsive **thermistors** or **Resistance Temperature Detectors (RTDs)**, typically housed in fan-aspirated or naturally ventilated radiation shields.
- **Humidity:Capacitive humidity sensors** are standard due to their good accuracy, fast response, and robust nature.
- **Pressure:** Compact and precise **electronic/piezoresistive barometers** are universally used.
- **Wind: Cup anemometers** and **wind vanes** are very common. Increasingly, **ultrasonic (sonic) anemometers** are being deployed

for their higher accuracy, lack of moving parts, and ability to measure turbulent fluxes.

- **Precipitation:Tipping bucket rain gauges** are the workhorse for liquid precipitation. Some advanced AWS might use **weighing rain gauges** for all precipitation types.
- **Solar Radiation:Pyranometers** (thermopile-based) or more affordable **photovoltaic (silicon photodiode) sensors**are often included.
- **Visibility:Forward scatter sensors** or **transmissometers** are used for automated visibility measurement, especially at airports.
- **Cloud:Ceilometers** (laser-based) automatically detect cloud base height and often provide information on multiple cloud layers.
- **Present Weather:** Automated sensors (e.g., **optical disdrometers** or **forward scatter sensors**) can identify precipitation type and intensity.

3.2.6 Specialized Weather Stations

These stations are deployed for specific research, environmental monitoring, or industry applications, and their sensor suites are tailored to those needs.

- **Agricultural Weather Stations**
  - Core AWS sensors (temperature, humidity, wind, rain).
  - Additional: **Soil moisture sensors** (e.g., TDR probes, capacitance probes), **leaf wetness sensors**, **pyranometers** (for evapotranspiration models), **soil temperature sensors** at various depths.
- **Hydrological Stations**
  - Core AWS sensors (temperature, precipitation).
  - Additional: **Snow depth sensors** (ultrasonic), **water level sensors** (for rivers/lakes), **streamflow velocity sensors**.
- **Air Quality Monitoring Stations**
  - Core AWS sensors (wind, temperature, humidity for dispersion modeling).
  - Primary focus: Specialized **gas analyzers** (e.g., for ozone, carbon monoxide, nitrogen oxides, sulfur dioxide), **particulate matter sensors** (PM2.5, PM10), **volatile organic compound (VOC) sensors**.
- **Research Flux Towers/Micrometeorological Stations:**
  - Highly sophisticated versions of AWS, often focused on energy and mass exchange between the surface and atmosphere.

- Employ high-frequency **ultrasonic anemometers** (3D sonic anemometers) for turbulent wind components, fast-response **infrared gas analyzers** for CO2 and water vapor fluxes, **net radiometers** (for all radiation components), and multiple **soil heat flux plates** and **soil temperature/moisture sensors**.
- **Marine Buoys/Oceanic Stations**
  - Sensors adapted for harsh marine environments.
  - Core AWS sensors (temperature, humidity, pressure, wind).
  - Additional: **Wave height/period sensors, sea surface temperature sensors, salinity sensors, ocean current meters, subsurface temperature chains**.

In summary, while basic meteorological parameters are measured by all types of weather stations, the specific sensor technologies, their precision, and the breadth of parameters measured are carefully selected to meet the operational requirements, research objectives, or specialized monitoring needs of each station type. The trend is towards more automated, electronic, and often non-mechanical sensors for improved accuracy, reliability, and continuous data transmission.

3.2.7 Siting Requirements

The effectiveness and representativeness of meteorological data depend not only on the accuracy of the sensors but also crucially on their **siting**—the physical location and installation environment of the weather station and its individual instruments. Proper siting ensures that observations are characteristic of the larger surrounding area and are not unduly influenced by localized anomalies. The World Meteorological Organization (WMO) provides comprehensive guidelines for siting to ensure global data comparability.

Here's a breakdown of siting requirements and the typical sensors used for various types of weather stations:

- **Manned / Traditional Weather Stations:** These stations are designed for comprehensive, often manual, observation and are typically located at airports, research facilities, or national meteorological centers. They adhere to very strict WMO guidelines to ensure data quality and international comparability.
- **Primary Purpose:** Provide full-suite, high-quality meteorological observations, often including visual observations, for synoptic forecasting, aviation, and climate records.

- **General Siting Requirements**
    - **Open, Level Area:** The station should be located in a flat, open area, preferably grass-covered, away from buildings, trees, and other obstructions.
    - **Representative Exposure:** The site should be representative of the general climate of the surrounding area, avoiding microclimates caused by urban heat islands, large bodies of water (unless specifically measuring coastal effects), or significant terrain features.
    - **Obstruction Clearance:** Instruments should be positioned such that obstructions (buildings, trees, fences) do not significantly impact measurements. A common rule for wind sensors is that the distance to any obstruction should be at least 10 times the height of the obstruction.
    - **Accessibility:** While needing an open area, it must also be accessible for observers and maintenance.
- **Typical Sensors & Siting Considerations**
    - **Air Temperature & Humidity (Liquid-in-Glass Thermometers, Psychrometers):** Housed in a **Stevenson screen** (a white, louvered box) mounted 1.25 to 2 meters above a natural, low-cut grass surface. The screen protects sensors from direct solar radiation, precipitation, and re-radiation from the ground, while allowing free airflow.
    - **Atmospheric Pressure (Mercury Barometer, Aneroid Barometer):** Typically located indoors in a stable environment, away from direct sunlight, drafts, and significant temperature fluctuations that could affect the instrument.
    - **Wind Speed & Direction (Cup Anemometer, Wind Vane):** Mounted on a mast at a standard height of 10 meters (33 feet) above ground level. The mast must be free of obstructions within a radius of at least 10 times the height of the obstruction.
    - **Precipitation (Standard Rain Gauge):** Placed in an open, level area, away from obstructions by a distance of at least twice the height of the obstruction. The rim of the gauge should be level and typically 30 cm (1 foot) above the ground to minimize splash-in and splash-out.
    - **Solar Radiation (Pyranometer):** Mounted on a level platform, typically at a height that ensures an unobstructed view of the entire sky (360-degree horizon) throughout the day and year.

- **Automated Weather Stations (AWS):** AWS are designed for continuous, unattended operation and are the most common type of weather station today, found in various environments from urban centers to remote wilderness (Fig. 3.1).

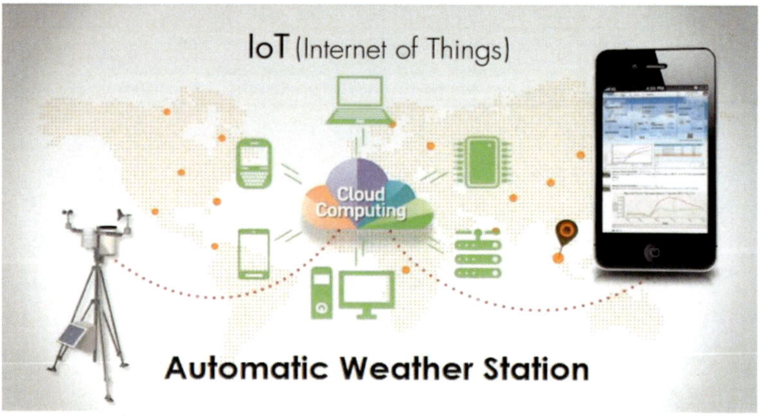

Fig. 3.1: Automated weather station

- **Primary Purpose:** Provide continuous, real-time, automated measurements of standard meteorological parameters for forecasting, climate monitoring, and data assimilation into models.
- **General Siting Requirements:** Similar to manned stations regarding open exposure and representativeness, but with added considerations for power, communication, and security in remote locations.
- **Typical Sensors & Siting Considerations**
  - **Air Temperature & Humidity (Thermistors, RTDs, Capacitive Humidity Sensors):** Housed in naturally aspirated or fan-aspirated radiation shields, mounted 1.25 to 2 meters above a natural surface (grass preferred, but adapted for other surfaces if necessary). The shield design is critical for accurate readings in an automated system.
  - **Atmospheric Pressure (Electronic/Piezoresistive Barometers):** Often integrated into the main data logger enclosure, which should be protected from extreme temperatures and vibrations.
  - **Wind Speed & Direction (Cup/Propeller Anemometers, Wind Vanes, Ultrasonic Anemometers):**Mounted at a standard height of 10 meters on a sturdy mast. Ultrasonic anemometers are increasingly preferred for their accuracy and lack of moving parts. Obstruction clearance rules (10x height) are strictly applied.

- **Precipitation (Tipping Bucket Rain Gauges, Weighing Rain Gauges):** Sited in an open, level area, with the gauge rim at 30 cm (1 foot) above ground. Weighing gauges are more robust for snow and provide intensity.
- **Solar Radiation (Pyranometers, Photovoltaic Sensors):** Mounted level with an unobstructed view of the sky, typically on the same mast as wind sensors but positioned to avoid shading from the mast itself.
- **Visibility (Forward Scatter Sensors, Transmissometers):** Requires a clear line of sight. For transmissometers, a specific baseline distance (e.g., 30-75 meters) is needed between the transmitter and receiver. Sited away from local exhaust or heat sources.
- **Cloud Base (Ceilometers):** Requires a clear, unobstructed view of the sky directly above. Often located near runways at airports.

- **Specialized Weather Stations**

These stations have specific objectives, leading to tailored sensor suites and unique siting considerations.

- *Agricultural Weather Stations*
- **Primary Purpose:** Monitor conditions relevant to crop growth, irrigation, pest management, and soil health.
- **General Siting Requirements:** Located within or adjacent to agricultural fields, representative of the specific crop or soil type being monitored. Avoid areas with unusual drainage or shading.
- **Typical Sensors & Siting Considerations**
    - **Standard AWS sensors:** Temperature, humidity, wind, precipitation (often at lower heights relevant to crop canopy).
    - **Soil Moisture Sensors:** Buried at various depths (e.g., 10 cm, 30 cm, 60 cm) within the crop root zone, ensuring good contact with the soil.
    - **Soil Temperature Sensors:** Buried at specified depths, often alongside soil moisture sensors.
    - **Leaf Wetness Sensors:** Placed within the crop canopy, mimicking the orientation and exposure of actual leaves.
    - **Evapotranspiration Sensors (e.g., Net Radiometers, Soil Heat Flux Plates):** Sited in the center of a large, uniform fetch of the crop, away from field edges or non-vegetated areas.

- *Hydrological Stations*
- **Primary Purpose:** Monitor water resources, flood potential, and drought conditions.
- **General Siting Requirements:** Located near rivers, lakes, reservoirs, or in specific watersheds to measure precipitation, snowpack, and water levels relevant to hydrological cycles.
- **Typical Sensors & Siting Considerations:**
  - **Precipitation (Weighing Rain Gauges, Snow Gauges):** Crucial, often located in open areas of the watershed. Snow gauges may have wind shields.
  - **Snow Depth Sensors (Ultrasonic):** Mounted above snow accumulation areas, clear of obstructions.
  - **Water Level Sensors (Pressure Transducers, Ultrasonic Sensors):** Installed in river channels, lakes, or wells, typically within stilling wells or protected structures to minimize turbulence and debris interference.
  - **Streamflow Velocity Sensors (Acoustic Doppler Current Profilers-ADCP):** Deployed in the river channel to measure water flow.
  - **Temperature:** Air temperature is needed for snowmelt calculations.
- *Air Quality Monitoring Stations*
- **Primary Purpose:** Measure concentrations of atmospheric pollutants.
- **General Siting Requirements:** Highly specific to the type of pollution being monitored (e.g., urban background, industrial impact, roadside). Sites must avoid immediate sources of pollution (e.g., direct exhaust, ventilation outlets) that would not be representative of ambient air quality.
- **Typical Sensors & Siting Considerations:**
  - **Standard AWS sensors:** Wind speed and direction are critical for understanding pollutant dispersion; temperature and humidity also affect pollutant behavior.
  - **Gas Analyzers:** Specialized instruments for specific gases (O_3, CO, NO_x, SO_2, etc.). Air intakes must be at a standard height (e.g., 3-10 meters) and positioned to avoid recirculation or localized sources.
  - **Particulate Matter Sensors (PM2.5, PM10):** Sample air through inlets designed to exclude larger particles and ensure representative sampling.

  - **Volatile Organic Compound (VOC) Sensors:** Similar intake requirements to gas analyzers.
- ***Research Flux Towers / Micrometeorological Stations***
- **Primary Purpose:** Measure high-frequency turbulent fluxes of momentum, heat, and trace gases (e.g., CO_2, water vapor) between the surface and the atmosphere.
- **General Siting Requirements:** Located within a large, homogeneous "fetch" area representative of the surface type (e.g., forest, grassland, urban canopy). Must be far from edges or changes in surface type.
- **Typical Sensors & Siting Considerations:**
  - **3D Ultrasonic Anemometers:** Mounted at specific heights (e.g., above canopy height for forests) on sturdy towers to measure turbulent wind components.
  - **Fast-Response Infrared Gas Analyzers:** Paired with sonic anemometers to measure high-frequency fluctuations of CO_2 and water vapor.
  - **Net Radiometers:** Positioned above the canopy to measure all incoming and outgoing radiation components.
  - **Soil Heat Flux Plates:** Buried in the soil to measure heat transfer into the ground.
  - **Multiple Temperature/Humidity Sensors:** At various heights on the tower to profile the atmospheric boundary layer.
- **Crucial Siting Principles Across All Station Types**
- **Representativeness:** The primary goal is for the measured data to be representative of the desired area (e.g., a city, a rural region, a specific field), not just the immediate point of measurement.
- **Exposure:** Instruments must have proper exposure to the element they are measuring, free from artificial obstructions or influences.
- **Standardization:** Adherence to WMO and national meteorological agency standards is essential for data comparability over time and across different locations.
- **Long-Term Stability:** The site should ideally be stable over long periods, minimizing changes in the surrounding environment that could affect observations (e.g., new buildings, tree growth).
- **Maintenance Access:** Sites must be accessible for regular maintenance, calibration, and troubleshooting of sensors.

Proper siting, coupled with accurate and well-maintained sensors, is fundamental to acquiring high-quality meteorological data, which in turn supports reliable weather forecasts, robust climate studies, and effective environmental management.

3.2.8 Data Transmission

Once meteorological data is collected by the sensors, its utility hinges on efficient and reliable transmission from the station to central processing facilities, forecasting centers, and data archives. This process is particularly critical for automated weather stations (AWS) that operate continuously and often in remote locations.

- **Primary Purpose:** To move observed data from the point of collection to where it can be processed, analyzed, disseminated, and assimilated into numerical models in a timely manner.
- **Key Components**
  - **Data Logger:** The central unit at the station that collects raw data from all connected sensors, performs initial processing (e.g., averaging, unit conversion), and stores the data temporarily.
  - **Communication Module (Modem/Transceiver):** Responsible for sending the processed data over a chosen communication medium.
  - **Antenna:** For wireless communication methods.
- **Methods of Transmission**
  - **Wired Connections (Ethernet, Fiber Optic)**
  - **Application:** Primarily used for stations located within a local area network (LAN), such as at airports, research campuses, or urban meteorological offices where direct cable infrastructure is feasible.
  - **Advantages:** High bandwidth, very reliable, low latency, secure.
  - **Disadvantages:** Limited by physical cable infrastructure, not suitable for remote or mobile stations.
- **Cellular Networks (GSM/3G/4G/5G)**
  - **Application:** Widely used for AWS in areas with cellular coverage. Data is transmitted via mobile data services.
  - **Advantages:** Relatively high bandwidth, widespread coverage in populated areas, cost-effective for moderate data volumes.
  - **Disadvantages:** Requires cellular signal, can be affected by network congestion, ongoing subscription costs.

- **Satellite Communication**
  - **Application:** Essential for remote or inaccessible locations (e.g., polar regions, oceans, mountains) where cellular or wired connections are unavailable (Fig. 3.2).

Fig. 3.2: Satellite communication

- **Types**
  - **Geostationary Operational Environmental Satellite (GOES) Meteorological Satellites:** Dedicated meteorological satellites that can receive data from ground-based Data Collection Platforms (DCPs).
  - **Low Earth Orbit (LEO) Satellites (e.g., Iridium, Argos, Starlink):** Provide global coverage, often used for buoys, remote AWS, and research stations.
  - **Advantages:** Global coverage, highly reliable for remote sites.
  - **Disadvantages:** Higher cost, potentially lower bandwidth, higher latency compared to cellular/wired.
- **Radio (VHF/UHF)**
  - **Application:** Used for short-to-medium range transmission, often for local networks (e.g., a network of stations around an airport or a research site) or for line-of-sight communication.
  - **Advantages:** Can be cost-effective for local networks, independent of public infrastructure.
  - **Disadvantages:** Limited range, requires line-of-sight, susceptible to interference.
- **LoRa/LoRaWAN (Long Range Wide Area Network)**
  - **Application:** Emerging technology for low-power, long-range Internet of Things (IoT) applications, suitable for transmitting small packets of meteorological data from distributed sensors.

  - **Advantages:** Extremely low power consumption (enabling long battery life), long range (several kilometers), low cost.
  - **Disadvantages:** Low bandwidth, not suitable for high-frequency or large data volumes.
- **Data Formats and Protocols**
  - **WMO Codes:** For international exchange, data is often encoded into standardized WMO formats like SYNOP (for surface observations) or METAR (for aviation routine weather reports).
  - **Common Data Formats:** Data loggers often output data in simple text formats (CSV), XML, or JSON for easier parsing and integration into databases.
  - **Communication Protocols:** Standard internet protocols (TCP/IP, FTP, HTTP/HTTPS) are used over wired and cellular networks. Satellite systems use their own proprietary protocols.
- **Challenges in Data Transmission**
  - **Power Consumption:** Especially critical for remote AWS relying on solar panels and batteries. Efficient communication modules are essential.
  - **Reliability in Harsh Environments:** Equipment must withstand extreme temperatures, moisture, wind, and lightning.
  - **Data Security:** Ensuring the integrity and confidentiality of transmitted data.
  - **Latency:** The time delay between observation and data reception, which is critical for real-time forecasting and warnings.
  - **Bandwidth:** Sufficient bandwidth is needed for the volume and frequency of data transmission.

The robust and timely transmission of observed data is the final, crucial step in the *in situ* observation process, transforming raw measurements into actionable information for meteorologists, researchers, and the public.

3.2.9 Importance and Applications

The observations from surface weather stations are indispensable for:

- **Local Weather Forecasting:** Providing real-time conditions that forecasters use to issue warnings, update short-range forecasts, and understand mesoscale weather phenomena.
- **Climate Monitoring:** Long-term records from these stations are crucial for tracking climate change, identifying trends, and assessing climate variability.

- **Aviation and Transportation:** Essential for flight planning, air traffic control, and road safety, providing critical information on visibility, wind, and precipitation.
- **Agriculture:** Data on temperature, precipitation, and soil moisture helps farmers make informed decisions about planting, irrigation, and harvesting.
- **Hydrology:** Monitoring precipitation and temperature influences water resource management, flood forecasting, and drought assessment.
- **Research:** Providing high-resolution data for validating atmospheric models, studying boundary layer processes, and understanding local climate interactions.

In essence, surface weather stations provide the foundational, highly localized data that integrates with broader remote sensing observations and numerical models to create a comprehensive picture of the Earth's atmosphere.

3.3 Upper Air Observations: Probing the Third Dimension of the Atmosphere

While surface weather stations provide invaluable data at ground level, the atmosphere is a dynamic, three-dimensional system where critical processes unfold high above our heads. To truly understand, predict, and model weather and climate, meteorologists must obtain detailed information about the vertical structure of the atmosphere. This is the domain of **upper air observations**, also known as **atmospheric soundings** or **upper-air soundings**, which involve direct, *in situ* measurements of meteorological parameters throughout the troposphere and into the stratosphere. This vertical profile data is not merely supplementary; it is indispensable for initializing numerical weather prediction models, diagnosing atmospheric stability, forecasting severe weather, and advancing atmospheric research.

3.3.1 The Imperative of Vertical Profiles

The atmosphere's behavior is fundamentally governed by how its properties change with height. Upper air observations are critical because they reveal:

- **Vertical Profiles of Key Parameters:** They provide precise measurements of temperature, humidity, pressure, and wind speed and direction at successive altitudes. This allows for the identification of distinct atmospheric layers, temperature inversions (where temperature increases with height), and regions of varying moisture content.
- **Atmospheric Stability:** The vertical temperature lapse rate (how temperature changes with height) directly determines atmospheric

stability. This dictates whether air parcels will rise or sink, profoundly influencing cloud formation, the potential for convection (e.g., thunderstorms), and the presence of turbulence.

- **Wind Shear:** Changes in wind speed or direction with altitude are crucial for understanding the organization and intensity of severe storms, the movement of weather systems, and significant hazards to aviation.
- **Moisture Transport and Distribution:** The vertical distribution of water vapor is a primary factor in cloud development, precipitation potential, and the overall energy budget of the atmosphere.
- **Initialization and Validation of Numerical Models:** Upper air data provides the essential three-dimensional "snapshot" of the atmosphere that forms the initial conditions for global and regional numerical weather prediction (NWP) models. Without these soundings, models would lack the necessary vertical detail to accurately simulate atmospheric evolution. They also serve as critical validation data for model outputs.

3.3.2 Primary Methods and Platforms for Upper Air Observations

The global upper-air observing network relies on a combination of ground-based launches and specialized mobile platforms.

- **Radiosondes and Rawinsondes (Weather Balloons):** These are the cornerstone of the global upper-air observation system, providing the most comprehensive and direct vertical profiles.
- **Radiosonde Operation:** A radiosonde is a compact, lightweight, and expendable instrument package containing highly sensitive sensors for temperature, relative humidity, and atmospheric pressure. It is attached to a large, unpiloted balloon, typically inflated with hydrogen or helium, and launched from a ground station. As the balloon ascends through the atmosphere (at an average rate of 5-6 meters per second, or about 1,000 feet per minute), the radiosonde continuously measures these parameters. The data is digitized and transmitted via a radio telemetry link (typically in the UHF or VHF band) back to a ground receiving station. The balloon continues to rise until it reaches an altitude where the atmospheric pressure is too low for the balloon material to withstand the internal pressure, causing it to burst (typically between 20-35 km or 65,000-115,000 feet). The radiosonde then descends slowly by a small parachute.
- **Rawinsonde Enhancement:** A rawinsonde is a radiosonde that is additionally tracked by ground-based radar, GPS, or other navigation systems as it ascends. By precisely tracking the sonde's horizontal

position over time, the system can calculate the **wind speed and direction** at various altitudes. Most modern radiosonde launches incorporate GPS tracking, making them rawinsonondes.

- **Global Network and Frequency:** Over 800 meteorological stations worldwide launch radiosondes, typically twice daily at internationally coordinated times (0000 UTC and 1200 UTC). This synchronized global network provides a consistent dataset for NWP models.
- **Output:** The raw data from a sounding is processed and often visualized on a thermodynamic diagram, such as a **Skew-T Log-P diagram**, which allows meteorologists to quickly assess atmospheric stability, identify inversions, and determine the potential for convection.
- **Siting Requirements (Launch Sites):** Radiosonde launch sites require a substantial, open, and level area free from significant obstructions (buildings, trees, power lines) to ensure the balloon's unobstructed ascent. Space is also needed for the inflation shelter, instrument preparation, and ground tracking equipment. Proximity to air traffic control is a consideration due to the balloon's flight path.
- **Sensors**
    - **Temperature:** Highly accurate thermistors or platinum Resistance Temperature Detectors (RTDs) for rapid response to changing conditions.
    - **Humidity:** Capacitive humidity sensors (polymer-based) or thin-film polymer sensors that change electrical properties based on water vapor absorption.
    - **Pressure:** Electronic piezoresistive barometers that measure pressure changes with extreme precision.
    - **Wind (Rawinsonde):** Integrated GPS receiver within the radiosonde package, which transmits its precise latitude, longitude, and altitude, allowing for the calculation of horizontal wind vectors by the ground station. Older systems used ground-based radar or radio direction finders to track a reflector or signal from the sonde.
- **Dropsondes**

  Dropsondes are specialized radiosondes deployed from an aircraft, providing vital vertical profiles in regions inaccessible to ground launches.

    - **Application:** Primarily used by "hurricane hunter" aircraft to probe the inner structure of tropical cyclones, providing critical data on wind, temperature, and moisture within the storm's core. They are also deployed during research campaigns over oceans or other data-sparse areas where ground-based soundings are not feasible.

- **Operation:** The dropsonde is released from the aircraft and descends by a small parachute. As it falls, it continuously measures and transmits data back to the deploying aircraft. The aircraft then relays this data to ground stations.
- **Sensors:** Similar to radiosondes, dropsondes measure temperature, humidity, pressure, and GPS-derived wind profiles during their descent.

- **Research Aircraft**

  Specially instrumented aircraft serve as mobile platforms for highly detailed and targeted *in situ* upper air observations.

  - **Application:** Ideal for studying specific atmospheric phenomena (e.g., cloud formation, atmospheric chemistry, severe storm dynamics, boundary layer processes, air pollution plumes). They offer the flexibility to fly specific patterns, target areas of interest, and collect data at very high spatial and temporal resolutions.
  - **Sensors:** Research aircraft carry an extensive suite of sophisticated, fast-response sensors mounted externally and internally:
  - **Standard Meteorological Instruments:** Highly accurate temperature probes (e.g., platinum resistance thermometers, thermistors in specialized housings), fast-response humidity sensors (e.g., chilled mirror hygrometers, Lyman-alpha hygrometers), and precision pressure transducers.
  - **Wind:** Multi-hole probes (e.g., five-hole probes) for measuring true airspeed and angles of attack/sideslip, combined with inertial navigation systems (INS) and GPS to calculate precise 3D wind vectors.
  - **Cloud Microphysics:** Optical array probes (e.g., FSSP, 2DC, 2DP) to measure the size, shape, and concentration of cloud droplets and ice crystals; hot-wire probes for liquid water content.
  - **Radiation:** Upward and downward looking radiometers for various wavelengths.
  - **Trace Gases & Aerosols:** Specialized instruments for measuring concentrations of atmospheric pollutants (e.g., ozone, CO, NOx), greenhouse gases (CO_2, CH_4), and aerosol properties (size distribution, chemical composition).

- **Siting:** As mobile platforms, they don't have fixed "siting" in the traditional sense. However, flight plans are meticulously designed to achieve research objectives while adhering to air traffic control regulations and safety protocols.
- **Complementary Upper Air Observation Techniques (Remote Sensing):** While the focus here is on *in situ* observations, it's important to note that other technologies provide crucial upper air data through remote sensing:
  - **Wind Profilers:** Ground-based Doppler radars that transmit pulses of electromagnetic energy vertically and at slight angles. By detecting the faint backscatter from atmospheric turbulence and measuring the Doppler shift of the returned signal, they continuously calculate wind speed and direction at various altitudes (typically up to 10-16 km). They offer excellent temporal resolution (e.g., every 10-30 minutes) but generally do not measure temperature or humidity.
  - **Lidars (Light Detection and Ranging):** Use laser pulses to measure profiles of aerosols, clouds, water vapor, temperature, and even wind (Doppler Lidars) by analyzing the scattered light.
  - **Weather Radars (Doppler Radar):** While primarily used for detecting precipitation, advanced Doppler weather radars can also infer wind speed and direction within precipitation areas, providing valuable wind shear information.
  - **Satellite Sounders:** Instruments on meteorological satellites (e.g., infrared and microwave sounders) measure radiation emitted or absorbed by atmospheric gases at different wavelengths. This data can be inverted to derive temperature and humidity profiles, especially crucial over data-sparse oceanic regions. While not direct *in situ* measurements, they provide global coverage and complement the *in situ* network.
- **Data Transmission and Dissemination**

  For radiosondes and dropsondes, the collected data is transmitted wirelessly to a ground station or aircraft. Once received, the data undergoes quality control checks and is then encoded into standardized formats (e.g., WMO TEMP code, BUFR) for efficient transmission. This data is rapidly disseminated globally via the **WMO Global Telecommunication System (GTS)**, making it available to all national meteorological services and research institutions for immediate use in forecasting and modeling. Data from research aircraft is often recorded onboard for post-flight analysis, but some real-time transmission capabilities exist for critical parameters.

- **Importance and Applications**

  Upper air observations are absolutely fundamental for

  - **Numerical Weather Prediction (NWP):** They provide the high-resolution, three-dimensional initial conditions that are critical for running sophisticated atmospheric models, leading to more accurate forecasts of temperature, precipitation, and severe weather events.
  - **Severe Weather Forecasting:** Meteorologists use sounding data to diagnose atmospheric instability (e.g., Convective Available Potential Energy - CAPE), identify capping inversions, assess wind shear, and determine the potential for thunderstorms, tornadoes, and heavy rainfall.
  - **Aviation Meteorology:** Crucial for flight planning, providing information on winds aloft, jet streams, turbulence, and potential icing conditions.
  - **Climate Monitoring and Research:** Long-term records of upper air conditions are essential for tracking climate variability, detecting trends in atmospheric temperature and moisture, and validating climate models.
  - **Atmospheric Research:** Provides the detailed observations needed to understand fundamental atmospheric processes, from boundary layer dynamics to stratospheric chemistry.

In essence, upper air observations provide the critical vertical dimension to our understanding of the atmosphere, synergistically combining with surface observations and satellite remote sensing to create a comprehensive, real-time picture of our planet's weather and climate system.

3.4 Specialized In Situ Observation Systems: Unveiling Atmospheric Intricacies

While the global networks of surface weather stations and routine radiosonde launches provide the foundational, synoptic-scale view of the atmosphere, many critical meteorological phenomena occur at finer scales, in challenging environments, or require highly specific measurements that standard systems cannot provide. This is where **specialized in situ observation systems** come into play. These advanced platforms are meticulously designed and instrumented to target particular atmospheric processes, monitor unique environments, or collect exceptionally high-resolution data, offering unparalleled insights into the atmosphere's most complex behaviors.

These systems are often at the forefront of meteorological research, pushing the boundaries of our understanding and significantly enhancing our ability to forecast and mitigate the impacts of severe weather and climate change.

3.4.1 Weather Reconnaissance Aircraft (e.g., «Hurricane Hunters»)

These are arguably the most iconic specialized *in situ* platforms, dedicated to flying directly into the heart of hazardous weather, particularly tropical cyclones (hurricanes, typhoons, cyclones).

- **Primary Purpose:** To provide real-time, high-resolution, direct measurements from within the core of tropical cyclones, which is crucial for improving forecasts of storm intensity, track, and structure. They are operated by agencies like the US Air Force Reserve's 53rd Weather Reconnaissance Squadron and NOAA's Hurricane Hunters.
- **Operation:** These robust, specially modified aircraft (e.g., WC-130J Super Hercules, WP-3D Orion) fly pre-determined patterns through the storm, including multiple penetrations of the eyewall and eye. They are equipped to withstand extreme turbulence and high winds.
- **Unique Advantages**
    - **Direct Core Penetration:** They are the only platforms capable of obtaining direct, *in situ* measurements from inside the most dangerous and critical parts of a hurricane, where satellite data can be obscured.
    - **Targeted Sampling:** Flight paths can be adjusted dynamically based on real-time data to target specific features like developing eyewalls or areas of rapid intensification.
    - **Real-time Operational Data:** The data collected is immediately transmitted to forecasting centers, providing vital input for urgent advisories and warnings, significantly impacting decisions related to evacuations and disaster preparedness.
- **Specialized Sensors & Siting/Deployment**
    - **Dropsondes:** The cornerstone of hurricane reconnaissance. These are sophisticated, parachute-borne instrument packages released from the aircraft. As they descend from flight level (e.g., 10,000 feet) to the ocean surface, they continuously measure and transmit vertical profiles of temperature, humidity, pressure, and GPS-derived wind speed and direction. Strategic deployment of dropsondes provides a 3D picture of the storm's inner core.

- **Flight-Level Sensors:** Standard meteorological sensors (temperature, humidity, pressure, wind) are mounted on the aircraft's exterior to provide continuous measurements at flight altitude.
- **Turbulence Probes:** Specialized sensors measure atmospheric turbulence experienced by the aircraft, providing insights into the storm's energy and structure.
- **Onboard Remote Sensing:** Often complemented by advanced Doppler radars (e.g., lower fuselage radar for precipitation, tail Doppler radar for detailed wind fields) that provides a broader context for the *in situ* measurements.
- **Siting:** No fixed "siting." Deployment is highly dynamic, driven by the location, movement, and intensity of the tropical cyclone.

3.4.2 Research Aircraft (Dedicated Scientific Platforms)

Distinct from reconnaissance aircraft, these are highly instrumented planes (e.g., NOAA P-3 Orion, NSF/NCAR C-130, NASA ER-2) used exclusively for scientific investigations across a vast range of atmospheric phenomena.

- **Primary Purpose:** To conduct detailed, high-resolution, and often multi-parameter *in situ* measurements for fundamental scientific research, addressing specific hypotheses about atmospheric processes.
- **Operation:** Flights are meticulously planned based on scientific objectives, often involving complex flight patterns (e.g., spirals, transects, formation flying with other aircraft) to sample specific atmospheric features or gradients. They can fly at various altitudes, from the boundary layer to the lower stratosphere.
- **Unique Advantages**
  - **Unparalleled Flexibility and Mobility:** Can access virtually any atmospheric region or phenomenon, including remote areas, specific cloud types, or pollution plumes.
  - **High Spatial and Temporal Resolution:** Capable of collecting data at very high frequencies (e.g., hundreds of samples per second) along the flight path, capturing fine-scale variations.
  - **Comprehensive Multi-Parameter Measurement:** Can carry a vast array of sophisticated instruments simultaneously, allowing for the co-located measurement of physical, chemical, and microphysical properties.

- **Specialized Sensors & Siting/Deployment**
- **High-Precision Meteorological Sensors:** Beyond standard measurements, these include fast-response temperature probes (e.g., platinum resistance thermometers with specialized housings to minimize heating), highly accurate humidity sensors (e.g., chilled-mirror hygrometers for dew point, Lyman-alpha hygrometers for fast response), and precise pressure transducers.
- **3D Wind Measurement Systems:** Advanced multi-hole probes (e.g., five-hole probes) measure true airspeed and angles of attack/sideslip, combined with highly accurate Inertial Navigation Systems (INS) and GPS to derive precise 3D wind vectors, essential for flux calculations.
- **Cloud Microphysics Probes:** A diverse suite of optical array probes (e.g., FSSP, CDP, 2DC, 2DP) to measure the size distribution, number concentration, and phase (liquid, ice, mixed) of cloud droplets and ice crystals. Hot-wire probes measure liquid water content.
- **Atmospheric Chemistry Instruments:** Highly sensitive gas analyzers for a wide range of trace gases (e.g., ozone, carbon monoxide, nitrogen oxides, sulfur dioxide, carbon dioxide, methane, volatile organic compounds), and aerosol spectrometers for particle size distribution and chemical composition.
- **Radiation Sensors:** Upward and downward looking radiometers for various spectral bands to study radiative transfer.
- **Siting:** No fixed siting; operations are based out of research airfields, with flight plans designed to meet scientific objectives within air traffic control regulations.

3.4.3 Uncrewed Aerial Systems (UAS) / Drones

UAS represent a rapidly evolving and transformative technology for *in situ* atmospheric observations, offering unprecedented flexibility and access to the lower atmosphere.

- **Primary Purpose:** To provide flexible, cost-effective, and often high-resolution atmospheric profiling and targeted measurements, particularly within the atmospheric boundary layer (ABL), urban environments, or for specific local phenomena (e.g., fog, low-level jets).
- **Operation:** Operated remotely by pilots, UAS can be programmed to fly specific vertical profiles (soundings), horizontal transects, or hover at specific altitudes. Their small size and maneuverability allow access to areas difficult or dangerous for manned aircraft.

- **Unique Advantages**
  - **Cost-Effectiveness:** Significantly cheaper to acquire and operate than manned aircraft, making them accessible for a wider range of research and operational uses.
  - **Flexibility and Rapid Deployment:** Can be deployed quickly and repeatedly from various locations, ideal for capturing transient phenomena.
  - **Access to Challenging Environments:** Can safely probe hazardous areas (e.g., volcanic plumes, immediate vicinity of wildfires, urban canyons, near-storm environments) or areas with complex terrain.
  - **Fine-Scale Resolution:** Their ability to fly slowly or hover allows for very high spatial and temporal resolution measurements.
- **Specialized Sensors & Siting/Deployment**
  - **Miniaturized Meteorological Sensors:** Custom-designed, lightweight versions of temperature (thermistors), humidity (capacitive), pressure (MEMS barometers), and wind (miniature sonic anemometers or pitot tubes).
  - **Air Quality Sensors:** Increasingly, miniaturized sensors for various pollutants (e.g., CO2, O3, PM2.5) are integrated for urban air quality mapping or plume tracking.
  - **Thermal Cameras:** Used for high-resolution mapping of surface temperatures, aiding in urban heat island studies.
  - **Small Lidars/Sodars:** Some larger UAS can carry compact remote sensing payloads for wind or aerosol profiling.
  - **Siting/Deployment:** Requires a clear launch and landing area. Operations are heavily governed by aviation regulations (e.g., FAA rules for line-of-sight, altitude limits, airspace restrictions), and careful consideration of local airspace and potential interference is crucial.

3.4.4 Instrumented Towers (Flux Towers, Communication Towers)

Tall towers provide continuous, multi-level, high-frequency *in situ* measurements, particularly valuable for studying the atmospheric boundary layer and land-atmosphere interactions.

- **Primary Purpose:** To measure vertical profiles of meteorological parameters and, critically, **turbulent fluxes** (the transfer of momentum, heat, water vapor, and trace gases like CO_2) between the Earth's surface and the atmosphere. This is fundamental for understanding energy and carbon cycles.

- **Operation:** Sensors are mounted at multiple levels along the tower's height, from just above the surface to several tens or hundreds of meters. Data is collected continuously and at very high frequencies (e.g., 10-20 Hz or more for turbulence measurements) to resolve rapid atmospheric fluctuations.
- **Unique Advantages**
  - **Oceanic Coverage:** Provide the only regular, direct *in situ* observations from vast stretches of the world's oceans, filling critical data gaps.
  - **Air-Sea Interface Data:** Measure parameters at the crucial boundary where the atmosphere and ocean interact, driving global weather and climate patterns.
  - **Long-Term Monitoring:** Many buoys are deployed for years, providing continuous time series data.
- **Specialized Sensors & Siting/Deployment**
  - **Standard AWS Sensors (Marine-Hardened):** Wind speed/direction, air temperature, relative humidity, and barometric pressure, all specifically designed to withstand the harsh marine environment (salt spray, corrosion, strong winds, wave motion).
  - **Sea Surface Temperature (SST) Sensors:** Thermistors or other probes specifically designed for accurate water temperature measurement at the surface.
  - **Wave Sensors:** Measure wave height, period, and direction using accelerometers or pressure transducers.
  - **Ocean Current Meters:** Measure current speed and direction at various depths.
  - **Salinity Sensors:** To measure the salt content of the seawater.
  - **Subsurface Sensors:** Temperature chains, pressure sensors, and other oceanographic instruments extending down into the water column.
  - **Siting:** Deployed in strategic locations to represent specific ocean basins, major shipping lanes, areas of significant air-sea interaction, or for specific research arrays (e.g., TAO/TRITON array in the Pacific for ENSO). Moorings must be incredibly robust to withstand severe storms and strong currents.

3.4.5 Mobile Meteorological Platforms (e.g., Storm Chaser Vehicles, Mobile Mesonets)

These are ground-based vehicles equipped with meteorological instrumentation, designed for rapid deployment and targeted observations of localized or transient weather phenomena.

- **Primary Purpose:** To collect high-resolution, localized *in situ* data within or immediately adjacent to severe weather events (e.g., tornadoes, supercells, winter storms) to understand their fine-scale structure, evolution, and interaction with the surface.
- **Operation:** Vehicles (often ruggedized or armored) drive into proximity of storms, deploying portable sensors or taking measurements from onboard equipment. They are a critical component of field campaigns like VORTEX-SE.
- **Unique Advantages**
    - **Extreme Proximity:** Can obtain measurements very close to or even within the immediate environment of hazardous weather phenomena, providing data that no other platform can.
    - **Rapid, Flexible Deployment:** Can respond quickly to developing situations and move with the storm, capturing its evolution.
    - **Ground-Truth for Remote Sensing:** Provide crucial ground-level data to validate and complement mobile Doppler radar observations and satellite imagery.
- **Specialized Sensors & Siting/Deployment**
    - **Mobile Mesonet Stations:** Portable meteorological stations deployed from the vehicle, often on extendable masts or tripods, to measure surface temperature, humidity, pressure, and wind in the immediate vicinity of a storm's inflow or outflow. Multiple vehicles can form a dense temporary network.
    - **Vehicle-Mounted Sensors:** Instruments on the vehicle's roof or extendable masts for continuous measurements while driving through different parts of a storm's environment.
    - **Rapid-Scan Radars:** Some vehicles carry small, rapidly scanning Doppler radars to complement the *in situ* measurements with remote sensing data.
    - **Siting/Deployment:** Highly dynamic and opportunistic, driven by the location and movement of the weather phenomenon. Safety of personnel is the paramount concern, requiring extensive training and real-time decision-making.

- These specialized *in situ* systems, often operating in concert with advanced remote sensing technologies, are indispensable for pushing the frontiers of meteorological understanding. They provide the unique, high-resolution, and targeted data necessary to unravel complex atmospheric processes, improve the accuracy of severe weather forecasts, and enhance our ability to predict and adapt to a changing climate.

3.5 Challenges and Limitations of In Situ Meteorological Data

While *in situ* observations form the bedrock of meteorological understanding, providing direct, high-fidelity measurements of atmospheric parameters, they are not without their inherent challenges and limitations. These issues can impact data quality, spatial and temporal coverage, and the overall utility of the observations for forecasting, research, and climate monitoring. Understanding these constraints is crucial for interpreting meteorological data accurately and for guiding the development of future observing systems.

3.5.1 Spatial Representativeness and Coverage

In situ observations are, by definition, **point measurements**. This fundamental characteristic leads to several spatial limitations:

- **Limited Spatial Coverage:** Even with extensive networks, individual stations provide data only for their immediate vicinity. Vast areas of the globe, particularly oceans, deserts, mountainous regions, and developing countries, remain data-sparse. This lack of coverage makes it challenging to accurately capture weather phenomena that span large regions or to track systems as they move into unobserved areas.
- **Difficulty in Capturing Mesoscale and Microscale Variability:** Weather phenomena often exhibit significant variability over short distances (mesoscale, e.g., thunderstorms, sea breezes) and very short distances (microscale, e.g., turbulence, urban heat islands). A sparse network of point observations struggles to resolve these fine-scale features, potentially missing crucial details that impact local forecasts or process understanding.
- **Representativeness of Site:** Even with strict siting guidelines (as discussed previously), no single point measurement can perfectly represent a larger area. Local topography, vegetation, nearby buildings, or even the type of ground cover immediately surrounding the station can create microclimates that differ from the broader regional conditions. This makes it challenging to generalize point observations to a wider domain without careful consideration.

3.5.2 Temporal Resolution and Continuity

The frequency and consistency of *in situ* observations also pose limitations

- **Discontinuous Observations:** Some critical *in situ* data, such as radiosonde launches, are typically performed only twice daily (0000 UTC and 1200 UTC) at most stations. This leaves significant temporal gaps, meaning that important atmospheric changes occurring between soundings might be missed. While AWS provide continuous data, their spatial density is still limited.
- **Gaps Due to Maintenance and Failure:** Instruments require regular maintenance, calibration, and can fail due to environmental factors or technical malfunctions. During these periods, data collection ceases, leading to gaps in the time series.
- **Challenges in Long-Term Homogeneity for Climate Studies:** For climate monitoring, consistent long-term records are paramount. However, over decades, changes in instrument technology, sensor types, manufacturers, calibration procedures, or even the immediate environment around a station (e.g., urban growth) can introduce artificial trends or biases into the data. This "inhomogeneity" requires complex statistical homogenization techniques to produce reliable climate records.

3.5.3 Logistical and Operational Challenges

Operating and maintaining *in situ* observing networks is a complex and resource-intensive endeavor:

- **High Cost:** The acquisition, installation, and ongoing maintenance of meteorological instruments, especially specialized platforms like research aircraft, buoys, or flux towers, are very expensive. This includes the cost of sensors, data loggers, communication systems, power infrastructure, and personnel.
- **Maintenance and Calibration:** Instruments degrade over time due to exposure to harsh weather elements (UV radiation, dust, salt, moisture). Regular calibration is essential to maintain accuracy, which can be labor-intensive, costly, and difficult in remote or inaccessible locations.
- **Power Requirements:** Automated weather stations and other electronic systems require a reliable power source. In remote areas, this often necessitates solar panels with battery backups, which can be bulky, expensive, and susceptible to weather conditions (e.g., prolonged cloudiness impacting solar charging).

- **Personnel Requirements:** Even for automated stations, trained personnel are needed for installation, troubleshooting, routine maintenance, data quality control, and, for manned stations, for manual observations.
- **Safety Concerns:** Deploying instruments in hazardous environments (e.g., storm chasing, hurricane reconnaissance flights, high-altitude balloon launches) or remote, rugged terrain poses significant safety risks to personnel.
- **Security and Vandalism:** Ground-based stations, particularly those in accessible or public areas, can be vulnerable to vandalism or theft, leading to data loss and costly repairs.

3.5.4 Environmental Factors and Instrument-Specific Limitations

The very environment being measured can negatively impact the instruments themselves:

- **Sensor Degradation and Drift:** Prolonged exposure to sunlight, extreme temperatures, humidity, dust, salt, and pollutants can cause sensors to degrade, leading to a gradual loss of accuracy (drift) over time. For example, humidity sensors can be particularly sensitive to contamination.
- **Icing:** In cold, moist conditions, instruments (especially wind sensors, temperature probes, and precipitation gauges) can become covered in ice. This can lead to erroneous readings, sensor damage, or complete data loss until the ice melts or is removed.
- **Precipitation Effects:** Rain and snow can directly impact sensor performance. For instance, precipitation on temperature or humidity sensors can cause evaporative cooling or wetting biases. Wind sensors can be affected by the physical impact of heavy rain or snow accumulation.
- **Radiation Errors:** Even with radiation shields, solar radiation can heat the instrument housing or the sensor itself, causing artificially high temperature readings, especially during calm, sunny conditions. Similarly, longwave radiation from the ground can affect nighttime temperature measurements.
- **Lag Errors (Response Time):** All sensors have a certain response time, meaning they don't instantly react to rapid changes in atmospheric conditions. This can lead to "lag errors," where the sensor underestimates the true variability or peaks/troughs of rapidly changing parameters (e.g., during turbulence or when a radiosonde passes through a sharp inversion).

- **Mechanical Wear and Tear:** Instruments with moving parts (e.g., cup anemometers, wind vanes, tipping bucket rain gauges) are susceptible to mechanical wear, friction, and damage from strong winds or debris.

3.5.5 Data Quality Control and Processing

Ensuring the accuracy and usability of *in situ* data requires extensive post-collection effort:

- **Rigorous Quality Control (QC):** Raw data from sensors often contains errors, outliers, or biases. Comprehensive automated and manual quality control procedures are necessary to identify and flag or correct these issues before the data is used. This involves checking for physical plausibility, consistency with other parameters, and adherence to expected ranges.
- **Data Formatting and Standardization:** For global exchange and assimilation into models, *in situ* data must be converted into standardized formats (e.g., WMO SYNOP, METAR, TEMP, BUFR). This conversion process requires careful handling to avoid errors.
- **Data Archiving and Accessibility:** Long-term storage and easy accessibility of vast amounts of *in situ* data are crucial for climate research and historical analysis. This requires robust data management systems.

Despite these challenges, *in situ* data remains irreplaceable. Its direct nature provides the ground truth necessary for calibrating and validating remote sensing observations, initializing and evaluating numerical weather prediction models, and understanding the fundamental physical processes that govern our atmosphere. Addressing these limitations through technological advancements, improved network design, and robust data management practices is an ongoing effort in meteorological science.

3.6 Conclusion

This chapter has explored the indispensable role of **in situ observations** in meteorology, delving into their fundamental nature as direct, on-site measurements of atmospheric parameters. We have seen how these observations, ranging from the comprehensive networks of surface weather stations providing ground-level insights to the vertical profiles obtained from radiosondes and specialized platforms like weather reconnaissance aircraft and flux towers, collectively build a detailed, three-dimensional picture of the atmosphere. Each system, with its unique suite of sensors and specific siting requirements, contributes vital pieces of information that are unattainable through other means.

Despite the inherent challenges and limitations—such as the point-based nature leading to spatial gaps, the logistical complexities of deployment and maintenance, and the susceptibility of sensors to environmental factors—*in situ* data remains absolutely critical. It serves as the essential "ground truth" for calibrating and validating remote sensing technologies, provides the crucial initial conditions for numerical weather prediction models, and offers the high-resolution detail necessary for understanding complex atmospheric processes. As meteorological science continues to advance, the synergy between *in situ* observations and remote sensing techniques will only deepen, with each complementing the other to provide an ever more complete and accurate understanding of our dynamic atmosphere. The ongoing evolution of sensor technology, communication systems, and the integration of novel platforms like UAS (uncrewed areal systems) ensures that *in situ* observations will continue to be an indispensable cornerstone of weather forecasting, climate monitoring, and atmospheric research for the foreseeable future.

4

Remote Sensing from Ground Based Platforms

4.1 Introduction

Having explored the direct, "hands-on" approach of *in situ* observations in the previous chapter, we now turn our attention to **remote sensing**. In contrast to *in situ* methods, remote sensing allows us to gather information about the atmosphere without physically touching it. Instead, it relies on the detection of electromagnetic radiation (or acoustic waves) that has been emitted, reflected, or scattered by the atmosphere or its constituents.

This chapter will focus specifically on **ground-based remote sensing platforms**. Positioned on or near the Earth's surface, these instruments offer unique advantages for continuous monitoring of the atmosphere above a fixed location, providing valuable insights into vertical profiles, temporal evolution, and localized phenomena. From radars that peer into storms to lidars that probe aerosols and winds, and radiometers that measure emitted radiation, ground-based remote sensing provides a vital link between highly localized *in situ* measurements and the broader spatial coverage offered by satellite and airborne systems.

We will delve into the fundamental principles behind several key ground-based remote sensing instruments used in meteorology. This will include an examination of how they interact with the atmosphere, the types of atmospheric parameters they can measure (such as precipitation, wind, temperature, humidity, aerosols, and trace gases), and the strengths and limitations of each technique. By understanding the capabilities of these ground-based "eyes" on the sky, we can better appreciate their contribution to weather forecasting, atmospheric research, and environmental monitoring.

4.2 Weather and Doppler Radar

Weather radar, particularly **Doppler weather radar**, stands as a pivotal ground-based remote sensing tool in modern meteorology. These active systems emit pulses of electromagnetic energy into the atmosphere and analyze the backscattered signals from hydrometeors (rain, snow, hail, drizzle) and other atmospheric targets (insects, dust). This analysis provides crucial insights into precipitation characteristics and, importantly with Doppler capability, the motion within the atmosphere (Fig. 4.1).

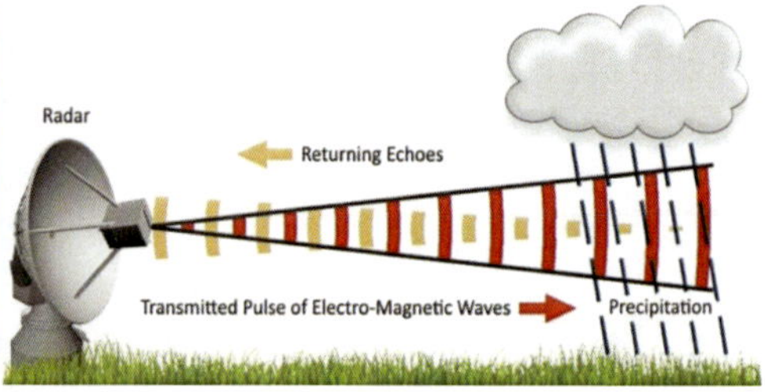

Fig. 4.1: Doppler weather radar system

4.2.1 Fundamental Principles

The operation of weather radar involves these key stages:.

- **Transmission:** The radar transmits short pulses of microwave energy (typically S-band or C-band) via a rotating antenna.
- **Interaction:** These pulses encounter atmospheric targets, and a portion of the energy is scattered back to the radar.
- **Reception:** The antenna receives these return echoes between the transmitted pulses.
- **Signal Processing:** The received signals are processed to extract:
  - **Range:** Determined by the time delay of the returning echo.
  - **Reflectivity:** The power of the returned signal, indicative of the intensity and type of targets.
  - **Doppler Velocity (Radial Velocity):** The frequency shift of the returned signal due to the motion of the targets towards or away from the radar.

The addition of **Doppler capability** is what distinguishes modern weather radar and provides the ability to measure the *velocity* of the targets along the radar's line of sight.

4.2.2 Evolution - From Conventional to Doppler and Dual-Polarization

- **Conventional (Non-Doppler) Radar:** Primarily measured reflectivity, offering information about the location and intensity of precipitation.
- **Doppler Radar:** Introduced the ability to measure the radial velocity of targets, revolutionizing the detection of wind patterns within storms, including rotation indicative of severe weather.

- **Dual-Polarization Doppler Radar:** The latest advancement, transmitting and receiving both horizontal and vertical polarized waves. This allows for the discrimination of precipitation types, estimation of rainfall rates, and the detection of non-meteorological targets with greater accuracy.

4.2.3 Information Provided by Doppler Weather Radar

Doppler weather radar provides a rich set of data products:

- **Reflectivity (Z):** (Measured in dBZ) Indicates precipitation intensity. Higher dBZ means heavier precipitation. Products include Base Reflectivity (lowest level) and Composite Reflectivity (maximum in the vertical column).
- **Radial Velocity (Vr):** (Measured in knots or m/s) Shows motion towards (negative, often green/blue) or away from (positive, often red/orange) the radar. Crucial for identifying rotation (mesocyclones) and wind shear.
- **Spectrum Width (SW):** Measures the variability of velocities within the radar sampling volume, indicating turbulence or mixed targets.

4.2.4 With Dual-Polarization Doppler Radar, additional products include

- **Differential Reflectivity (ZDR):** (Measured in dB) Indicates the difference in reflectivity between horizontal and vertical polarizations, providing information about the shape of hydrometeors (e.g., flattened raindrops have positive ZDR).
- **Correlation Coefficient (CC):** (Dimensionless, 0-1) Measures the uniformity of hydrometeors within the sampling volume. Low CC can indicate mixed precipitation types or non-meteorological targets like tornado debris.
- **Specific Differential Phase (KDP):** (Measured in degrees per kilometer) The difference in phase shift between horizontal and vertical pulses as they propagate through precipitation, related to liquid water content and useful for rainfall estimation.

4.2.5 Applications of Doppler Weather Radar

The data from Doppler weather radar is vital for:

- **Severe Weather Warnings:** Detecting rotation in thunderstorms (leading to tornado warnings), identifying hail, and tracking damaging winds. The addition of dual-polarization helps confirm the presence of tornado debris.

- **Precipitation Estimation:** Providing quantitative precipitation estimates (QPE) for hydrology and flash flood warnings. Dual-polarization improves the accuracy of these estimates by accounting for precipitation type.
- **Aviation Safety:** Detecting wind shear and microbursts near airports, as well as tracking hazardous precipitation.
- **Research:** Studying the dynamics and microphysics of storms, the structure of the atmospheric boundary layer, and other meteorological phenomena.

4.2.6. Advantages of Doppler Weather Radar

- **Large Area Coverage:** Monitors weather conditions over a wide radius.
- **Timely Updates:** Scans are typically completed every few minutes, providing near real-time information.
- **Internal Storm Structure:** Reveals details about the motion and composition within storms.
- **Remote Measurement:** Observes hazardous weather safely from a distance.
- **Enhanced Capabilities with Doppler and Dual-Pol:** Doppler allows for wind velocity measurements, and dual-polarization aids in precipitation typing and non-meteorological target identification.

4.2.7 Limitations of Doppler Weather Radar

- **Beam Blockage:** Obstructions can prevent the radar from seeing lower levels of the atmosphere.
- **Beam Broadening:** Resolution decreases with distance.
- **Earth's Curvature:** The radar beam height increases with distance, limiting low-level detection at far ranges.
- **Attenuation:** Heavy precipitation can weaken the radar signal.
- **Anomalous Propagation (AP):** Can lead to false echoes from non-precipitation targets under certain atmospheric conditions.
- **Bright Band:** Can overestimate precipitation intensity at the melting layer.
- **Ground Clutter:** Echoes from stationary objects need to be filtered.
- **Non-Meteorological Echoes:** Insects, birds, etc., can sometimes be detected.

4.2.8 Siting Considerations for Doppler Weather Radar

Optimal locations aim to:

- Minimize beam blockage.
- Provide broad, unobstructed coverage.
- Ensure accessibility for maintenance.

In conclusion, **Doppler weather radar** represents a significant advancement in our ability to remotely sense the atmosphere. By measuring not only the intensity of precipitation but also its motion, and with dual-polarization, its type and characteristics, it provides invaluable data for weather forecasting, severe weather warnings, and atmospheric research, despite its inherent limitations.

4.2.9 Data Products from Weather Radars -Base and Derived, and Their Uses

Weather radars, especially modern Doppler and dual-polarization systems, are sophisticated instruments that collect vast amounts of raw atmospheric data. This raw data is then processed into various **data products**, which are visual representations or numerical outputs that meteorologists use to interpret atmospheric conditions, diagnose weather phenomena, and issue forecasts and warnings. These products are broadly categorized into **base products** (representing the direct, fundamental measurements from the radar) and **derived products** (which are computed from the base products to provide more specific, interpretable information).

- **Base Products (Level II Data)**

  Base products are the fundamental, unprocessed measurements directly obtained from the radar's electromagnetic pulses. They represent the raw, foundational information for each elevation angle scanned by the radar.

- **Reflectivity (Z)**
  - **What it measures:** The power of the energy scattered back to the radar from targets (hydrometeors like rain, snow, hail, as well as insects, dust, or debris). It's a measure of how "reflective" these targets are.
  - **Units:** Expressed in decibels of Z (dBZ).
  - **Interpretation:** Higher dBZ values indicate a greater concentration of targets or larger targets, generally correlating with heavier precipitation. For example, light rain might be 20-30 dBZ, moderate rain 30-40 dBZ, and heavy rain or hail can exceed 50-60 dBZ.

- **Use in Meteorology**
  - **Detecting Precipitation:** The most basic use is to show where precipitation is occurring.
  - **Estimating Intensity:** Provides a first-order estimate of precipitation rate (e.g., light, moderate, heavy).
  - **Tracking Storms:** Allows forecasters to monitor the movement and evolution of rainbands, thunderstorms, and snow events.
  - **Identifying Strong Storms:** Very high reflectivity values (e.g., >50 dBZ) can indicate the presence of large hail or very heavy rainfall, prompting closer investigation for severe weather.
- **Common Displays**
  - **Base Reflectivity:** Shows reflectivity at the lowest elevation angle (e.g., 0.5 degrees), providing a near-surface view of precipitation.
  - **Composite Reflectivity:** Displays the maximum reflectivity value observed at any elevation angle above a specific point, giving an overview of the strongest echoes in the entire vertical column of the atmosphere. Useful for identifying the overall intensity and vertical extent of storms.
- **Radial Velocity (Vr)**
  - **What it measures:** The component of target motion directly towards or away from the radar. This is derived from the **Doppler effect** – the shift in frequency of the returned radar pulse caused by target movement.
  - **Units:** Typically in knots or meters per second (m/s).
- **Interpretation**
  - **Negative values (often blue/green):** Indicate motion *towards* the radar.
  - **Positive values (often red/orange):** Indicate motion *away* from the radar.
  - **Velocity Couple/Shear:** Adjacent areas of strong inbound and outbound velocities indicate rotation or strong wind shear.
- **Use in Meteorology**
  - **Detecting Rotation (Mesocyclones):** The primary tool for identifying mesocyclones (rotating updrafts within thunderstorms), which are strong indicators of potential tornado formation. Forecasters look for distinct "couplets" of inbound and outbound velocities side-by-side.

- **Identifying Wind Shear:** Crucial for aviation safety (e.g., microbursts, low-level wind shear near airports) and understanding storm dynamics.
- **Estimating Wind Fields:** Provides insights into the overall wind patterns within a storm or atmospheric layer.
- **Tracking Storms:** Helps determine the true movement of a storm, not just its reflectivity pattern.

- **Spectrum Width (SW)**
  - **What it measures:** A measure of the variability or dispersion of velocities within a single radar pulse volume. It quantifies how much the individual targets within that volume are moving at different radial velocities.
  - **Units:** Typically in knots or m/s.
  - **Interpretation**
  - **Low SW:** Indicates uniform motion (e.g., laminar flow, or all targets moving at roughly the same speed).
  - **High SW:** Suggests turbulent air, strong wind shear, or a mixture of different types of targets moving at varying speeds (e.g., rain and hail tumbling, or debris).
- **Use in Meteorology**
  - **Identifying Turbulence:** High values can indicate areas of strong turbulence within a storm or clear air.
  - **Locating Tornadoes:** Extremely high spectrum width values, especially when co-located with a tight velocity couplet, can be an indicator of a tornado's presence due to the intense, chaotic winds.
  - **Detecting the Melting Layer:** Often shows higher values in the bright band due to the coexistence of melting snow and rain, which have different fall speeds.

4.2.10 Dual-Polarization Base Products

Modern dual-polarization radars transmit and receive both horizontally and vertically polarized electromagnetic waves, providing additional base products that offer deeper insights into the characteristics of the targets.

- **Differential Reflectivity (ZDR)**
- **What it measures:** The ratio of the reflectivity of the horizontally polarized pulse to the reflectivity of the vertically polarized pulse.
- **Units:** Decibels (dB).

- **Interpretation:** Provides information about the *shape* and *orientation* of the precipitation particles.
  - **Positive ZDR:** Indicates targets that are wider horizontally than vertically (e.g., large, oblate raindrops that flatten as they fall; horizontally oriented ice crystals).
  - **Near-zero ZDR:** Indicates more spherical targets (e.g., small raindrops, hailstones, tumbling ice crystals).
  - **Negative ZDR:** Very rare, indicates targets taller than wide (e.g., vertically oriented ice crystals in strong electric fields).
- **Use in Meteorology**
  - **Distinguishing Precipitation Types:** Helps differentiate between rain (positive ZDR), hail (near-zero ZDR), and snow (variable ZDR depending on crystal type).
  - **Estimating Rainfall Rates:** Large, oblate raindrops have high ZDR, and incorporating ZDR improves the accuracy of rainfall estimates, especially for heavy rain.
  - **Identifying Hail:** A "hail core" will often show high reflectivity (Z) but near-zero ZDR.
  - **Detecting Updrafts:** Strong updrafts can loft large raindrops, leading to areas of high ZDR above the surface.
- **Correlation Coefficient (CC)**
- **What it measures:** A dimensionless value (between 0 and 1) that indicates how uniformly the horizontal and vertical pulses are behaving within the pulse volume. It's a measure of the diversity of shapes and sizes of targets.
- **Interpretation**
  - **High CC (close to 1.0):** Indicates a uniform population of targets (e.g., all rain, all snow, all insects).
  - **Low CC (below 0.8-0.9):** Suggests a mixture of different types of targets (e.g., rain mixed with hail), or the presence of highly irregular, non-meteorological targets.
- **Use in Meteorology**
  - **Filtering Non-Meteorological Echoes:** Very low CC values are excellent for identifying ground clutter, biological targets (insects, birds), and anomalous propagation, allowing forecasters to remove them from precipitation displays.

- **Detecting Tornado Debris Signatures (TDS):** A rapid drop in CC values (often below 0.8) co-located with high reflectivity and a strong velocity couplet is a near-certain indicator that a tornado is on the ground and lofting irregular debris. This is a critical warning tool.
- **Identifying Melting Layer (Bright Band):** Often shows a slight dip in CC as snow melts into rain, creating a mixture of hydrometeor types.

- **Specific Differential Phase (KDP)**
  - **What it measures:** The rate of change of the differential phase shift between the horizontal and vertical pulses as they propagate through precipitation.
  - **Units:** Degrees per kilometer (°/km).
  - **Interpretation:** Directly related to the concentration of liquid water in the radar beam. It increases with the amount of liquid water the beam passes through.
- **Use in Meteorology**
  - **Estimating Heavy Rainfall Rates:** KDP is particularly robust for estimating very heavy rainfall, especially in areas with large raindrops or mixed-phase precipitation, as it is less affected by attenuation than reflectivity.
  - **Identifying Heavy Rain/Hail Cores:** High KDP values often pinpoint areas of extremely heavy rainfall or where large hail is present.
  - **Distinguishing Hail from Heavy Rain:** When combined with Z and ZDR, KDP helps differentiate between very heavy rain (high Z, high ZDR, high KDP) and large hail (high Z, near-zero ZDR, but often low KDP unless mixed with heavy rain).

4.2.11 Derived Products (Level III Data)

Derived products are created by applying algorithms and mathematical computations to the base products. They are designed to provide more specific, actionable information for forecasters and are often presented in a more user-friendly format.

- **Vertical Integrated Liquid (VIL)**
- **Derivation:** Calculated by integrating reflectivity values vertically through the entire atmospheric column above a specific point.
- **Significance:** Represents the total amount of liquid water contained in a vertical column of air.

- **Use in Meteorology**
    - **Assessing Storm Intensity:** High VIL values indicate a deep column of intense precipitation, suggesting a strong potential for heavy rainfall and/or large hail.
    - **Identifying Severe Storms:** Often used as an indicator of severe thunderstorm potential, especially when values are rapidly increasing.
- **Storm Relative Velocity (SRM)**
    - **Derivation:** Radial velocity data with the mean motion of the storm subtracted.
    - **Significance:** Removes the storm's overall translation, allowing forecasters to see the *true* rotation or motion within the storm more clearly, as if they were moving with the storm.
    - **Use in Meteorology**
    - **Enhanced Mesocyclone Detection:** Makes it easier to identify and analyze mesocyclones and other rotational features within thunderstorms without the confusion of the storm's bulk motion.
    - **Diagnosing Storm Structure:** Helps understand inflow and outflow patterns relative to the storm.
- **Mesocyclone Detection Algorithm (MDA) / Tornado Vortex Signature (TVS)**
- **Derivation:** Algorithms analyze radial velocity data for specific patterns of inbound and outbound velocities in close proximity, indicative of rotation.
- **Significance**
    - **MDA:** Identifies potential mesocyclones (rotating updrafts), which are strong precursors to tornadoes. It often outputs a "strength" and "depth" of the rotation.
    - **TVS:** A stronger, tighter, and often deeper rotation signature identified by the algorithm, indicating a high probability of a tornado or strong low-level mesocyclone.
- **Use in Meteorology**
    - **Issuing Tornado Warnings:** These are primary automated tools that alert forecasters to dangerous rotation within thunderstorms, directly supporting the issuance of tornado warnings.
- **Hail Detection Algorithm (HDA):**

- **Derivation:** Uses a combination of reflectivity (high values) and dual-polarization parameters (e.g., near-zero ZDR, low CC) to estimate the probability of hail and its size.
- **Significance:** Provides automated alerts for potential hail.

- **Use in Meteorology**
  - **Issuing Severe Thunderstorm Warnings:** Helps forecasters determine if a storm is producing or is likely to produce damaging hail.
  - **Assessing Hail Risk:** Provides estimated maximum hail size.
- **Quantitative Precipitation Estimates (QPE) / Accumulation**
  - **Derivation:** Converts reflectivity (and increasingly, dual-polarization data like KDP and ZDR) into estimated rainfall rates, which are then accumulated over time (e.g., 1-hour, 3-hour, 24-hour totals).
  - **Significance:** Provides spatial maps of estimated rainfall.
- **Use in Meteorology**
  - **Flood Forecasting:** Crucial for identifying areas of heavy rainfall accumulation that could lead to flash floods or river flooding.
  - **Water Resource Management:** Used for tracking water input into reservoirs and watersheds.
  - **Drought Monitoring:** Long-term QPEs help assess rainfall deficits.
- **Echo Tops**
  - **Derivation:** Identifies the highest altitude at which a detectable reflectivity echo (e.g., 18.5 dBZ) is observed.
  - **Significance:** Provides an estimate of the maximum vertical extent of a storm.
- **Use in Meteorology**
  - **Assessing Storm Intensity:** Taller storms often indicate stronger updrafts and potentially more severe weather.
  - **Aviation:** Provides information on the vertical extent of hazardous weather for flight planning.
- **VAD Wind Profile (VWP)**
  - **Derivation:** Uses radial velocity data from a full 360-degree scan at a single elevation angle (Velocity Azimuth Display - VAD) to derive a profile of horizontal wind speed and direction with height.

  - **Significance:** Provides a sounding-like profile of winds aloft from a ground-based radar.
- **Use in Meteorology**
  - **Supplementing Radiosondes:** Fills in gaps between radiosonde launches by providing continuous wind profiles.
  - **Diagnosing Wind Shear:** Helps identify layers of strong wind shear, which are important for storm organization and aviation.
  - **Model Initialization:** Provides valuable wind data for numerical weather prediction models.
- **Hydrometeor Classification Algorithm (HCA)**
  - **Derivation:** Combines all dual-polarization base products (Z, ZDR, CC, KDP) with temperature information (often from numerical models or nearby soundings) to classify the type of precipitation or atmospheric target.
  - **Significance:** Provides a much clearer picture of what the radar is "seeing."
- **Use in Meteorology**
  - **Differentiating Precipitation:** Automatically classifies areas as light rain, heavy rain, snow, graupel, hail, biological targets, or ground clutter.
  - **Improving Warnings:** Enables more specific and accurate warnings (e.g., "heavy rain" vs. "hail").
  - **Understanding Storm Microphysics:** Helps researchers study the processes of precipitation formation.
- **Tornado Debris Signature (TDS) / Debris Ball**
  - **Derivation:** A specific pattern in dual-polarization data, characterized by very low Correlation Coefficient (CC), often co-located with high Reflectivity (Z) and sometimes high Differential Reflectivity (ZDR), within a strong velocity couplet.
  - **Significance:** This is a strong indicator that a tornado is on the ground and lofting irregular debris, providing near-certain confirmation of a tornado's presence.
- **Use in Meteorology**
  - **Confirming Tornadoes:** The most definitive radar signature of a tornado, allowing forecasters to issue or extend tornado warnings with high confidence.

The interplay between these base and derived products allows meteorologists to rapidly assess complex atmospheric situations, from widespread rain to rapidly developing severe thunderstorms, providing crucial information for public safety and scientific advancement.

4.3 Wind Profilers and Sodars

Wind profilers and SODARs are both ground-based remote sensing instruments used to measure wind speed and direction at various heights above the ground. They are valuable tools for atmospheric research, weather forecasting, and various applications like wind energy and aviation (Fig. 4.2). Here's a breakdown of each:

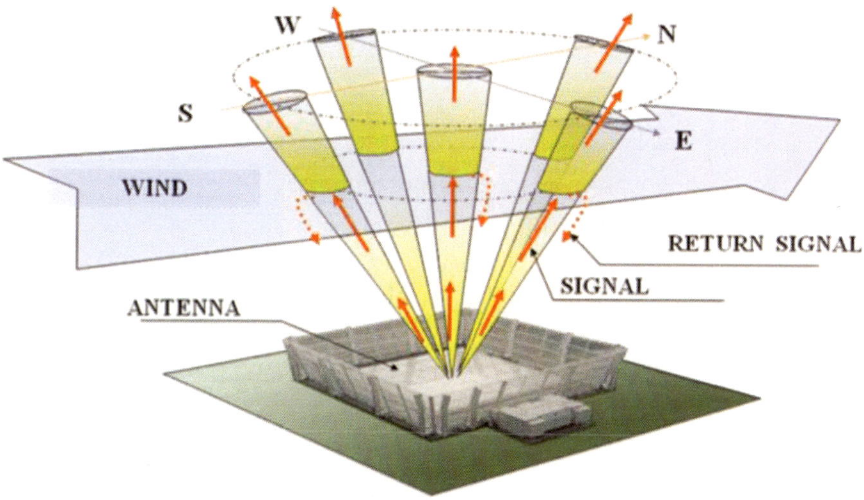

Fig. 4.2: Wind profilers and sodars

- **Wind Profilers**
- **Operating Principle:** Wind profilers typically use radar (electromagnetic waves) to detect wind.They transmit pulses of electromagnetic energy into the atmosphere, and these pulses are scattered back by changes in the refractive index of the air, which are often caused by turbulence. The Doppler shift of the returned signal is analyzed to determine the wind speed along the radar beam's direction. By using multiple beams pointed in different directions (usually one vertical and two or more oblique), the three-dimensional wind vector (horizontal and vertical components) can be calculated.
- **Frequency:** Radar wind profilers often operate in the VHF (30-300 MHz) or UHF (300-1000 MHz) frequency bands.

- **Range:** Radar wind profilers generally have a longer range, capable of measuring winds up to several kilometers in altitude, extending through the boundary layer and into the troposphere. Some specialized types, like boundary-layer radar wind profilers, focus on the lower atmosphere with higher resolution.
- **Weather Dependency:** Radar wind profilers using longer wavelengths are less affected by clouds and precipitation, allowing for all-weather operation.
- **Applications**
  - Weather forecasting
  - Climate research
  - Aviation safety (detecting wind shear)
  - Air quality monitoring
  - Severe weather studies
  - Wind energy assessment at higher altitudes
- **SODARs (Sonic Detection and Ranging)**
- **Operating Principle:** SODARs use sound waves instead of electromagnetic waves. They transmit acoustic pulses into the atmosphere, and these waves are scattered back by temperature inhomogeneities and turbulence. Similar to radar profilers, the Doppler shift of the returning sound waves is used to determine the wind speed along the acoustic beam. Multiple beams (typically one vertical and two oblique) allow for the calculation of the 3D wind vectors.
- **Frequency:** SODARs operate with acoustic frequencies, typically ranging from below 1000 Hz to over 4000 Hz.14
- **Range:** SODARs typically have a shorter vertical range compared to radar wind profilers, generally measuring winds up to a few hundred meters (e.g., 15 m to 500 m, sometimes up to 1-2 km under favorable conditions).15 They are particularly useful for studying the atmospheric boundary layer with high vertical resolution.
- **Weather Dependency:** The performance of SODARs can be affected by environmental noise (like traffic or industrial sounds) and strong winds at the surface, which can interfere with the acoustic signals.16Heavy precipitation can also affect the measurements.
- **Applications**
  - Wind energy assessment, especially for turbine height winds (50m to 200m)

- Air quality studies (monitoring mixing height & atmospheric stability)
- Boundary layer research
- Airport site evaluation (e.g., monitoring low-level wind shear)
- Turbulence studies

- **Key Differences Summarized**

Feature	Wind Profiler (Radar)	SODAR (Acoustic)
Wave Type	Electromagnetic waves (radar)	Sound waves (acoustic)
Frequency	VHF/UHF (MHz range)	Audible/Ultrasound (kHz range)
Scattering by	Refractive index fluctuations	Temperature inhomogeneities, turbulence
Typical Range	Higher (up to several km)	Lower (up to a few hundred meters)
Weather	Less affected by clouds/light rain	More susceptible to noise, heavy rain
Applications	Upper atmosphere, general meteorology	Boundary layer, wind energy at lower altitudes

In some cases, radar and SODAR wind profilers can be combined to provide detailed wind information across a larger vertical range with good resolution at lower levels and coverage at higher altitudes.

4.4 LIDARS

Okay, let's elaborate further and rewrite the description of LiDAR, focusing more on the nuances of its atmospheric measurement capabilities alongside its surface mapping applications.

4.4.1 LIDAR: Illuminating the World and the Atmosphere

- **LIDAR (Light Detection and Ranging)** is a versatile remote sensing technology that employs laser light to probe the environment. It excels not only at creating high-resolution three-dimensional maps of the Earth's surface and objects but also at characterizing the composition and dynamics of our atmosphere. By analyzing the interaction of laser light with its targets, LiDAR provides valuable insights into both the physical structure and the constituents of the world around us and the air above.
- **Dual Capabilities: Surface Mapping and Atmospheric Profiling**

  LiDAR's strength lies in its ability to perform two primary, yet often distinct, types of measurements:

  - **Surface and Object Mapping:** This is achieved by precisely measuring the time it takes for laser pulses to travel to a target and return. This **time-of-flight** (t) is used to calculate the distance (d) using the speed of light (c): d=ct/2

- By rapidly scanning a laser across a scene, millions of these distance measurements are collected, forming a dense **point cloud**, a 3D representation of the surveyed area or object. The intensity of the reflected light at each point can also provide information about the surface properties.
- **Atmospheric Profiling:** Here, LiDAR analyzes the light that is backscattered by atmospheric particles (aerosols, cloud droplets/ice crystals) and molecules. The characteristics of this backscattered light – its intensity, the change in its wavelength (Doppler shift), and its polarization – reveal crucial information about the atmosphere's composition and dynamics at different altitudes.

- **Atmospheric Measurement Techniques in Detail**

  Several specialized LiDAR techniques are used for atmospheric studies:

  - **Elastic Backscatter LiDAR:** This is the most common type for atmospheric work. It measures the intensity of the backscattered light as a function of altitude. Changes in backscatter intensity indicate the presence of aerosol layers or cloud boundaries. The range-resolved backscatter signal can be used to study the vertical distribution and optical properties of aerosols and clouds.
  - **Raman LiDAR:** This technique exploits the Raman scattering effect, where the wavelength of the backscattered light is shifted due to interaction with molecules. By analyzing these wavelength shifts, Raman LiDAR can provide information about the type and concentration of specific molecules (e.g., water vapor, nitrogen, oxygen) as a function of altitude, even in the presence of aerosols and clouds.
  - **Differential Absorption Lidar (DIAL):** As mentioned before, DIAL uses two closely spaced laser wavelengths. One wavelength is strongly absorbed by the target gas, while the other is weakly absorbed. By measuring the difference in backscattered signals at these two wavelengths, the concentration of the gas (e.g., ozone, water vapor, carbon dioxide, methane) along the laser path can be determined with high sensitivity and range resolution.
  - **Doppler LiDAR:** This technique measures the Doppler shift of the backscattered light, which is caused by the motion of the scattering particles (aerosols, cloud droplets). This allows for the measurement of the radial velocity of these particles along the laser beam's direction. By using multiple beams pointed in different directions, the three-dimensional wind vector (wind speed and direction) can be

retrieved as a function of height. Doppler LiDAR is crucial for wind profiling, studying atmospheric turbulence, and detecting wind shear.

4.4.2 Applications Revisited (with more emphasis on atmosphere)

- **Surface Mapping Applications:** (These remain largely as described before, focusing on creating 3D models for various fields like autonomous vehicles, surveying, forestry, etc.)
- **Atmospheric Monitoring Applications**
    - **Aerosol Studies:** Monitoring the vertical distribution, transport, and optical properties of pollutants, dust, volcanic ash, and other particulate matter, crucial for air quality and climate studies.
    - **Cloud Research:** Investigating cloud structure, height, phase (water/ ice), and their impact on radiative transfer and precipitation.
    - **Trace Gas Monitoring:** Measuring the concentration and distribution of greenhouse gases (e.g., CO2, CH4), pollutants (e.g., ozone, nitrogen dioxide), and water vapor, essential for understanding climate change and atmospheric chemistry.
    - **Wind Profiling:** Providing high-resolution vertical profiles of wind speed and direction, vital for weather forecasting, aviation safety (wind shear detection), and wind energy applications.
    - **Boundary Layer Studies:** Examining the structure and dynamics of the lower atmosphere, including mixing height and turbulence.

In essence, LiDAR acts as a "laser radar," providing detailed information about both the solid Earth and the dynamic atmosphere above it. The specific type of LiDAR and the way the backscattered light is analyzed determine whether the focus is on creating spatial maps or on characterizing the atmospheric constituents and their motion

4.5 GPS Meteorology: Sensing the Atmosphere with Satellites

GPS Meteorology is a branch of atmospheric science that uses the signals from the Global Positioning System (GPS) satellite constellation to gather information about the Earth's atmosphere. Instead of primarily focusing on navigation and positioning, GPS meteorology leverages the fact that atmospheric conditions affect the propagation of GPS radio signals. These effects, while a source of error for precise positioning, contain valuable information about the state of the atmosphere.

- **How it Works:** GPS satellites transmit radio signals at two main frequencies (L1 and L2) as they orbit the Earth. Ground-based GPS receivers, typically part of permanent geodetic networks, continuously track these signals. As the signals travel through the atmosphere, they are:

- **Delayed:** The speed of the GPS signals is slightly reduced by the presence of the atmosphere, particularly by water vapor in the troposphere.
- **Refracted (bent):** The path of the signals is bent as they pass through layers of the atmosphere with varying refractive indices (due to changes in temperature, pressure, and humidity).

GPS meteorology exploits these atmospheric effects by

- **Ground-based GPS Meteorology:** Uses networks of ground-based GPS receivers. By precisely measuring the arrival time of the signals from multiple satellites, the **total delay** caused by the atmosphere can be estimated. Since the delay due to the well-mixed "dry" atmosphere (primarily nitrogen and oxygen) can be modeled using surface pressure measurements, the remaining delay is primarily due to **water vapor**. This "wet delay" is then used to estimate the **integrated precipitable water vapor (IPWV)**, which is the total amount of water vapor in a vertical column of the atmosphere above the receiver.
- **Space-based GPS Meteorology (Radio Occultation):** Involves Low Earth Orbiting (LEO) satellites receiving signals from the GPS satellites as they rise or set behind the Earth's limb. As the GPS signals pass through different layers of the atmosphere on their way to the LEO satellite, they are refracted (bent) by varying amounts depending on the atmospheric density. By precisely measuring this bending angle as a function of the ray's tangent height, vertical profiles of **refractivity** can be retrieved. These refractivity profiles can then be used to derive profiles of temperature, pressure, and humidity in the troposphere and stratosphere, especially when combined with other data sources.
- **What it Can Measure in Meteorology**

  GPS meteorology provides valuable measurements for various meteorological applications
- **Integrated Precipitable Water Vapor (IPWV):** Ground-based GPS is particularly effective at measuring the total column amount of water vapor. This is crucial for:
  - Improving short-term weather forecasts, especially for precipitation events. Increased IPWV can be an indicator of potential heavy rainfall.
  - Monitoring atmospheric moisture, which plays a significant role in the Earth's climate system.

- **Atmospheric Refractivity Profiles:** Space-based GPS radio occultation provides vertical profiles of atmospheric refractivity, which can be used to derive:
  - **Temperature profiles:** Especially in the upper troposphere and stratosphere.
  - **Pressure profiles:** Particularly in the upper troposphere and stratosphere.
  - **Humidity profiles:** Primarily in the troposphere, although this retrieval is more complex and often requires constraints from other data.
- **Zenith Total Delay (ZTD) and Zenith Wet Delay (ZWD):** These are intermediate products in ground-based GPS meteorology. ZTD represents the total atmospheric delay in the vertical direction, while ZWD isolates the delay caused by water vapor. These can be directly assimilated into weather models.
- **Ionospheric Effects:** While often treated as noise in standard GPS positioning, the dispersive nature of the ionosphere (its effect on signal speed depends on frequency) allows dual-frequency GPS receivers to estimate the **Total Electron Content (TEC)** in the ionosphere, which is relevant for space weather studies.

In summary, GPS meteorology offers a cost-effective, all-weather, and continuous way to monitor key atmospheric variables, particularly water vapor and upper atmospheric conditions, contributing significantly to weather forecasting, climate monitoring, and atmospheric research.

4.6 Conclusion

The exploration of ground-based remote sensing platforms underscores their indispensable role in modern meteorology and atmospheric science. Instruments such as **Wind Profilers** (both radar and SODAR variants), **SODARs**, and **GPS Meteorology** systems collectively provide a continuous, high-resolution window into the dynamic processes of the lower atmosphere.

These platforms offer significant advantages, including their ability to operate in various weather conditions (especially radar-based systems), provide real-time data, and fill critical observational gaps that traditional in-situ measurements or satellite observations alone cannot address. They offer unparalleled temporal and vertical resolution, particularly within the atmospheric boundary layer, which is crucial for understanding local weather phenomena, air quality, and the exchange of energy and moisture between the Earth's surface and the atmosphere.

From precisely measuring wind profiles and turbulence with Wind Profilers and SODARs to quantifying the vital atmospheric water vapor content through GPS signal delays, these ground-based technologies enhance our ability to:

- **Improve weather forecasting:** Especially for short-term and severe weather events, by providing crucial moisture and wind data.
- **Advance climate research:** By contributing to a more comprehensive understanding of atmospheric composition and processes.
- **Support various applications:** Including aviation safety, wind energy assessment, and environmental monitoring.

As meteorological demands grow for more detailed and timely atmospheric information, ground-based remote sensing platforms will continue to evolve and become even more integral to our observational networks, solidifying their position as foundational tools for both operational meteorology and scientific discovery.

5

Satellite Remote Sensing

5.1 Introduction

The Earth's atmosphere is a vast, dynamic, and interconnected system, constantly in motion and influencing every aspect of our lives. Accurately observing and understanding this complex system is fundamental to predicting weather, monitoring climate change, and responding to environmental challenges. While ground-based platforms provide invaluable high-resolution data at specific locations, theitr spatial coverage is inherently limited.

This is where **satellite remote sensing** emerges as a revolutionary tool in meteorology. By observing the Earth and its atmosphere from space, satellites offer an unparalleled vantage point, providing global and continuous coverage that is impossible to achieve with ground-based instruments alone. They can capture data from remote oceans, deserts, and polar regions, offering a truly holistic view of atmospheric processes.

Satellite remote sensing enables meteorologists to measure a wide array of atmospheric parameters, including temperature and humidity profiles, cloud properties (type, height, movement), wind fields, precipitation rates, and the distribution of atmospheric gases and aerosols. This wealth of data is crucial for initializing and validating numerical weather prediction models, tracking severe storms, monitoring long-term climate trends, and conducting fundamental atmospheric research. This chapter will delve into the fascinating world of satellite remote sensing in meteorology, exploring the principles behind how these orbiting observatories collect data, the various types of meteorological satellites and their specialized instruments, and the diverse applications of the information they provide for understanding and forecasting our planet's weather and climate.

5.2 Orbits: A Celestial Dance for Meteorological Insights

The path a satellite takes around the Earth, known as its **orbit**, is a critical design choice that fundamentally determines its observational capabilities in remote sensing, especially for meteorology. Different orbits offer distinct advantages in terms of spatial coverage, temporal resolution (how often a location is observed), and the types of atmospheric phenomena that can be effectively monitored. For Earth-observing meteorological satellites, two primary categories of orbits are predominantly utilized: **Geostationary Earth Orbit (GEO)** and **Low Earth Orbit (LEO)**, with a significant emphasis on **Polar Orbits** within the LEO category.

5.2.1 Geostationary Earth Orbit (GEO)

- **Characteristics**
  - **Altitude:** Approximately 35,786 km (22,236 miles) directly above the Earth's Equator.
  - **Orbital Period:** Exactly 24 hours, precisely matching the Earth's rotation period.
  - **Apparent Position:** From the ground, a geostationary satellite appears to remain fixed over a single point on the Equator.
  - **Coverage:** Each GEO satellite provides a continuous view of roughly one-third of the Earth's surface. A network of typically three to four geostationary satellites can provide near-global coverage of the Earth's disk, excluding the extreme polar regions.
- **Advantages for Meteorology**
  - **Continuous Monitoring:** Offers constant, real-time imagery and data for a vast geographical area. This is invaluable for tracking rapidly evolving weather phenomena such as hurricanes, severe thunderstorms, and frontal systems.
  - **High Temporal Resolution:** Allows for very frequent updates (e.g., every 10–15 minutes, or even faster for specific "rapid scan" regions), which is essential for "nowcasting" (short-term forecasting) and monitoring hazardous weather.
  - **Wind Derivation:** The continuous sequence of images enables meteorologists to estimate wind speeds and directions by tracking the movement of clouds or water vapor features.
- **Disadvantages for Meteorology**
  - **Coarser Spatial Resolution:** Due to the immense altitude, the ground resolution of images is significantly lower compared to LEO satellites.
  - **Limited Polar Coverage:** Cannot effectively observe areas near the poles (typically above 60 o latitude) due to the low viewing angle and the curvature of the Earth.

- **Oblique Viewing Angles:** At higher latitudes within their coverage area, the satellite's oblique viewing angle can distort observations and complicate the retrieval of certain atmospheric profiles.

- **Examples:** GOES (Geostationary Operational Environmental Satellite - USA), Meteosat (Europe), Himawari (Japan), Fengyun (China).

5.2.2 Low Earth Orbit (LEO)

LEO satellites orbit much closer to the Earth's surface, typically between 200 km and 1,000 km (120 miles to 620 miles). Their lower altitude means they orbit the Earth much faster, usually completing an orbit in about 90–100 minutes. Within the LEO category, **Polar Orbits** are of particular importance for meteorological remote sensing.

- **Polar Orbits**
- **Characteristics**
  - **Path:** These satellites travel in an orbit that passes close to or directly over the Earth's geographical poles on each revolution.
  - **Global Coverage:** As the Earth rotates beneath the satellite's orbital plane, a polar-orbiting satellite eventually covers the entire globe over a series of successive orbits.
  - **Revisit Time:** Any given point on Earth is typically observed a few times a day, with more frequent coverage closer to the poles where the orbital paths converge.
- **Advantages for Meteorology**
  - **True Global Coverage:** Provides observations over the entire planet, including the crucial polar regions which are vital for climate monitoring and are inaccessible to GEO satellites.
  - **High Spatial Resolution:** Their proximity to Earth allows for much finer spatial resolution, enabling detailed observations of smaller-scale weather features, surface characteristics, and atmospheric vertical structures.
  - **Vertical Atmospheric Sounding:** Polar orbiters are ideal platforms for instruments designed to measure vertical profiles of temperature, humidity, and atmospheric gases (e.g., through infrared/microwave sounders or radio occultation techniques), providing detailed "slices" of the atmosphere.

- **Sun-Synchronous Orbits (SSO) - A Special Type of Polar Orbit**
- **Characteristics**
  - **Consistent Local Time:** A sun-synchronous orbit is a specific type of polar orbit where the satellite crosses the Equator at roughly the same local mean solar time on each pass. This means that the angle of illumination from the sun on the Earth's surface is nearly constant each time the satellite passes over a given latitude.
  - **Orbital Plane Precession:** This is achieved by carefully tuning the orbit's inclination and altitude so that the orbital plane precesses (rotates) at the same rate as the Earth revolves around the Sun.
- **Advantages for Meteorology (especially for SSO)**
  - **Consistent Lighting Conditions:** The constant solar illumination angle is highly beneficial for optical instruments, ensuring consistent data collection conditions and simplifying the comparison of images and measurements over days, months, or even years. This is crucial for long-term climate monitoring and detecting subtle environmental changes.
  - **Optimized for Passive Sensors:** Many passive remote sensing instruments (e.g., radiometers measuring reflected sunlight or emitted thermal radiation) perform best under consistent illumination.
- **Disadvantages for LEO/Polar/SSO**
  - **Lower Temporal Resolution:** While providing global coverage, any single location is observed far less frequently than by a geostationary satellite. This limits their utility for real-time tracking of rapidly evolving, short-lived weather systems.
  - **Narrow Swath:** Each orbital pass covers a relatively narrow strip of the Earth's surface, requiring many orbits to achieve global coverage.
- **Examples:** NOAA POES (Polar-orbiting Operational Environmental Satellites - USA), MetOp (Europe), Aqua, Terra, Suomi NPP (USA).

5.2.3 Complementary Roles of Orbits

In conclusion, Geostationary and Low Earth Orbiting satellites play complementary, rather than competing, roles in meteorological remote sensing. GEO satellites provide the continuous, high-temporal resolution "big picture" for tracking large-scale, rapidly changing weather systems over fixed regions. LEO satellites, particularly those in polar and sun-synchronous orbits, offer comprehensive global coverage with high spatial and vertical resolution, making them indispensable for detailed atmospheric soundings, long-term climate monitoring, and observing the critical polar regions. Together, these diverse orbital strategies form the robust and multi-layered observational backbone of modern weather forecasting and climate science.

5.3 Sensors in Satellite Remote Sensing: The Eyes, Ears, and Profilers in Space

In the realm of satellite remote sensing for meteorology, **sensors** are the sophisticated instruments onboard spacecraft that are designed to detect and measure electromagnetic radiation. These instruments act as the "eyes and ears" of the satellite, capturing specific wavelengths of energy reflected from the Sun or emitted by the Earth's surface, atmosphere, and clouds. The design and capabilities of these sensors—whether they are primarily **imagers** or **sounders**—determine the type of meteorological information they can provide.

Sensors are broadly categorized into two main types based on their energy source: **Passive Sensors** and **Active Sensors**.

5.3.1 Passive Sensors

Passive sensors detect naturally occurring electromagnetic radiation. They do not generate their own energy. This category includes both imagers and sounders (Fig. 5.1).

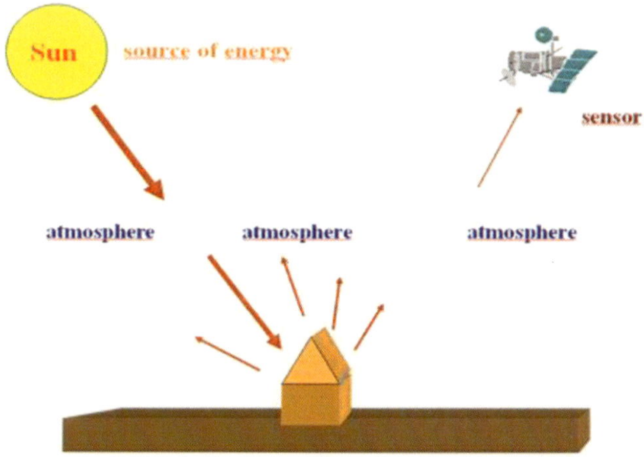

Fig. 5.1: Passive sensor

- **Imagers (Radiometers/Scanners for Spatial Views)**
- **Principle:** Imagers typically operate across multiple spectral bands (visible, near-infrared, and infrared). They sweep across a wide area to capture a "picture" or a broad spatial view of the Earth and its

atmosphere. They measure the intensity of reflected sunlight (in visible/near-infrared) or emitted thermal radiation (in infrared).

- **What they measure in Meteorology**
  - **Cloud Properties:** Provide detailed imagery of cloud cover, cloud type (e.g., distinguishing between low stratus and high cirrus), cloud texture, and cloud movement. By tracking cloud features over time, wind vectors can be derived.
  - **Surface Features:** Monitor snow and ice cover, vegetation health, land-use changes, and sea surface characteristics (e.g., sea surface temperature, ocean currents, chlorophyll concentration).
  - **Atmospheric Phenomena:** Detect and track large-scale aerosol events like smoke plumes from wildfires, dust storms, and volcanic ash clouds.
- **Limitations:** Visible imagers cannot operate at night. Infrared imagers can operate day and night but can be obscured by thick clouds when trying to view the surface below.
- **Examples:** Advanced Baseline Imager (ABI) on GOES, Visible Infrared Imaging Radiometer Suite (VIIRS) on Suomi NPP.
- **Sounders (Radiometers for Vertical Profiles)**
  - **Principle:** Sounders are specialized radiometers designed to measure radiation at very specific, narrow spectral bands that are sensitive to the absorption and emission characteristics of atmospheric gases (like water vapor, carbon dioxide, and oxygen) at different altitudes. By analyzing the radiation from these different layers, they can construct vertical profiles of atmospheric properties.
- **Types of Sounders**
  - **Infrared (IR) Sounders:** Detect thermal infrared radiation. By selecting wavelengths where specific gases absorb and emit, they can derive:
  - **Atmospheric Temperature Profiles:** From the surface up to the stratosphere.
  - **Water Vapor Profiles:** The distribution of moisture at different altitudes.
  - **Trace Gas Concentrations:** For certain gases with strong infrared absorption features.

- **Microwave (MW) Sounders:** Detect microwave radiation. Crucially, microwave radiation can penetrate most non-precipitating clouds, making them ideal for:
  - **All-Weather Temperature and Humidity Profiles:** Providing profiles even when clouds are present.
  - **Cloud Liquid Water and Ice Water Content:** Insights into cloud microphysics.
  - **Precipitation Rate Estimation:** Especially over oceans.
- **Advantages:** Provide critical vertical information about the atmosphere, essential for numerical weather prediction models. Microwave sounders offer all-weather capability.
- **Examples:** Cross-track Infrared Sounder (CrIS) on Suomi NPP, Infrared Atmospheric Sounding Interferometer (IASI) on MetOp, Advanced Technology Microwave Sounder (ATMS) on Suomi NPP.

5.3.2 Active Sensors

Active sensors generate their own electromagnetic radiation (e.g., laser pulses or radar beams) and then measure the portion of that radiation that is reflected or scattered back to the sensor (Fig. 5.2).

- **Radar (e.g., Precipitation Radar):**
  - **Principle:** Transmits microwave pulses and measures the strength and time of the returned echoes from precipitation particles (rain, snow, hail). The Doppler shift of the echoes can also be used to measure the velocity of these particles.
- **What they measure in Meteorology**
  - **Precipitation Intensity and Distribution:** Provides a 3D structure of precipitation within clouds.
  - **Cloud Structure:** Information about the vertical extent of precipitating clouds.
  - **Storm Dynamics:** Insights into the internal structure and movement of storm systems.
- **Examples:** Precipitation Radar (PR) on TRMM (Tropical Rainfall Measuring Mission), CloudSat's Cloud Profiling Radar (CPR).

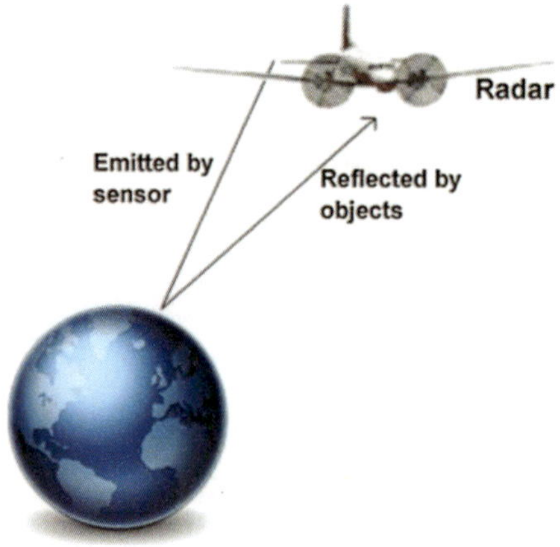

Fig. 5.2: Active sensor

- **Lidar (Light Detection and Ranging)**
- **Principle:** Transmits laser pulses and measures the backscattered light from atmospheric aerosols, cloud particles, and molecules. The time-of-flight provides distance, and the characteristics of the backscattered signal (intensity, polarization, wavelength shift) provide information about the scattering medium. Doppler LiDAR can also measure wind velocity.
- **What they measure in Meteorology**
  - **Vertical Profiles of Aerosols:** Distribution, concentration, and type of dust, smoke, and pollution layers.
  - **Cloud Layering and Height:** Precise determination of cloud base and top altitudes, and detection of thin cloud layers often missed by passive sensors.
  - **Trace Gases:** Using techniques like Differential Absorption Lidar (DIAL), concentrations of specific gases like water vapor, ozone, and carbon dioxide can be measured.
  - **Wind Profiles:** Doppler LiDAR can provide vertical wind profiles.
- **Examples:** Cloud-Aerosol Lidar with Orthogonal Polarization (CALIOP) on CALIPSO, Atmospheric Dynamics Mission Aeolus (ADM-Aeolus) with its Doppler Wind Lidar (DWL).

5.3.3 Conclusion on Sensors

The sophisticated suite of sensors on meteorological satellites, encompassing both passive (imagers and sounders) and active (radar and lidar) technologies, allows for a comprehensive and multi-spectral observation of the Earth's atmosphere and surface. Imagers provide the broad spatial context and track visible phenomena, while sounders delve into the vertical structure of the atmosphere, providing crucial profiles. Active sensors offer unique capabilities for probing through clouds and precisely measuring particles and winds. This continuous stream of diverse data is fundamental for advancing numerical weather prediction models, improving our understanding of atmospheric processes, & monitoring the intricate dynamics of our planet's climate system.

5.4 Visible and Infrared Imagery

Visible and Infrared (IR) imagery are two of the most fundamental and widely used types of data collected by meteorological satellites. They provide complementary views of the Earth's atmosphere and surface, allowing meteorologists to observe a wide range of phenomena crucial for weather forecasting and climate monitoring. Both types of imagery are typically acquired by **passive sensors** known as **imagers** or **radiometers**.

5.4.1 Visible Imagery

- **What it is:** Visible imagery captures the portion of the electromagnetic spectrum that is visible to the human eye, typically wavelengths between approximately 0.4 and 0.7 micrometers. It essentially measures **sunlight reflected**by the Earth's surface, clouds, and atmospheric particles.
- **How it Works:** The satellite sensor detects the intensity of sunlight scattered back towards space. Brighter objects reflect more sunlight, while darker objects absorb more.
- **What it Measures/Shows in Meteorology**
  - **Clouds:** Clouds appear white because they are excellent reflectors of sunlight. Thicker clouds (like cumulonimbus) appear brighter white, while thinner clouds (like cirrus) might appear grayer. This allows meteorologists to:
  - Identify cloud cover and patterns.
  - Distinguish between different cloud types based on their texture and brightness.
  - Track cloud movement to estimate wind direction and speed.
  - Locate fog and low stratus clouds (which appear bright).
- **Surface Features:** Land surfaces appear in varying shades of gray, depending on their reflectivity (albedo). Snow and ice appear very bright. Vegetated areas are typically darker than barren land.
- **Aerosols:** Dense plumes of smoke from wildfires, dust storms, and volcanic ash clouds can appear as distinct, often hazy, features.

- **Advantages**
- **High Spatial Detail:** Can provide very sharp images, especially from LEO satellites, allowing for detailed observation of cloud structures and surface features.
- **Intuitive:** The images closely resemble what a human eye would see from space, making them easy to interpret.
- **Limitations**
  - **Daylight Only:** Requires sunlight to work, so it cannot provide data during nighttime or in polar winter.
  - **Cannot "See Through" Clouds:** Only shows the top of the cloud layer; information about what's beneath is obscured.
  - **Affected by Sun Glint:** Over water bodies, direct reflection of the sun (sun glint) can create bright spots that might be misinterpreted.

5.4.2 Infrared (IR) Imagery

- **What it is:** Infrared imagery detects **thermal infrared radiation** emitted by objects, typically in wavelengths between approximately 3 and 15 micrometers. All objects with a temperature above absolute zero (0 K or −273.15°C) emit infrared radiation. The amount of radiation emitted is directly related to the object's temperature (warmer objects emit more radiation).
- **How it Works:** The satellite sensor measures the intensity of the emitted IR radiation. This intensity is then converted into a temperature reading. In meteorological images, a common convention is to display warmer temperatures as darker shades and colder temperatures as brighter shades (white).
- **What it Measures/Shows in Meteorology**
  - **Cloud Top Temperature and Height:** This is a primary application. Colder temperatures indicate higher altitudes in the atmosphere. Therefore, brighter white clouds in an IR image are typically high-level clouds (like cumulonimbus anvils or cirrus), while darker clouds are lower. This allows meteorologists to:
  - Determine the vertical extent and structure of clouds.
  - Identify areas of strong convection (tall, cold cloud tops often associated with severe weather).
  - Track cloud movement day and night.
- **Surface Temperature:** When there are no clouds, IR imagery can measure land surface temperature and sea surface temperature. This

is vital for understanding heat distribution, ocean currents, and land-atmosphere interactions.

- **Atmospheric Features**
  - **Water Vapor Channels:** Specific IR wavelengths are sensitive to water vapor absorption/emission. These "water vapor channels" show the distribution of moisture in the middle and upper troposphere, even in clear skies, revealing atmospheric circulation patterns and dry air intrusions.
  - **Volcanic Ash and Dust:** These plumes have unique IR signatures that can be used for detection and tracking, especially at night.
- **Advantages**
  - **Day and Night Capability:** Does not rely on sunlight, making it invaluable for continuous monitoring 24 hours a day.
  - **Cloud Height Information:** Directly provides an estimate of cloud top temperature, which can be correlated to cloud height.
  - **Surface Temperature Measurement:** Essential for various environmental and oceanographic applications.
- **Limitations**
  - **Lower Spatial Detail:** Generally has coarser spatial resolution compared to visible imagery due to the longer wavelengths.
  - **Cannot "See Through" Clouds:** Like visible imagery, it only shows the temperature of the cloud top, not what's underneath.
  - **Temperature Inversions:** Can be tricky to interpret in cases of temperature inversions (where temperature increases with height), as a low, warm cloud might appear similar to the warm ground.

5.4.3 Complementary Nature

Visible and infrared imagery are often used together to provide a more complete picture of atmospheric conditions. For instance, a meteorologist might use a visible image during the day to see the detailed texture of a thunderstorm, and then switch to an infrared image at night to track its cold cloud tops and estimate its intensity. The combination of these two fundamental types of satellite data forms the backbone of operational weather monitoring and forecasting globally.

5.5 Microwave Remote Sensing

Microwave remote sensing is a crucial technique in satellite meteorology, offering unique capabilities that complement visible & infrared observations.

Unlike visible and infrared radiation, microwave radiation has longer wavelengths (typically ranging from 1 mm to 1 m), which allows it to penetrate most clouds, light rain, and even some vegetation. This "all-weather" capability makes microwave sensors indispensable for continuous monitoring of atmospheric and surface conditions, especially in cloudy or stormy environments where other sensors are limited.

Microwave remote sensing can be performed using both **passive** and **active** sensors.

5.5.1 Passive Microwave Remote Sensing (Microwave Radiometers/ Sounders)

- **What it is:** Passive microwave sensors detect naturally emitted microwave radiation from the Earth's surface, atmosphere, and clouds. All objects with a temperature above absolute zero emit microwave radiation, and the intensity of this emission is related to their temperature and emissivity.
- **How it Works:** The sensor measures the microwave energy (brightness temperature) at specific frequencies. Different atmospheric gases (like oxygen and water vapor) absorb and emit microwave radiation at characteristic frequencies. By selecting these frequencies, scientists can infer the temperature and humidity of different atmospheric layers. The presence of liquid water (rain, clouds) or ice (snow, hail) also significantly affects the microwave signal.
- **What it Measures/Shows in Meteorology**
    - **Atmospheric Moisture (Water Vapor):** This is a primary strength. By tuning to frequencies sensitive to water vapor absorption (e.g., around 22 GHz and 183 GHz), microwave sounders can derive:
    - **Vertical Humidity Profiles:** Providing crucial moisture information at different altitudes, even through non-precipitating clouds.
    - **Total Precipitable Water Vapor (TPW):** The total amount of water vapor in a column of the atmosphere, vital for understanding the hydrological cycle and forecasting heavy rainfall.
- **Precipitation:** Microwave radiometers are excellent for estimating rainfall rates, particularly over oceans, by detecting the emission from raindrops. Over land, the signal is more complex due to varying surface emissivity, but the scattering by ice particles in convective clouds can still be used to estimate precipitation.
- **Sea Surface Temperature (SST):** Microwave radiation emitted from the ocean surface is influenced by its temperature. While less precise

than infrared for clear-sky conditions, microwave sensors can provide SST data even through clouds, offering continuous global coverage for oceanographic and climate studies.

- **Temperature Profiles:** By tuning to frequencies where oxygen absorbs strongly (e.g., around 60 GHz), microwave sounders can derive vertical temperature profiles from the surface to the stratosphere, even through non-precipitating clouds.
- **Cloud Liquid Water Content:** The amount of liquid water in non-precipitating clouds, important for understanding cloud formation and evolution.
- **Sea Surface Wind Speed:** The roughness of the ocean surface (due to wind) affects its microwave emissivity, allowing for wind speed estimation.
- **Sea Ice Concentration and Type:** Different types of ice and open water have distinct microwave signatures, making these sensors vital for polar monitoring.
- **Snow Depth and Water Equivalent:** Over land, microwave signals are affected by snow cover, allowing for estimates of snow properties.
- **Advantages**
  - **All-Weather Capability:** Can "see through" most non-precipitating clouds, smoke, and dust, providing data where visible and infrared sensors are blind.
  - **Day and Night Operation:** Does not rely on sunlight.
  - **Direct Measurement of Atmospheric Constituents:** Provides direct information on temperature and water vapor profiles.
- **Limitations**
  - **Coarser Spatial Resolution:** Due to the longer wavelengths, microwave sensors generally have much coarser spatial resolution compared to visible and infrared imagers.
  - **Complex Signal Interpretation over Land:** The highly variable emissivity of land surfaces makes retrievals over land more challenging than over oceans.
  - **Scattering by Heavy Precipitation:** Very heavy rain or large ice particles can scatter microwave radiation, making accurate retrievals difficult in intense storms.
- **Examples:** Advanced Technology Microwave Sounder (ATMS) on Suomi NPP/JPSS, Microwave Humidity Sounder (MHS) on MetOp, Special Sensor Microwave Imager/Sounder (SSMIS) on DMSP.

5.5.2 Active Microwave Remote Sensing (Radar)

- **What it is:** Active microwave sensors (radars) transmit their own microwave pulses and then measure the characteristics of the signal that is reflected or scattered back from targets.
- **How it Works:** The sensor sends out a pulse of microwave energy. When this pulse encounters objects like precipitation particles or the Earth's surface, a portion of the energy is scattered back to the sensor. By measuring the time it takes for the pulse to return, the distance to the target is determined. The strength of the returned signal provides information about the size and concentration of the targets, and the Doppler shift can reveal their velocity.
- **What it Measures/Shows in Meteorology**
  - **Precipitation Intensity and 3D Structure:** Spaceborne precipitation radars (like those on TRMM and GPM) provide unparalleled detailed vertical profiles of rain, snow, and hail within clouds, showing the internal structure and evolution of storm systems. This is crucial for quantitative precipitation estimation.
  - **Cloud Profiling:** Cloud radars (like on CloudSat) can detect even non-precipitating clouds and provide their vertical extent, layering, and information about cloud particle sizes, offering insights into cloud microphysics.
  - **Surface Topography (Radar Altimetry):** Measures the precise height of the ocean surface (for sea level rise, ocean currents) and ice sheets.
  - **Surface Roughness/Wind (Scatterometers):** Measures the backscatter from the ocean surface, which is related to wind speed and direction near the surface.
- **Advantages**
  - **Direct and Quantitative Measurement of Precipitation:** Provides highly accurate precipitation data, including its vertical distribution.
  - **Penetration Capability:** Can probe through clouds and into the interior of storm systems.
  - **Vertical Profiling:** Offers unique vertical information on cloud and precipitation structure.
- **Limitations**
  - **Narrow Swath:** Typically have a much narrower observation swath compared to passive imagers, meaning they don't provide continuous wide-area coverage.

- **Power Intensive:** Require significant power to transmit pulses.
- **Data Volume:** Can generate large volumes of data.

- **Examples:** Precipitation Radar (PR) on TRMM, Cloud Profiling Radar (CPR) on CloudSat, Ku/Ka-band Precipitation Radar (DPR) on GPM, Scatterometers (e.g., ASCAT on MetOp).

5.5.3 Conclusion on Microwave Remote Sensing

Microwave remote sensing, encompassing both passive (radiometers/ sounders) and active (radar) technologies, provides an invaluable complement to visible and infrared observations. Its **all-weather capabilities** and ability to penetrate clouds make it essential for continuous monitoring of **precipitation**, **atmospheric moisture (water vapor)**, and **sea surface temperature**, even in challenging cloudy or stormy conditions. This makes microwave data critical for improving global weather forecasts, tracking severe storms, and advancing our understanding of the Earth's water cycle and climate system.

5.6 Hyperspectral and Multispectral Sounders in Satellite Meteorology

In satellite remote sensing, **sounders** are specialized passive instruments designed to measure electromagnetic radiation (typically infrared or microwave) emitted by the Earth's atmosphere at various wavelengths. By carefully selecting these wavelengths, sounders can infer the vertical distribution, or "profile," of atmospheric properties like temperature and humidity. The key distinction among sounders lies in the number and narrowness of the spectral channels they employ: **multispectral** and **hyperspectral**.

5.6.1 Multispectral Sounders

- **What it is:** Multispectral sounders measure radiation in a relatively small number of distinct, broad spectral bands (typically tens of channels). Each channel is designed to capture energy from a specific part of the electromagnetic spectrum, often targeting absorption or emission features of key atmospheric gases.
- **How it Works:** These sounders use filters or gratings to isolate radiation within a few predefined, wider wavelength ranges. For example, some channels might be chosen to be highly sensitive to water vapor, others to carbon dioxide (for temperature sounding), and others to "window" regions (for surface temperature). The measured radiance in each channel is then used in sophisticated retrieval algorithms to infer atmospheric profiles.

- **What it Measures in Meteorology**
  - **Atmospheric Temperature Profiles:** By observing infrared or microwave radiation emitted by well-mixed gases like carbon dioxide or oxygen, multispectral sounders can derive vertical profiles of atmospheric temperature. Different channels are designed to be sensitive to radiation originating from different altitudes, allowing for a layered temperature measurement.
  - **Atmospheric Humidity Profiles:** Channels sensitive to water vapor absorption allow for the retrieval of vertical profiles of atmospheric humidity (moisture content).
  - **Total Precipitable Water (TPW):** The integrated amount of water vapor in a column of the atmosphere.
  - **Cloud Top Pressure/Height/Temperature:** By analyzing radiance in channels affected by clouds, these properties can be estimated.
- **Advantages**
  - **Simpler Design:** Generally less complex and less expensive to build and operate than hyperspectral sounders.
  - **Higher Temporal Resolution (often):** Can sometimes achieve faster scan rates or cover wider swaths, providing more frequent updates.
  - **Robustness:** Less sensitive to minor calibration errors due to broader channels.
- **Limitations**
  - **Limited Vertical Resolution:** Due to the broadness and limited number of channels, the vertical resolution of the retrieved profiles is coarser. They provide less detailed information about fine atmospheric layering.
  - **Less Information Content:** Cannot resolve subtle spectral features that might indicate the presence of other trace gases or provide more precise information about atmospheric state.
- **Examples**
  - **Infrared:** High-resolution Infrared Sounder (HIRS) on NOAA POES satellites.
  - **Microwave:** Advanced Technology Microwave Sounder (ATMS) on Suomi NPP/JPSS, Microwave Humidity Sounder (MHS) on MetOp.

5.6.2 Hyperspectral Sounders

- What it is: Hyperspectral sounders measure radiation in a very large number of very narrow, contiguous spectral bands (typically hundreds to thousands of channels). This creates a virtually continuous spectrum of the emitted or reflected radiation.

- How it Works: These instruments use interferometers or advanced spectrometers to split the incoming radiation into a vast number of very fine spectral bins. This exceptionally high spectral resolution allows them to resolve minute details in the absorption and emission spectra of atmospheric gases. Each narrow channel provides a slightly different "view" into the atmosphere, allowing for a much more precise determination of temperature and moisture at numerous thin layers.
- **What it Measures in Meteorology**
  - **Highly Detailed Atmospheric Temperature Profiles:** Provides significantly improved vertical resolution for temperature profiles, crucial for accurately representing atmospheric stability, inversions, and frontal boundaries.
  - **Highly Detailed Atmospheric Humidity Profiles:** Offers very fine vertical resolution for water vapor, essential for understanding moisture transport, the initiation of convection, and the potential for heavy precipitation.
  - **Trace Gas Concentrations:** The ability to resolve subtle absorption features allows for the retrieval of vertical profiles of various trace gases (e.g., ozone, carbon monoxide, methane, sulfur dioxide), which is vital for air quality monitoring and atmospheric chemistry studies.
  - **Cloud Microphysical Properties:** Can provide more detailed information about cloud phase and particle size distributions.
- **Advantages**
  - **Superior Vertical Resolution:** Offers significantly more detailed vertical profiles of temperature and humidity, providing a much clearer picture of atmospheric layering and fine-scale structures.
  - **Enhanced Trace Gas Detection:** Enables the profiling of a wider range of trace gases with greater accuracy and sensitivity.
  - **Improved Cloud Characterization:** Better discrimination of cloud properties and their impact on atmospheric processes.
  - **Higher Information Content:** Provides more comprehensive data for assimilation into numerical weather prediction models, leading to more accurate and skillful forecasts.

- **Limitations**
  - **Complex and Expensive:** More technologically complex and costly to build, operate, and process data from.
  - **Large Data Volume:** Generates a massive amount of raw data, requiring significant processing power and storage infrastructure.
  - **Slower Scan Rates (sometimes):** The need for high spectral resolution can sometimes lead to slower scan rates or narrower swaths compared to some multispectral imagers.
- **Examples**
  - **Infrared:** Atmospheric Infrared Sounder (AIRS) on Aqua, Infrared Atmospheric Sounding Interferometer (IASI) on MetOp, Cross-track Infrared Sounder (CrIS) on Suomi NPP/JPSS.

5.6.3 Complementary Roles

Both multispectral and hyperspectral sounders play vital roles in satellite meteorology. While multispectral sounders offer broad coverage and robustness for operational monitoring, hyperspectral sounders provide the exquisite vertical detail necessary for advanced atmospheric research and for pushing the boundaries of numerical weather prediction. The continuous development and deployment of these sophisticated instruments are fundamental to our ever-improving understanding and forecasting of the Earth's dynamic atmosphere.

5.7 Satellite Data Products: Types, Formats, and Uses in Meteorology

Satellite remote sensing generates a vast amount of raw data, which is then processed and transformed into various **satellite data products**. These products are derived measurements or interpretations of the raw sensor data, designed to be directly useful for meteorological analysis, forecasting, and research. The specific product, its format, and its intended use are all interconnected.

5.7.1 Types of Satellite Data Products

Satellite data products can be broadly categorized based on the atmospheric or surface parameter they represent. Here are some of the most common and important meteorological products:

- **Imagery (Visible, Infrared, Water Vapor):**
  - **Visible Imagery:** Shows reflected sunlight, indicating cloud cover, cloud texture, snow/ice, and land features during daylight. Used for tracking cloud movement, identifying fog, and monitoring surface changes.

- **Infrared (IR) Imagery:** Shows emitted thermal radiation, providing cloud top temperature and height (colder = higher), surface temperature (when clear), and tracking cloud movement day and night.
- **Water Vapor Imagery:** Derived from specific IR or microwave channels sensitive to water vapor absorption/emission. Reveals the distribution of moisture in the middle and upper troposphere, even in clear skies, indicating atmospheric flow and potential for convection.

- **Atmospheric Profiles**
  - **Temperature Profiles:** Vertical distribution of air temperature from the surface to the stratosphere, crucial for understanding atmospheric stability, inversions, and frontal zones. Derived from IR and microwave sounders.
  - **Humidity Profiles:** Vertical distribution of atmospheric moisture, essential for forecasting precipitation, fog, and understanding the hydrological cycle. Derived from IR and microwave sounders.
- **Cloud Products**
  - **Cloud Mask:** Identifies pixels as either cloudy or clear.
  - **Cloud Top Height/Pressure/Temperature:** Precise measurements of the top of cloud layers.
  - **Cloud Optical Thickness/Effective Radius:** Information about the physical properties of clouds, impacting radiative transfer.
  - **Cloud Phase:** Distinguishing between liquid water clouds and ice clouds.
- **Precipitation Products**
  - **Rainfall Rate:** Estimates of instantaneous or accumulated precipitation over land and oceans, derived from microwave radiometers and active radars.
  - **Snow Cover/Depth:** Mapping the extent and depth of snow.
- **Wind Products**
  - **Atmospheric Motion Vectors (AMVs):** Wind speed and direction derived by tracking the movement of cloud features or water vapor patterns in successive satellite images (from geostationary imagers).
  - **Scatterometer Winds:** Near-surface wind speed and direction over oceans, derived from active microwave scatterometers.
  - **Lidar Winds:** Vertical profiles of wind speed and direction derived from Doppler lidars.

- **Surface Products**
  - **Sea Surface Temperature (SST):** Temperature of the ocean's top layer, vital for oceanography, climate monitoring, and hurricane intensity forecasting. Derived from IR and microwave radiometers.
  - **Land Surface Temperature (LST):** Temperature of the land surface.
  - **Vegetation Indices (e.g., NDVI):** Reflectivity measurements used to assess vegetation health and density.
  - **Snow and Ice Cover:** Extent and properties of snow and ice on land and sea.
- **Atmospheric Composition Products**
  - **Aerosol Optical Depth (AOD):** A measure of how much aerosols (dust, smoke, pollution) block light, indicating air quality and aerosol loading.
  - **Trace Gas Concentrations/Profiles:** Vertical profiles or total column amounts of gases like ozone (O3), carbon monoxide (CO), methane (CH4), sulfur dioxide (SO2), etc., derived from hyperspectral sounders.
  - **Volcanic Ash Detection:** Identification and tracking of volcanic ash plumes.

5.7.2 Data Formats

Satellite data products are stored and distributed in various standardized formats to ensure interoperability and ease of use by different meteorological and scientific communities. Common formats include:

- **NetCDF (Network Common Data Form):** A widely used, self-describing, machine-independent data format for array-oriented scientific data. It's highly flexible and efficient for storing meteorological variables (e.g., temperature, humidity, wind) along with their dimensions (latitude, longitude, altitude, time).
- **HDF (Hierarchical Data Format):** A set of file formats (HDF4, HDF5) designed to store and organize large amounts of complex data. Popular in NASA missions, it supports various data types and structures, making it suitable for multi-dimensional satellite products.
- **GRIB (GRIdded Binary):** A highly compact binary format used by meteorological centers worldwide for storing and exchanging gridded meteorological data, including output from numerical weather prediction models and satellite-derived products.

- **BUFR (Binary Universal Form for the Representation of meteorological data):** A flexible binary format for the exchange of observational data, including satellite radiance measurements and derived products.
- **GeoTIFF:** A standard TIFF image file format that includes embedded geospatial information (e.g., map projection, coordinates), making it suitable for satellite imagery that needs to be displayed on a map.
- **ASCII/CSV:** For smaller datasets or specific parameters, plain text formats like ASCII (American Standard Code for Information Interchange) or CSV (Comma Separated Values) might be used, though they are less efficient for large, multi-dimensional data.

5.7.3 Uses of Satellite Data Products in Meteorology

Satellite data products are fundamental to almost every aspect of modern meteorology:

- **Numerical Weather Prediction (NWP)**
  - **Data Assimilation:** Temperature and humidity profiles, wind vectors, and radiance measurements are assimilated into NWP models to improve the initial conditions, which is crucial for accurate forecasts.
  - **Model Validation:** Satellite products are used to verify and improve the performance of NWP models by comparing model outputs with actual observations.
- **Weather Forecasting and Nowcasting**
  - **Real-time Monitoring:** Imagers provide continuous updates on cloud cover, storm development, and frontal systems, enabling forecasters to track weather evolution.
  - **Severe Weather Prediction:** Tracking of hurricane intensity (e.g., using cloud top temperatures), thunderstorm development, and the movement of severe weather systems.
  - **Fog and Aviation Hazards:** Identification of fog, low clouds, and volcanic ash plumes for aviation safety.
- **Climate Monitoring and Research**
  - **Long-term Trends:** Consistent, global records of temperature, humidity, cloud properties, sea ice extent, and trace gases are essential for monitoring climate change and its impacts.
  - **Atmospheric Processes:** Studying cloud formation, atmospheric circulation, energy balance, and the global water cycle.

- **Aerosol and Air Quality Studies:** Monitoring the transport and impact of pollutants and natural aerosols.

- **Hydrology and Water Resources**
  - **Precipitation Estimation:** Crucial for flood forecasting, drought monitoring, and water resource management, especially in regions with sparse ground-based rain gauges.
  - **Snow and Ice Monitoring:** Assessing snowpack for water supply predictions and monitoring changes in glaciers and ice sheets.
- **Oceanography**
  - **Sea Surface Temperature:** Tracking ocean currents, El Niño/La Niña events, and marine heatwaves.
  - **Sea Surface Winds:** Important for marine forecasting and understanding ocean-atmosphere interactions.
  - **Sea Ice Extent:** Monitoring arctic and antarctic ice cover for shipping and climate studies.
- **Environmental Applications**
  - **Volcanic Ash Advisories:** Tracking ash plumes to warn aviation.
  - **Wildfire Monitoring:** Detecting active fires and tracking smoke plumes.
  - **Dust Storms:** Monitoring the initiation and transport of dust.

In essence, satellite data products transform raw measurements into actionable information, forming the backbone of global weather and climate services and enabling a deeper understanding of our planet's complex systems.

5.8 Conclusion

This chapter has illuminated the pivotal role of satellite remote sensing in modern meteorology. Orbiting platforms, whether geostationary or in low Earth orbit, equipped with a diverse array of sophisticated sensors, provide an indispensable global perspective on our planet's atmosphere and surface.

From the continuous, synoptic views offered by visible and infrared imagers, allowing us to track weather systems as they evolve, to the detailed vertical profiles of temperature and humidity provided by multispectral and hyperspectral sounders, satellites furnish critical data for weather analysis and forecasting. Furthermore, microwave sensors overcome the limitations of clouds, offering insights into precipitation, atmospheric moisture, and sea surface conditions under various weather scenarios. Active sensors like radar and lidar add another dimension, enabling the direct measurement of precipitation structure, cloud heights, aerosol distributions, and even wind profiles.

The transformation of raw satellite measurements into a multitude of data products, disseminated in standardized formats, underscores the utility of this technology across a broad spectrum of meteorological applications. These products are fundamental to initializing and validating numerical weather prediction models, monitoring severe weather, studying climate change, and supporting various environmental and oceanographic disciplines.

In conclusion, satellite remote sensing has revolutionized our ability to observe and understand the Earth's atmosphere. Its global reach, continuous monitoring capabilities, and the wealth of information derived from its sensors have become integral to operational meteorology, climate science, and environmental stewardship. As technology continues to advance, we can anticipate even more sophisticated satellite systems providing increasingly detailed and accurate data, further enhancing our ability to predict and prepare for the ever-changing dynamics of our planet.

6

Data Formats, Storage and Access

6.1 Introduction

Our journey through remote sensing has, thus far, focused on the remarkable capabilities of various platforms and sensors to observe Earth's atmosphere and surface. From the localized, high-resolution insights provided by ground-based instruments like Wind Profilers and SODARs, to the expansive, global coverage offered by an array of satellite-borne imagers, sounders, radars, and lidars, we've seen how a continuous stream of invaluable meteorological data is acquired.

However, the mere act of collecting this data, no matter how sophisticated the technology or comprehensive the observations, represents only the initial phase of its lifecycle. The true power and utility of remote sensing data are unlocked only when it can be efficiently organized, securely preserved, and readily available to the diverse community of meteorologists, climate scientists, researchers, and operational forecasters who rely on it daily. This necessitates a profound understanding of the principles and practices governing **data formats**, **data storage**, and **data access**.

The sheer volume, velocity, and variety of remote sensing data pose significant challenges. Petabytes of information are generated annually, requiring robust infrastructure and standardized methodologies to ensure its integrity, interoperability, and long-term usability. Without effective strategies for managing this deluge of information, the potential of remote sensing to enhance weather prediction, advance climate science, and inform critical decision-making would remain largely untapped.

Chapter 6 will delve into these foundational aspects of data management. We will explore the various **standardized and specialized data formats** that have been developed to efficiently represent complex meteorological and remote sensing observations, ensuring consistency and ease of exchange across different systems and institutions. Subsequently, we will examine the evolving landscape of **data storage solutions**, from traditional archival systems to modern cloud-based paradigms, considering the requirements for scalability, reliability, and cost-effectiveness. Finally, we will investigate the diverse **mechanisms and platforms for data access**, discussing how users can discover, retrieve, and effectively utilize this invaluable information to drive scientific discovery, improve forecasting capabilities, and address pressing environmental challenges. Understanding these elements is paramount for transforming raw observations into actionable knowledge in the meteorological domain.

6.2 Data Formats: General, Standardized, and Specialized in Remote Sensing

In the intricate world of remote sensing and meteorology, **data formats** are the structured ways in which digital information is organized, stored, and exchanged. They define how raw measurements, derived products, and associated metadata are encoded into files. The choice of data format is critical as it impacts storage efficiency, processing speed, ease of sharing, and the overall usability of the data. Given the vast volume and complexity of remote sensing data, various formats have evolved, falling broadly into general, standardized, and specialized categories (Fig. 6.1).

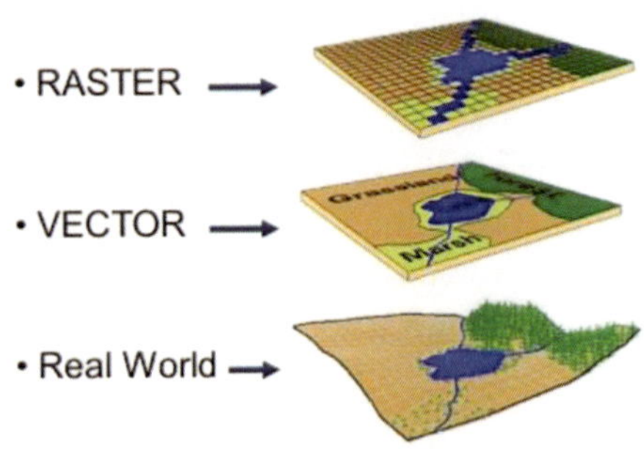

Fig. 6.1: Data format in remote sensing

6.2.1 General Data Formats

General data formats are typically simple, widely compatible, and often human-readable. While versatile for basic data, they are generally not optimized for the complex, multi-dimensional, and metadata-rich nature of remote sensing data.

- **Characteristics**
  - **Simplicity:** Easy to understand and parse with common programming tools.
  - **Human-Readable:** Often plain text, allowing direct inspection.
  - **Broad Compatibility:** Can be opened and read by almost any text editor or basic software.

- **Examples**
  - **ASCII (American Standard Code for Information Interchange):** A foundational character encoding. Data is stored as plain text characters. While individual values are readable, large scientific datasets in pure ASCII are inefficient.
  - **CSV (Comma Separated Values):** A very common plain-text format where data values are separated by commas (or other delimiters) and rows are separated by newlines. Often used for tabular data, such as lists of station observations or simple time series.
  - **Plain Text Files:** Any file where data is simply written as human-readable text, often with custom delimiters or fixed column widths.
- **Pros:** Easy to create, share, and debug for small datasets; highly interoperable across different operating systems and basic software.
- **Cons:** Inefficient for large datasets (large file sizes); lacks built-in metadata capabilities (requiring external documentation); difficult to represent multi-dimensional arrays or complex hierarchical structures; prone to parsing errors without strict adherence to custom rules.

6.2.2 Standardized Data Formats

Standardized data formats are designed to overcome the limitations of general formats by providing a robust structure for complex scientific data, often incorporating metadata directly within the file. They promote interoperability and long-term archival.

- **Characteristics**
  - **Self-Describing:** Contains metadata (e.g., variable names, units, dimensions, coordinate systems, creation date) within the file, making it understandable without external documentation.
  - **Machine-Independent/Platform-Agnostic:** Can be read and written across different computer systems and operating systems.
  - **Support for Complex Data Structures:** Capable of storing multi-dimensional arrays, hierarchical groups, and various data types.
  - **Community-Driven Standards:** Developed and maintained by scientific communities to ensure broad adoption and consistency.
- **Examples**
  - **NetCDF (Network Common Data Form)**
  - **Description:** A widely adopted interface for array-oriented data. It is self-describing, allowing users to understand the data's structure and

content without external information. It's highly efficient for storing time series and gridded data (e.g., temperature fields over time).

- **Use in Meteorology:** A cornerstone format for climate model output, reanalysis datasets, and many gridded satellite products (e.g., sea surface temperature maps, precipitation estimates).

- **HDF (Hierarchical Data Format)**
  - **Description:** A set of file formats (HDF4, HDF5) designed to store and organize extremely large and complex collections of heterogeneous data. HDF files can contain multiple datasets, groups, and metadata, forming a hierarchical structure.
  - **Use in Meteorology:** Extensively used by NASA and other space agencies for storing raw satellite radiance data, atmospheric profiles (e.g., from AIRS, CrIS), aerosol optical depth, and various Earth system science products.
- **GeoTIFF**
  - **Description:** An extension of the standard TIFF (Tagged Image File Format) that embeds geospatial metadata (e.g., map projection, coordinate system, georeferencing information) directly within the image file.
  - **Use in Meteorology:** Commonly used for satellite imagery (visible, infrared) that needs to be displayed and analyzed within Geographic Information Systems (GIS) or other mapping software.
- **Pros:** Excellent for data sharing and long-term archiving; rich metadata support; efficient for large, complex datasets; widely supported by scientific software libraries.
- **Cons:** Can be more complex to work with initially than general formats; requires specific libraries for reading/writing.

6.2.3 Specialized Data Formats

Specialized data formats are developed for very specific types of data or applications within a particular domain. They are often highly optimized for compactness and efficiency for their intended purpose, but may be less general-purpose (Fig. 6.2).

- **Characteristics**
  - **Domain-Specific:** Tailored to the unique characteristics of data from a particular instrument or application.
  - **Highly Optimized:** Designed for maximum storage efficiency and/or processing speed for their specific data type.

- **Often Binary:** Frequently use binary encoding for compactness, making them non-human-readable without specialized tools.

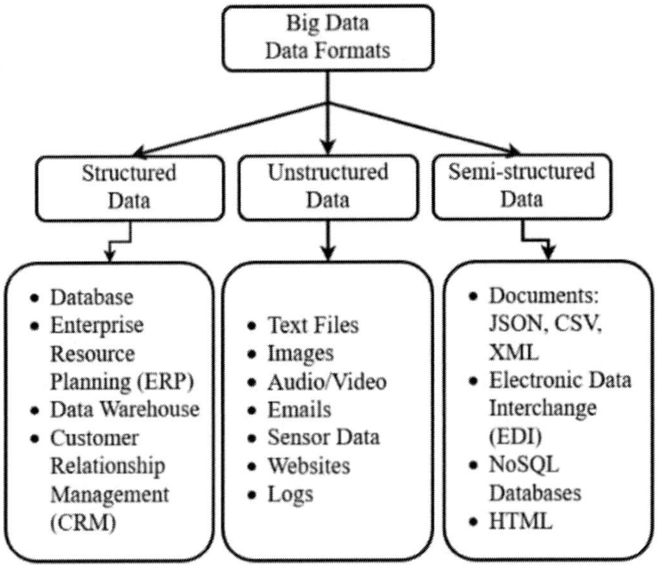

Fig. 6.2: Big data formats

- **Examples**
  - **GRIB (GRIdded Binary)**
  - **Description:** A highly compact binary format developed by the World Meteorological Organization (WMO) for the efficient storage and exchange of gridded meteorological data. It uses complex compression schemes.
  - **Use in Meteorology:** The de-facto standard for numerical weather prediction (NWP) model output, reanalysis data, and many operational gridded forecasts distributed by national meteorological centers (e.g., NOAA, ECMWF).
  - BUFR (Binary Universal Form for the Representation of meteorological data):
  - **Description:** Another WMO-standardized binary format designed for the flexible and efficient exchange of observational data. It can encode a wide variety of meteorological observations, including surface measurements, radiosonde soundings, and raw satellite radiance data.

    - **Use in Meteorology:** Widely used for transmitting real-time observational data from various sources to NWP centers for data assimilation.
- **LAS (LASer)**
    - **Description:** A public file format for the exchange of 3D point cloud data, specifically designed for LiDAR data. It stores XYZ coordinates, intensity, return number, classification, and other attributes for each point.
    - **Use in Remote Sensing:** The standard format for storing and exchanging airborne and terrestrial LiDAR data, used for terrain modeling, forestry, and urban planning.
- **JPEG 2000 (for imagery):** Offers superior compression and quality compared to traditional JPEG, often used for high-resolution satellite imagery where efficient storage and transmission are critical.
- **Pros:** Extremely compact and efficient for their specific data types; designed to meet the rigorous demands of operational systems.
- **Cons:** Require specialized software and libraries for decoding; less flexible for general-purpose data storage; can be more complex to implement.

In conclusion, the choice of data format is a strategic decision in remote sensing and meteorology. While general formats offer simplicity for basic tasks, standardized formats like NetCDF and HDF provide the robustness and metadata capabilities needed for complex scientific data. Specialized formats like GRIB and BUFR are crucial for the efficient operational exchange of highly specific meteorological information. Together, these formats form the backbone of the global data infrastructure that supports weather forecasting, climate research, and a myriad of environmental applications.

6.3 Data Archiving: The Custodians of Meteorological History and Future Research

Data archiving is the systematic process of storing and preserving meteorological data for extended periods, safeguarding its integrity and ensuring its usability for future generations. For a field like meteorology, where historical context is vital for understanding climate variability and validating long-term models, robust archiving is absolutely essential.

- **Core Purpose**
    - **Perpetual Preservation:** To ensure that invaluable observational records, model outputs, and derived products remain available and

usable for decades, centuries, or even longer, serving as an unbroken historical record of Earth's atmospheric and climatic state.

- **Climate Change Monitoring and Research:** Provides the foundational, consistent, and long-term datasets necessary for detecting subtle climate trends, validating climate models, and understanding the complex dynamics of a changing planet.
- **Scientific Reproducibility and Validation:** Enables researchers to verify past findings, conduct reanalysis studies (integrating historical observations with modern models), and build upon existing knowledge, fostering scientific rigor and progress.
- **Forecasting Model Development:** Historical data is crucial for training and validating new generations of numerical weather and climate models, improving their accuracy and skill.
- **Legal and Regulatory Compliance:** Adherence to national and international mandates for data retention and accessibility, particularly for public safety and environmental monitoring.

- **Key Characteristics**
  - **Absolute Data Integrity:** Paramount emphasis on preventing data loss or corruption through rigorous measures such as checksums, redundant storage (e.g., RAID configurations, geographically distributed copies), and robust error correction codes.
  - **Cost-Effectiveness at Scale:** Prioritizes high-capacity, lower-cost storage media optimized for infrequent access, such as magnetic tapes or cloud cold storage tiers, to manage petabytes to exabytes of data economically.
  - **Comprehensive Metadata:** Meticulous preservation of all associated metadata (e.g., instrument type, observation time, location, processing levels, quality flags, units, data provenance) is critical to ensure future interpretability and usability of the data by diverse users.
  - **Format Migration Strategies:** Proactive planning and execution of data migration to newer, more sustainable, and widely supported data formats over time to combat technological obsolescence and ensure long-term readability.
- **Common Methods and Technologies**
  - **Magnetic Tape Libraries (e.g., LTO - Linear Tape-Open):** A traditional and still highly prevalent method for massive, offline, long-term data storage. Tapes offer excellent density, low cost per terabyte, and a long shelf life, though access times can be slower.

- **Optical Disc Archives (e.g., M-DISC, specialized archival Blu-ray):** Offer very long physical durability and resistance to environmental factors, making them suitable for ultra-long-term cold storage.
- **Cloud Archival Storage Services:** Major cloud providers (e.g., Amazon Glacier, Google Cloud Archive, Azure Archive Storage) offer highly durable, scalable, and cost-effective storage tiers specifically designed for infrequently accessed archival data, often with retrieval times ranging from minutes to hours.
- **On-Premise Disk Storage Arrays (NAS/SAN):** While more expensive per terabyte than tape, large disk arrays are used for "warm" archives—historical data that is accessed more frequently than true cold archives.
- **Specialized Archival Systems:** Large meteorological and climate data centers develop custom-built data management systems that integrate hardware, software, and policies for robust data ingestion, validation, versioning, and long-term preservation.

- **Persistent Challenges**
  - **Exorbitant Data Volume:** The continuous influx of data from global observing systems and high-resolution models creates immense logistical and technical hurdles for storage and management.
  - **Technological Obsolescence:** The rapid evolution of storage hardware and software formats demands ongoing migration efforts to ensure data readability across decades.
  - **Data Curation and Discoverability:** Ensuring that archived data remains well-documented, properly indexed, and easily discoverable by future users, even those unfamiliar with the original data source or mission.
  - **Sustainable Funding:** The long-term nature of archiving requires consistent and substantial financial commitment for infrastructure, maintenance, and expert personnel.
- **Prominent Archival Institutions:** NOAA's National Centers for Environmental Information (NCEI), NASA's Earth Observing System Data and Information System (EOSDIS), the European Centre for Medium-Range Weather Forecasts (ECMWF) archives, the European Organisation for the Exploitation of Meteorological Satellites (EUMETSAT) archives, and national meteorological services worldwide.

6.3.1 Databases: The Dynamic Gateways to Meteorological Information

Databases are structured collections of data, meticulously organized for efficient storage, rapid retrieval, dynamic management, and complex querying. In meteorology, databases are primarily employed to manage metadata, index vast datasets, store frequently accessed derived products, and facilitate near real-time data flow for operational systems.

- **Core Purpose**
    - **Efficient Data Retrieval and Querying:** Optimized for quick searching, filtering, and retrieval of specific data records based on various criteria (e.g., time range, geographic area, observation type, model run, processing level).
    - **Metadata Management:** Centralized storage of comprehensive metadata (e.g., instrument ID, observation time, location, processing level, quality flags, data provenance, file locations) that enables users to precisely discover and understand relevant data.
    - **Data Discovery and Cataloging:** Powering user-facing catalogs, portals, and Application Programming Interfaces (APIs) that allow users to browse, search, and select available meteorological products from diverse sources.
    - **Operational Data Flow:** Supporting the rapid ingestion, processing, and distribution of near real-time observational data (e.g., satellite imagery, radar scans, surface station reports) and model outputs for operational weather forecasting and monitoring systems.
    - **Data Integration:** Facilitating the seamless integration and fusion of different types of meteorological data (e.g., satellite, in-situ, model) for comprehensive analysis.
- **Relevant Database Types**
    - **Relational Databases (RDBMS):** (e.g., PostgreSQL, MySQL, Oracle) Excel at managing highly structured, tabular data with predefined schemas. They are ideal for storing metadata (e.g., observation times, station locations, product availability, processing logs) and support powerful SQL (Structured Query Language) for complex filtering, sorting, and joining of information. Geospatial extensions like PostGIS for PostgreSQL are crucial for spatial indexing and querying of meteorological data.
    - **NoSQL Databases:** (e.g., MongoDB for document-oriented, Apache Cassandra for wide-column, Elasticsearch for search/indexing) Offer flexible schemas and horizontal scalability, making them suitable

for managing large volumes of semi-structured metadata, rapidly changing data streams (e.g., real-time sensor feeds), or building powerful search capabilities over data catalogs.

- **Time-Series Databases:** (e.g., InfluxDB, TimescaleDB, OpenTSDB) Specifically optimized for storing and querying time-stamped data, which is inherent to almost all meteorological observations and model outputs. They are highly efficient for handling high-volume, high-frequency sensor readings or derived time-series products.

- **How Databases Interact with Meteorological Data**
  - **Metadata Catalogs:** Databases serve as the backend for sophisticated data catalogs. When a user submits a query (e.g., "all surface temperature observations in Europe for the last 24 hours," or "all GRIB files from the latest global model run"), the database rapidly queries its indexed metadata tables to identify the relevant data records or file pointers.
  - **Indexing:** Databases create and maintain indexes on key attributes (e.g., acquisition time, geographic coordinates, instrument ID, model run time, variable name) to dramatically speed up data searches, even across petabytes of underlying data.
  - **Pointers to Data Files:** Crucially, databases typically do not store the raw, massive meteorological data files (e.g., large NetCDF, HDF, GRIB files) directly. Instead, they store "pointers" or file paths (URIs) to where the actual data files are located in an archive, object storage, or a distributed file system.
  - **Operational Data Stores:** For near real-time applications, databases manage the rapid ingestion, processing, and distribution of recently acquired and processed data, making it immediately available to forecasters, automated systems, and public dissemination channels.
  - **Data Provenance and Lineage:** Databases can track the origin and processing history of data products, linking derived products back to their raw observational inputs or model runs.
- **Persistent Challenges**
  - **Schema Design and Evolution:** Designing flexible and scalable database schemas that can accommodate the diverse and evolving nature of meteorological data and metadata.
  - **Performance Tuning:** Optimizing database performance for massive data ingestion rates and complex, multi-dimensional queries across heterogeneous data types.

- **Data Consistency and Synchronization:** Ensuring data consistency across distributed database systems and synchronization with underlying archives.
- **Integration Complexity:** Integrating databases with various data acquisition systems, processing pipelines, archival systems, and user access portals.

6.3.2 The Symbiotic Relationship: Archives & Databases in the Data Lifecycle

Data archiving and databases are not isolated components but rather form a synergistic and tiered ecosystem essential for comprehensive meteorological data management. They represent different stages and access patterns within the data lifecycle:

- **Data Ingestion & Processing:** Raw data from all sources (sensors, models) is acquired and undergoes initial processing.
- **Cataloging (Database):** Metadata for the newly processed data products is ingested into databases, making the data discoverable and searchable.
- **Operational Access (Database/Warm Storage):** Recently acquired and frequently accessed data may reside in faster, database-backed storage tiers for near real-time operational use by forecasters and automated systems.
- **Long-Term Preservation (Archive):** As data ages or becomes less frequently accessed, it is migrated to cost-effective, high-durability archival storage. The database retains the metadata and a pointer to the archived location, ensuring its long-term discoverability.

This tiered approach, often referred to as **data tiering** (e.g., hot, warm, cold storage), ensures that data is stored in the most appropriate and cost-effective manner based on its access frequency and criticality. Databases provide the intelligent front-end for discovery, management, and operational access, while archives serve as the secure, enduring back-end repository. Adherence to **metadata standards** (e.g., ISO 19115, WMO Core Metadata Profile) is crucial for seamless integration between these components, ultimately supporting the **FAIR principles** (Findable, Accessible, Interoperable, Reusable) for scientific data. This comprehensive approach is vital for maximizing the scientific and societal impact of all meteorological data.

6.4 NoSQL Databases

NoSQL stands for "Not only SQL" or "non-relational." These databases are designed to handle large volumes of unstructured, semi-structured, and structured data with flexible schemas, differing from traditional relational databases (which use SQL and have rigid, table-based schemas).

Here's a breakdown of NoSQL databases in the context of meteorological data:

6.4.1 Characteristics and Benefits for Meteorological Data

- **Flexibility:** Meteorological data comes in various forms (gridded model outputs, point observations, satellite imagery). NoSQL databases can often handle these diverse formats more naturally than strictly structured relational databases. For example, you could store each model grid point as a document with associated metadata.
- **Scalability:** Meteorological datasets are often very large and continuously growing. NoSQL databases are generally designed for horizontal scaling, meaning you can add more servers to handle increased data volume and traffic, which is crucial for big meteorological data.
- **Performance:** For certain types of queries and data access patterns common in meteorology (e.g., time series analysis, spatial lookups), some NoSQL databases can offer better performance than traditional SQL databases.
- **Schema-less Design:** The evolving nature of meteorological models and data formats can be accommodated more easily by the schema-less nature of many NoSQL databases. You don't need to pre-define the structure of your data rigidly.

6.4.2 Types of NoSQL Databases Relevant to Meteorology

- **Document Databases (e.g., MongoDB, Amazon DocumentDB):** Store data as JSON-like documents. This can be useful for storing metadata alongside gridded data or for representing complex observational datasets.
- **Key-Value Stores (e.g., Redis, Amazon DynamoDB):** Store data as key-value pairs. Could be used for caching frequently accessed meteorological data or for session management in web-based meteorological portals.
- **Wide-Column Stores (e.g., Apache Cassandra, Apache HBase):** Store data in columns rather than rows, which can be efficient for analytical queries across large datasets, potentially useful for climate data analysis.
- **Time-Series Databases (e.g., InfluxDB, TimescaleDB):** Optimized for storing and querying time-stamped data, which is fundamental to almost all meteorological observations and model outputs. These are particularly well-suited for visualizing trends and performing temporal analysis.

- **Graph Databases (e.g., Neo4j, Amazon Neptune):** Focus on relationships between data points. While perhaps less directly applicable to raw meteorological data, they could be used for network analysis of weather patterns or for managing complex metadata relationships.

6.4.3 Use Cases in Meteorology

- Storing and querying large ensembles of climate model simulations.
- Real-time ingestion & analysis of sensor data from distributed networks.
- Building scalable web applications for visualizing meteorological data.
- Managing metadata for vast archives of satellite and radar data.
- Analyzing trends in long-term observational datasets.

6.5 Cloud Storage Solutions

Cloud storage solutions refer to services that allow users to store data on remote servers, typically maintained by a third-party provider, rather than directly on local hardware. This data is accessible over a network (usually the internet) from various devices. For the vast and ever-growing datasets in meteorology, cloud storage has become an increasingly vital component of data management strategies, offering significant advantages over traditional on-premise infrastructure.

- **What are Cloud Storage Solutions?**

  At its core, cloud storage involves storing digital data in logical pools across multiple virtual servers, rather than on a single dedicated server. The physical storage spans multiple data centers, and the environment is typically managed by a cloud service provider (CSP) like Amazon Web Services (AWS), Google Cloud Platform (GCP), or Microsoft Azure. Users pay for the storage consumed, often on a "pay-as-you-go" model, and can scale their storage capacity up or down as needed.

- **Key Benefits for Meteorological Data**

  Cloud storage offers several compelling advantages that are particularly beneficial for the unique demands of meteorological data:

  - **Scalability:** Meteorological data volumes are enormous and constantly growing (petabytes to exabytes). Cloud storage can scale almost infinitely and on-demand, eliminating the need for organizations to continuously invest in and manage their own physical storage hardware.
  - **Accessibility:** Data stored in the cloud can be accessed from anywhere in the world with an internet connection. This facilitates collaboration

among geographically dispersed researchers, forecasters, and international meteorological organizations.

- **Durability and Reliability:** Cloud providers design their storage services for extreme durability, often replicating data across multiple physical devices and geographical regions. This significantly reduces the risk of data loss due to hardware failures or localized disasters.
- **Cost-Effectiveness:** While initial setup might seem complex, the pay-as-you-go model and reduced need for in-house hardware, maintenance, and IT personnel can lead to significant cost savings, especially for large archives or variable data loads.
- **Disaster Recovery and Business Continuity:** Data is inherently distributed and backed up across multiple locations, providing built-in disaster recovery capabilities. In the event of a local outage, data remains accessible from other regions.
- **Integration with Cloud Computing:** Cloud storage seamlessly integrates with cloud-based computing resources (e.g., virtual machines, serverless functions, big data analytics services). This allows for "data proximity"– processing data where it lives, reducing transfer times and costs, which is critical for computationally intensive meteorological models and analyses.
- **Security:** Cloud providers invest heavily in security infrastructure, protocols, and compliance certifications, often exceeding what individual organizations can achieve on their own.

6.5.1 Common Types of Cloud Storage Solutions

Cloud storage services are typically offered in different tiers, optimized for various access patterns and cost requirements:

- **Object Storage**
    - **Description:** This is the most common and flexible type of cloud storage for unstructured data (e.g., files, images, videos, large datasets). Data is stored as "objects" within "buckets," each identified by a unique key. It's highly scalable and durable.
    - **Tiers:** Cloud providers offer different tiers within object storage, optimized for access frequency:
    - **Hot Storage (Standard/Frequent Access):** For data that is accessed frequently (e.g., current weather observations, recent model outputs, active research datasets). Offers low latency and high throughput.

- **Warm Storage (Infrequent Access):** For data accessed less frequently but still needing quick retrieval (e.g., historical reanalysis data, older model runs). Lower storage cost, but slightly higher retrieval fees.
- **Cold Storage (Archive/Deep Archive):** For long-term archival of rarely accessed data (e.g., very old observational records, raw satellite radiance data). Offers the lowest storage cost, but retrieval times can range from minutes to hours, with higher retrieval fees. This tier is particularly relevant for meteorological data archiving.
- **Examples:** Amazon S3 (Simple Storage Service) with its various storage classes (Standard, Infrequent Access, Glacier, Glacier Deep Archive), Google Cloud Storage (Standard, Nearline, Coldline, Archive), Azure Blob Storage (Hot, Cool, Archive).

- **File Storage**
  - **Description:** Provides shared file system interfaces (like NFS or SMB) that allow multiple users or applications to access and share data simultaneously, similar to network-attached storage (NAS) on-premises.
  - **Use in Meteorology:** Useful for collaborative workspaces, shared model input/output directories, or for applications that require a traditional file system interface.
  - **Examples:** Amazon EFS (Elastic File System), Google Cloud Filestore, Azure Files.
- **Block Storage**
  - **Description:** Provides raw, unformatted storage volumes that can be attached to virtual servers (e.g., cloud-based virtual machines). It's ideal for high-performance applications that require low-latency access to data, like databases or operating system disks.
  - **Use in Meteorology:** Primarily used for the underlying storage of cloud-based virtual machines running specific meteorological applications, databases (e.g., a PostgreSQL database for metadata), or high-performance computing (HPC) workloads.
  - **Examples:** Amazon EBS (Elastic Block Store), Google Persistent Disk, Azure Managed Disks.

6.5.2 Cloud Storage and Meteorological Data Management

Meteorological organizations leverage cloud storage solutions in various ways:

- **Data Lakes:** Building massive data lakes using object storage to store all raw and processed meteorological data in its native format, enabling flexible analytics and future processing.
- **Operational Data Pipelines:** Integrating cloud storage into real-time data ingestion and processing pipelines for immediate access by forecasting systems.
- **Archival and Long-Term Preservation:** Migrating vast historical datasets from tape libraries to cloud cold storage tiers for cost-effective, highly durable archiving.
- **Data Distribution:** Using cloud storage as a robust and scalable platform to distribute meteorological data products to users worldwide.
- **Research and Development:** Providing researchers with easy access to large datasets and co-located computing power for complex analyses without needing to download data.
- **Disaster Recovery:** Creating redundant copies of critical operational data in different cloud regions to ensure business continuity.

In essence, cloud storage solutions are transforming how meteorological data is managed, stored, and accessed, enabling greater flexibility, scalability, and collaboration, which are all critical for advancing weather forecasting and climate science in the era of big data.

6.6 Data Access APIs and Protocols

Once meteorological data is formatted and stored, the crucial next step is enabling efficient **data access**. This is primarily achieved through the use of **Application Programming Interfaces (APIs)** and standardized **protocols**, which define how different systems and users can request, retrieve, and interact with the data.

6.6.1 Data Access Protocols

Data access protocols are the fundamental rules and standards that govern how data is transmitted and retrieved over a network. They ensure that clients (users or applications) and servers (data repositories) can communicate effectively.

- **Description:** The foundation of data communication on the World Wide Web. It's a stateless protocol where clients send requests to servers, and servers respond. HTTPS adds encryption for secure transmission.
- **Use in Meteorology:** Widely used for accessing web-based data portals, downloading files (e.g., NetCDF, GRIB), and interacting with web services (often built as RESTful APIs).

- **FTP/SFTP (File Transfer Protocol/Secure)**
  - **Description:** Protocols designed specifically for transferring files between a client and a server. SFTP adds security through SSH encryption.
  - **Use in Meteorology:** Still used by some organizations for bulk data transfer of large meteorological datasets.
- **OPeNDAP (Open-source Project for a Network Data Access Protocol)**
  - **Description:** A protocol specifically designed for accessing scientific data, particularly gridded data like that from climate models or satellite products. It allows clients to request subsets of data (e.g., a specific variable over a certain region and time range) without downloading the entire file.
  - **Use in Meteorology:** Popular for accessing climate and oceanographic datasets stored in formats like NetCDF and HDF.
- **THREDDS Data Server (TDS)**
  - **Description:** A web server that provides metadata and data access for scientific datasets using various protocols, including OPeNDAP, HTTP, and WMS/WCS (for geospatial data).
  - **Use in Meteorology:** Many meteorological data providers use TDS to serve their datasets to the community.
- **Zarr**
  - **Description:** A format for the storage of chunked, compressed, N-dimensional arrays. It's designed for parallel access and can be efficiently accessed over HTTP using byte-range requests.
  - **Use in Meteorology:** Increasingly being adopted for large-scale meteorological datasets in the cloud, enabling efficient partial reads.

6.6.2 Data Access APIs (Application Programming Interfaces)

APIs are sets of rules and specifications that allow different software applications to communicate and exchange data. For meteorological data, APIs provide a more structured and often more user-friendly way to access information compared to directly working with protocols.

- **RESTful APIs (Representational State Transfer)**
  - **Description:** A widely adopted architectural style for building web services. RESTful APIs typically use HTTP methods (GET, POST, PUT, DELETE) to interact with resources (data) identified by URLs.

They often return data in formats like JSON or XML, which are easy for programs to parse.

- **Use in Meteorology:** Many modern meteorological data providers offer RESTful APIs to access a variety of data, from real-time observations to model forecasts and historical climate data. Users can often specify parameters (e.g., location, time, variables) in the API request to get exactly the data they need.

- **Python Libraries (e.g., xarray, rioxarray, MetPy)**
  - **Description:** While not strictly APIs in the web service sense, these libraries provide programmatic interfaces within Python to access and manipulate meteorological data in specific formats (e.g., NetCDF, GRIB, Zarr). They abstract away the complexities of file reading and protocol handling.
  - **Use in Meteorology:** Extremely popular among researchers & operational users who work with meteorological data programmatically.
- **GraphQL**
  - **Description:** A query language for your API and a server-side runtime for executing those queries. GraphQL allows clients to request exactly the data they need and nothing more, making data retrieval more efficient.
  - **Use in Meteorology:** While less common than REST, some newer meteorological data services might adopt GraphQL for more flexible data querying.
- **OGC Web Services (WMS, WCS, WFS)**
  - **Description:** Standards defined by the Open Geospatial Consortium (OGC) for serving geospatial data over the web.
  - **WMS (Web Map Service):** Produces map images of geospatial data.
  - **WCS (Web Coverage Service):** Provides access to coverage data (e.g., gridded data) with its original values.
  - **WFS (Web Feature Service):** Serves vector data (e.g., weather stations as points).
  - **Use in Meteorology:** Used for accessing and visualizing gridded meteorological data (like model outputs or satellite imagery) in a geospatial context.
- **Examples of Data Access in Meteorology**
  - A weather app on your phone might use a RESTful API provided by a national meteorological service to fetch the latest temperature and precipitation data for your location.

- A climate researcher might use the xarray Python library with an OPeNDAP URL to access a subset of variables from a large climate model dataset stored on a THREDDS server.
- A forecasting center might use SFTP to automatically download the latest GRIB files from a global model run.
- A web portal visualizing satellite imagery might use HTTP to fetch GeoTIFF tiles.

6.7 Downloads: Acquiring Meteorological Data

In the context of meteorological data management, **downloads** refer to the process by which users or automated systems retrieve data files from remote servers or archives to their local computing environment. This is a fundamental step for anyone wishing to analyze, visualize, or integrate meteorological data into their own applications or workflows. Given the often massive size and continuous nature of meteorological datasets, efficient and reliable download mechanisms are crucial.

6.7.1 What "Downloads" Entail

A download operation typically involves a client (your computer, a script, or an application) initiating a request to a server where the data resides. The server then transmits the data file(s) over a network connection to the client. The success and speed of a download depend on several factors, including:

- **File Size:** Meteorological datasets can range from small text files of a single observation to multi-terabyte model outputs or satellite archives.
- **Network Bandwidth:** The speed of the internet connection between the client and the server.
- **Server Load:** The capacity and current demand on the data provider's server.
- **Protocol Efficiency:** The underlying data transfer protocol used.

6.7.2 Common Methods and Protocols for Downloads

While we discussed data access protocols and APIs previously, here we focus specifically on their application for the *action* of downloading data files:

- **HTTP/HTTPS Direct Downloads**
  - **Mechanism:** This is the most common and user-friendly method. Data providers often host files on web servers, allowing users to simply click a link in a web browser to initiate a download. For programmatic access, tools like wget or curl (command-line utilities) or libraries in programming languages (e.g., Python's requests library, urllib) can be used to fetch files via HTTP/HTTPS.

- **Use in Meteorology:** Widely used for downloading individual or small collections of meteorological data files (e.g., a specific satellite image, a single day's model run, a reanalysis subset). Many data portals provide direct download links.
- **Advantages:** Universally supported, simple for manual downloads, secure with HTTPS.
- **Limitations:** Less efficient for very large volumes of data or when only a small subset of a large file is needed (often requires downloading the entire file).

- **FTP/SFTP (File Transfer Protocol/Secure File Transfer Protocol)**
  - **Mechanism:** Dedicated protocols for file transfer. FTP is unencrypted, while SFTP provides secure, encrypted transfers over SSH. Users typically use an FTP client software or command-line tools.
  - **Use in Meteorology:** Still employed by some meteorological centers for bulk data distribution, especially for operational data feeds or large archival datasets where security and reliability are paramount.
  - **Advantages:** Optimized for file transfer, robust for large files, SFTP offers strong security.
  - **Limitations:** Requires a dedicated client or command-line interface; less flexible for data subsetting than some newer protocols.
- **OPeNDAP (Open-source Project for a Network Data Access Protocol)**
  - **Mechanism:** While a data access protocol, OPeNDAP facilitates "virtual downloads" or subsetting. Instead of downloading an entire large NetCDF or HDF file, users can specify a spatial subset, a time range, or specific variables. The server then delivers only the requested data, effectively acting as a highly efficient download mechanism for partial data.
  - **Use in Meteorology:** Extremely valuable for researchers working with large gridded datasets (e.g., climate model outputs, reanalysis products, satellite composites) where downloading the full dataset is impractical. Libraries like Python's xarray can seamlessly interact with OPeNDAP servers.
  - **Advantages:** Highly efficient for subsetting, reduces bandwidth and storage requirements on the client side.
  - **Limitations:** Requires data to be hosted on an OPeNDAP-enabled server; not all data types are suitable.

- **Cloud Storage APIs/SDKs (Software Development Kits)**
  - **Mechanism:** When data is stored in cloud object storage (e.g., Amazon S3, Google Cloud Storage, Azure Blob Storage), cloud providers offer robust APIs and SDKs for programmatic access. These allow developers to integrate download capabilities directly into their applications, often with advanced features like multi-part downloads for very large files, resumable transfers, and fine-grained access control.
  - **Use in Meteorology:** Increasingly common for accessing data from cloud-based meteorological data lakes and archives. This is the preferred method for cloud-native applications and workflows.
  - **Advantages:** Highly scalable, durable, optimized for large files, integrated with cloud computing resources.
  - **Limitations:** Requires familiarity with cloud provider's specific APIs/SDKs.
- **Web Portals and Data Services**
  - **Mechanism:** Many meteorological data centers provide user-friendly web portals that act as a front-end to their archives and databases. Users can browse, search, visualize, and then select data for download. These portals often abstract the underlying protocols, providing a simplified user experience. Some portals also offer custom ordering or subsetting services.
  - **Use in Meteorology:** The primary interface for many users to discover and acquire data from major meteorological data providers (e.g., NOAA NCEI, ECMWF, EUMETSAT).
  - **Advantages:** User-friendly, often provides metadata and visualization tools, can handle complex data requests.
  - **Limitations:** Can be slower for bulk programmatic downloads compared to direct API/protocol access.

6.7.3 Considerations for Downloading Meteorological Data

- **Data Volume:** Be mindful of file sizes. Downloading global, high-resolution datasets can consume significant bandwidth and local storage.
- **Subsetting vs. Full Download:** Always consider if you need the entire file or if subsetting (e.g., via OPeNDAP or a data service) is possible and more efficient.
- **Automation:** For recurring data needs, automate downloads using scripts (e.g., Python, Bash) that leverage the appropriate protocols or APIs.

- **Authentication:** Many valuable datasets require user authentication (API keys, tokens, login credentials) for access.
- **Data Usage Policies:** Be aware of any terms of use or attribution requirements from data providers.
- **Error Handling:** Implement robust error handling in automated download scripts to manage network interruptions, server errors, or corrupted files.

In conclusion, "downloads" are the practical means by which meteorological data moves from its storage location to the user's workspace. The methods range from simple web browser clicks to sophisticated programmatic interactions with cloud storage APIs and specialized scientific protocols, all designed to facilitate the efficient acquisition of the vast and vital information that drives meteorological science and operations.

In conclusion, a diverse set of data access protocols and APIs are essential for making meteorological data usable. Protocols provide the underlying communication mechanisms, while APIs offer more structured and often more user-friendly ways for applications and users to interact with the data. The choice of protocol or API often depends on the type of data, the needs of the user or application, and the capabilities provided by the data server. The trend is towards more modern, web-friendly APIs (like REST) and protocols that allow for efficient partial data access (like OPeNDAP and Zarr), especially as meteorological datasets continue to grow in size and complexity.

6.8 Metadata Standards in Remote Sensing and Meteorology

In the world of remote sensing and meteorology, **metadata** is "data about data." It provides essential descriptive information about a dataset, making it understandable, discoverable, accessible, and reusable. Without robust metadata, even the most valuable scientific data can become meaningless or difficult to use.

Metadata standards are formal rules and guidelines for creating, organizing, and documenting metadata. They define a common set of elements, their definitions, and how they should be structured. The adoption of these standards is crucial for ensuring interoperability, discoverability, and long-term usability of meteorological and Earth observation data across different systems, organizations, and scientific disciplines.

6.8.1 Why Metadata Standards are Essential

- **Discoverability:** Allows users to find relevant data by searching for specific attributes (e.g., sensor type, geographic area, time range, measured variable, data quality). Without standards, searching across different data providers would be chaotic.

- **Interoperability:** Enables different software systems and applications to understand and process data from various sources. A standardized description means a tool designed to read data from one source can also understand data from another, provided both adhere to the same metadata standard.
- **Usability and Interpretability:** Provides crucial context for data users. This includes information about units, coordinate systems, processing history, sensor calibration, data quality, and any limitations, preventing misinterpretation or misuse.
- **Long-Term Archiving and Preservation:** Ensures that data remains understandable and usable decades into the future, even as the original creators or systems become obsolete. Metadata acts as a permanent "instruction manual" for the data.
- **Data Provenance and Lineage:** Documents the origin of the data, how it was collected, and any transformations it underwent, which is vital for scientific credibility and reproducibility.
- **Data Sharing and Exchange:** Facilitates seamless exchange of data between different organizations, national meteorological services, and international research initiatives.

6.8.2 Key Metadata Standards in Remote Sensing and Meteorology

Several international and domain-specific metadata standards are widely adopted in the meteorological and Earth observation communities:

- **ISO 19115 (Geographic Information — Metadata)**
  - **Description:** This is a comprehensive international standard for describing geographic information and services. It defines a wide range of metadata elements covering identification, extent, quality, spatial and temporal properties, lineage, and more. It's a foundational standard for geospatial data.
  - **Use in Meteorology:** Widely used for describing satellite imagery, gridded model outputs, and other spatially referenced meteorological datasets. Many national and international data centers use ISO 19115 or profiles derived from it (like ISO 19139 for XML encoding) to describe their holdings.
- **WMO Core Metadata Profile (WCMP)**
  - **Description:** A profile (subset and extension) of ISO 19115 specifically tailored by the World Meteorological Organization

(WMO) for meteorological and climatological data. It focuses on elements most relevant to atmospheric science, such as observation types, instrument characteristics, and meteorological variables.

- **Use in Meteorology:** Promotes consistent metadata for all WMO-related data, including observations from surface stations, radiosondes, radar, and satellite products, facilitating data exchange within the global meteorological community.

- **FGDC CSDGM (Federal Geographic Data Committee Content Standard for Digital Geospatial Metadata)**
  - **Description:** A U.S. national standard for geospatial metadata, predating ISO 19115 but still in use, particularly within U.S. government agencies. It provides a detailed set of elements for describing geospatial data.
  - **Use in Meteorology:** Used by some U.S. agencies (e.g., NOAA) for describing their geospatial meteorological datasets, though there's a general trend towards ISO standards.
- **NetCDF Conventions (e.g., CF Conventions - Climate and Forecast Metadata Conventions)**
  - **Description:** While NetCDF is a data format, the **CF Conventions** are a set of metadata conventions that specify how to describe data stored in NetCDF files, particularly for climate and forecast models. They define standard names for variables, units, coordinate systems, and other attributes.
  - **Use in Meteorology:** Absolutely critical for making NetCDF files self-describing and interoperable within the climate and weather modeling communities. They ensure that different models and analysis tools can correctly interpret the data stored in NetCDF files.
- **GRIB and BUFR Templates/Tables**
  - **Description:** For specialized binary formats like GRIB and BUFR, metadata is embedded directly within the binary structure using predefined "templates" and "tables" (code lists). These tables define the meaning of specific bits and bytes within the message, including parameters, units, levels, and instrument types.
  - **Use in Meteorology:** Essential for the efficient, real-time exchange of operational meteorological observations (BUFR) and model outputs (GRIB) globally. The WMO maintains these tables to ensure universal interpretation.

- **Dublin Core Metadata Initiative (DCMI)**
  - **Description:** A simpler, more general-purpose metadata standard consisting of 15 core elements (e.g., title, creator, date, subject). While not as comprehensive for scientific data as ISO standards, it's often used for basic resource discovery in broader digital libraries.
  - **Use in Meteorology:** May be used in conjunction with other standards for high-level cataloging of meteorological resources or documents.

6.8.3 Benefits of Adhering to Metadata Standards

- **Enhanced Data Discovery:** Researchers can easily find relevant datasets from diverse sources using standardized search queries.
- **Increased Data Interoperability:** Data from different instruments or models can be combined and analyzed more seamlessly.
- **Improved Data Quality and Trust:** Clear documentation of data lineage, processing steps, and quality flags builds confidence in the data's reliability.
- **Facilitated Data Sharing:** Promotes efficient exchange of data between national and international meteorological centers, supporting global forecasting efforts.
- **Support for Open Science:** Makes data more accessible and reusable, fostering transparency and collaboration in scientific research.
- **Long-Term Value:** Ensures that current investments in data collection continue to yield scientific and societal benefits for decades to come.

In essence, metadata standards are the "Rosetta Stone" for meteorological data. They provide the common language and structure necessary to unlock the full potential of the vast and complex information collected from remote sensing platforms and other sources, transforming raw data into actionable knowledge for understanding and predicting Earth's atmosphere.

6.9 Data Storage in Meteorology

The sheer volume, diversity, and continuous influx of meteorological data necessitate robust and sophisticated **data storage** solutions. From real-time observations pouring in every second to massive archives spanning centuries, effective data storage is fundamental to the entire meteorological enterprise, supporting everything from immediate weather warnings to long-term climate research.

6.9.1 The Imperative of Meteorological Data Storage

Data storage in meteorology serves several critical purposes:

- **Operational Forecasting:** Real-time observations and current model outputs must be stored for immediate access by forecasters to issue warnings, update forecasts, and monitor evolving weather.
- **Numerical Weather Prediction (NWP):** Model initialization requires access to the latest global and regional observations. Model outputs themselves generate vast amounts of gridded data that need to be stored for subsequent analysis, dissemination, and verification.
- **Climate Monitoring and Research:** Long-term, consistent archives of observational data are essential for detecting climate trends, understanding climate variability, and validating climate models. This requires data to be preserved for decades, even centuries.
- **Atmospheric Research:** Scientists need access to specialized, high-resolution datasets for detailed studies of atmospheric processes, requiring flexible and often high-performance storage.
- **Verification and Reanalysis:** Stored data is used to verify the accuracy of past forecasts and to create "reanalysis" datasets, which combine historical observations with modern models to produce consistent, long-term atmospheric records.

6.9.2 Characteristics of Meteorological Data Influencing Storage

Meteorological data presents unique challenges for storage due to its inherent characteristics:

- **Volume (Big Data):** Modern observing systems (e.g., high-resolution radars, geostationary satellites with frequent scans, high-resolution NWP models) generate petabytes of data annually. Historical archives already contain exabytes of information.
- **Velocity (High Ingest Rate):** Data streams in continuously and rapidly, especially from real-time operational networks, requiring storage systems capable of high-speed ingestion.
- **Variety (Heterogeneous Data):** Data comes in many different formats (e.g., raw sensor outputs, WMO codes, NetCDF, GRIB, HDF5), from diverse sources (surface stations, balloons, aircraft, satellites, radar, models), and with varying spatial and temporal resolutions.
- **Long-Term Preservation:** Much of the data, particularly climate records, must be stored indefinitely, requiring highly stable and durable storage media.

- **Access Patterns:** Data needs to be accessed in various ways: real-time streaming, batch processing for research, random access for specific queries, and sequential access for large-scale analysis.

6.9.3 Types of Meteorological Data Stored

- **Observational Data**
    - **Surface Observations:** METARs, SYNOPs, automated weather station data.
    - **Upper Air Observations:** Radiosonde soundings (TEMP), aircraft reports (AIREP), wind profiler data.
    - **Radar Data:** Base products (Reflectivity, Velocity, Spectrum Width, Dual-Pol products) at various elevation angles, and derived products (VIL, QPE).
    - **Satellite Data:** Raw radiance data, imagery (visible, infrared, water vapor), derived products (cloud properties, sea surface temperature, atmospheric profiles).
    - **Specialized Data:** From flux towers, buoys, UAS, research aircraft, lightning networks.
- **Numerical Weather Prediction (NWP) Model Data**
    - **Initial Conditions:** The global and regional analyses used to start model runs.
    - **Forecast Outputs:** Gridded fields of atmospheric variables (temperature, pressure, wind, humidity, precipitation) at various forecast hours and vertical levels. This includes deterministic forecasts, ensemble forecasts (multiple model runs to quantify uncertainty), and reanalysis datasets.
    - **Climatological Data:** Long-term averages, extremes, indices, and homogenized datasets derived from observations.

6.9.4 Storage Technologies and Architectures

Meteorological data storage has evolved significantly, moving from traditional tape archives to modern distributed and cloud-based systems:

- **Tape Libraries (Hierarchical Storage Management - HSM)**
    - Description: Historically, magnetic tapes were the primary medium for long-term archiving due to their low cost per terabyte and high durability. HSM systems automatically migrate less frequently accessed data to tape.
    - **Use:** Still used for deep archives of historical data where immediate access is not critical.

- Advantages: Very cost-effective for long-term, cold storage; high data integrity.
- **Disadvantages:** Slow retrieval times (latency), requires mechanical systems.

- **Direct-Attached Storage (DAS) & Network-Attached Storage (NAS)**
  - **Description**: Hard disk drives directly connected to servers (DAS) or shared over a network (NAS).
  - **Use:** For active, frequently accessed data, local processing, and smaller datasets.
  - **Advantages:** Relatively fast access, simple to manage for smaller scales.
  - **Disadvantages:** Limited scalability, performance bottlenecks as data grows.
- **Storage Area Networks (SANs)**
  - **Description:** Dedicated high-speed networks that connect servers to shared pools of storage devices.
  - **Use:** For mission-critical applications requiring high performance and shared access, such as real-time NWP data.
  - **Advantages:** High performance, good scalability for enterprise environments.
  - **Disadvantages:** Complex to set up and manage, expensive.
- **Distributed File Systems (e.g., Hadoop HDFS, Lustre)**
  - **Description:** Designed to store and manage very large datasets across clusters of commodity hardware. Data is broken into blocks and distributed across multiple nodes.
  - **Use:** Increasingly used by large meteorological centers for storing massive NWP model outputs and large observational archives.
  - **Advantages:** Highly scalable, fault-tolerant, cost-effective for petabyte-scale data.
  - **Disadvantages:** Can be complex to administer, not optimized for small, random reads.
- **Object Storage (e.g., Amazon S3, Google Cloud Storage, Ceph)**
  - **Description:** Data is stored as "objects" (files) along with rich metadata, accessed via unique identifiers over HTTP/HTTPS. It's inherently distributed and highly scalable.

  - **Use:** Ideal for cloud-based storage of large, static datasets like satellite imagery, radar archives, and NWP model outputs. Increasingly popular for data lakes.
  - **Advantages:** Massive scalability, high durability, cost-effective, easily accessible via APIs.
  - **Disadvantages:** Not designed for high-performance transactional workloads.
- **Databases (Relational, NoSQL, Time-Series)**
- **Description**
  - **Relational Databases (e.g., PostgreSQL, MySQL):** Structured data in tables.
  - **NoSQL Databases (e.g., MongoDB, Cassandra):** Flexible schema, suitable for unstructured or semi-structured data.
  - **Time-Series Databases (e.g., InfluxDB, TimescaleDB):** Optimized for storing and querying time-stamped data.
  - **Use:** For storing metadata, station configurations, quality control flags, specific point observations (e.g., METARs), and time-series of sensor readings.
  - **Advantages:** Efficient querying, data integrity, good for structured data.
  - **Disadvantages:** Can struggle with the sheer volume and complexity of gridded model data or raw satellite imagery.

6.9.5 Challenges in Meteorological Data Storage

- **Volume Management:** Managing petabytes to exabytes of data requires constant scaling, optimization, and lifecycle management (moving data between hot, warm, and cold storage tiers).
- **Data Integrity and Durability:** Ensuring that data remains uncorrupted and accessible over decades, despite hardware failures, software bugs, and media degradation. Redundancy (RAID, replication) is crucial.
- **Cost:** The financial burden of acquiring, maintaining, and powering massive storage infrastructure is significant.
- **Performance vs. Cost Trade-offs:** Balancing the need for rapid access to real-time data with the cost-effectiveness of long-term archiving.
- **Metadata Management:** Storing and linking comprehensive metadata (information about the data, such as sensor type, calibration, processing

history, geographic coordinates, time stamps) is crucial for data discovery and usability.

- **Data Migration:** As storage technologies evolve, periodically migrating vast archives to newer, more efficient systems is a complex and resource-intensive task.
- **Security:** Protecting sensitive data from unauthorized access or cyber threats.

6.9.6 Emerging Trends in Meteorological Data Storage

- **Cloud Adoption:** National meteorological services and research institutions are increasingly leveraging public and private cloud platforms for scalable, flexible, and cost-effective data storage and processing.
- **Data Lakes:** Building centralized repositories that store raw, unprocessed data in its native format, allowing for flexible analysis later. This is often built on object storage and distributed file systems.
- **Big Data Technologies:** Utilizing frameworks like Apache Hadoop and Spark for processing and analyzing massive datasets stored in distributed systems.
- **Containerization and Microservices:** Packaging applications and services (including data storage interfaces) into portable containers for easier deployment and management.
- **FAIR Data Principles:** Emphasis on making data Findable, Accessible, Interoperable, and Reusable, which heavily influences storage strategies and metadata practices.

In conclusion, the storage of meteorological data is a dynamic and complex field, driven by the ever-increasing volume and diversity of atmospheric information. Effective storage strategies are not just about holding data; they are about ensuring its long-term preservation, integrity, and efficient accessibility, thereby enabling the continuous advancement of weather forecasting, climate science, and environmental understanding.

6.10 Data Retrieval in Meteorology

Once meteorological data has been collected, formatted, and stored, the crucial next step is **data retrieval**. This involves accessing and obtaining the specific data needed for a particular purpose, whether it's for real-time weather forecasting, in-depth climate research, or any application in between. The methods and complexities of data retrieval in meteorology are as varied as the data itself, depending on factors like the data type, storage location, user needs, and the urgency of access.

6.10.1 Real-Time Data Retrieval for Operational Forecasting

For operational meteorologists, timely access to the latest observations and model outputs is paramount. This often involves automated systems that continuously stream data:

- **Global Telecommunication System (GTS):** A coordinated global network for the rapid exchange of meteorological data. National Meteorological Centers (NMCs) around the world contribute and receive data in near real-time via dedicated telecommunication links. Retrieval here is often a continuous process, with systems constantly ingesting and processing incoming data.
- **Direct Broadcast from Satellites:** Some users, particularly in remote locations or specialized centers, directly receive real-time data streams from meteorological satellites as they pass overhead. Specialized software is used to capture and process this data.
- **Real-Time Data Portals and APIs:** Many meteorological organizations provide web-based portals or Application Programming Interfaces (APIs) that allow users to programmatically access the latest observations (surface, upper air, radar, satellite) and model analyses/forecasts. These often allow filtering by location, parameter, and time.
- **Specialized Networks:** For specific applications like aviation, data from airport weather stations (METARs, TAFs) is often retrieved through dedicated networks and displayed in specialized formats.

6.10.2 Data Retrieval for Research and Analysis

Researchers often require access to historical datasets, sometimes spanning decades, and may need to perform complex queries based on various criteria:

- **Data Archives:** Meteorological agencies and research institutions maintain vast archives of historical data. Retrieval from these archives can involve:
- **Web-based interfaces:** Allowing users to select data based on parameters, time periods, and geographical areas.
- **Command-line tools:** For more advanced users to directly interact with the archive systems.
- **Programming APIs:** Enabling researchers to write scripts (e.g., in Python, R) to automate the download and processing of large datasets.

- **Climate Data Records (CDRs):** Carefully processed and homogenized datasets specifically designed for climate studies. Retrieval often involves accessing dedicated portals or data centers that curate these records.
- **Model Reanalysis Data:** Datasets that combine historical observations with modern weather models to create a consistent, gridded representation of the atmosphere over past decades. Retrieval typically involves accessing large scientific datasets through specific data portals.

6.10.3 Methods and Technologies for Data Retrieval

Various technologies and methods are employed for data retrieval:

- **File Transfer Protocol (FTP) and Secure FTP (SFTP):** Common protocols for downloading files, often used for accessing large datasets from archives.
- **Hypertext Transfer Protocol (HTTP) and HTTPS:** The foundation of web-based data access, used by many portals and APIs.
- **Open Geospatial Consortium (OGC) Standards:** Standards like Web Map Service (WMS) and Web Coverage Service (WCS) allow for the retrieval and visualization of geospatial meteorological data.
- **Specialized Data Formats and Libraries:** Data is often stored in formats like NetCDF or GRIB, and specific programming libraries (e.g., xarray, MetPy in Python) are used for efficient reading and manipulation of these formats.
- **Database Queries:** For data stored in databases, Structured Query Language (SQL) or similar query languages are used to retrieve specific subsets of information based on defined criteria.

6.10.4 Challenges in Data Retrieval

Despite advancements, several challenges remain in meteorological data retrieval:

- **Data Volume:** The sheer volume of meteorological data can make retrieval of large datasets time-consuming and resource-intensive.
- **Data Heterogeneity:** Data from different sources (satellites, radars, stations, models) can be in different formats and have varying metadata, making integrated retrieval and analysis complex.
- **Network Bandwidth:** Especially for users in remote locations or when dealing with large files, network bandwidth can be a limiting factor.

- **Metadata Quality:** Accurate and comprehensive metadata (information about the data) is crucial for effective retrieval. Inconsistent or incomplete metadata can hinder the ability to find and use the desired data.
- **Accessibility and Cost:** While much meteorological data is publicly available, some specialized datasets may have access restrictions or associated costs.

6.10.5 Trends in Data Retrieval

- **Cloud-Based Data Access:** Increasingly, meteorological data is being stored and accessed through cloud platforms, offering scalability and easier access via APIs.
- **FAIR Data Principles:** The principles of Findability, Accessibility, Interoperability, and Reusability (FAIR) are guiding efforts to improve data management and retrieval.
- **Standardized APIs:** Efforts are underway to develop more standardized APIs for accessing meteorological data across different providers.

In summary, data retrieval is a critical link in the chain from observation to application. Efficient and user-friendly retrieval mechanisms are essential to unlock the full potential of the vast amounts of meteorological information available.

6.11 Conclusion

This chapter has underscored that the true value of meteorological observations and model outputs extends far beyond their initial collection or generation. It critically depends on how this vast and diverse array of information is **formatted, stored, and retrieved**. We've explored the necessity of standardized data formats, from raw instrument outputs to highly processed and internationally exchanged codes, ensuring interoperability and consistency across the global meteorological community. The discussion also highlighted the immense challenges associated with storing petabytes of historical and real-time data efficiently and securely, emphasizing the shift towards robust archival systems and scalable cloud-based solutions. Finally, we examined the diverse strategies for data retrieval, from real-time streams for operational forecasting to complex queries for climate research, all designed to make the right data available to the right user at the right time.

In essence, effective data management—encompassing meticulous formatting, secure and scalable storage, and agile retrieval mechanisms—is the invisible backbone of modern meteorology. It transforms raw numbers into actionable intelligence for weather warnings, fuels the continuous improvement of numerical weather prediction models, underpins long-term climate monitoring, and enables groundbreaking atmospheric research. As the volume and complexity of meteorological data continue to grow exponentially with new observing systems and higher-resolution models, the challenges in these areas will only intensify. Therefore, continued innovation in data science, robust infrastructure development, and unwavering commitment to international standards will be paramount to unlocking the full potential of atmospheric information for the benefit of society.

7

Meteorological Data Quality Control and Preprocessing

7.1 Introduction

The preceding chapters have detailed the myriad ways in which meteorological data is collected, from direct *in situ* observations to various ground-based and space-borne remote sensing platforms. We have also explored the crucial aspects of data formatting, storage, and retrieval, recognizing the immense volume and diversity of atmospheric information. However, the journey from raw measurement to actionable insight is not complete without a critical intermediate step: **data quality control (QC) and preprocessing**.

No observational system, regardless of its sophistication, is entirely free from errors, biases, or inconsistencies. Sensors can drift, communication links can suffer interference, environmental factors can contaminate readings, and automated systems can malfunction. Consequently, raw meteorological data often contains spurious values, missing records, or systematic biases that, if left unaddressed, can severely compromise the accuracy of weather forecasts, the reliability of climate analyses, and the validity of scientific research.

This chapter will delve into the essential processes of **quality control**, which involves rigorously identifying, flagging, and, where appropriate, correcting or removing erroneous data points. We will explore various automated and manual techniques employed to ensure the integrity and reliability of meteorological observations. Furthermore, we will examine **preprocessing** steps, which prepare the quality-controlled data for its intended use, including tasks such as unit conversions, interpolation to common grids, and handling of missing values. By understanding the meticulous procedures involved in data quality control and preprocessing, we gain a profound appreciation for how raw atmospheric signals are transformed into the trustworthy information that underpins all of modern meteorology.

7.2 Importance of Data Quality in Meteorology

The integrity and reliability of meteorological data are not merely desirable attributes; they are absolutely paramount to the entire discipline. In a field where decisions can impact lives, economies, and long-term environmental strategies, the **quality of data** serves as the bedrock upon which all subsequent analyses, forecasts, and scientific conclusions are built. Compromised data quality can lead to a cascade of negative consequences, undermining the very purpose of meteorological observation (Fig. 7.1).

Here's why data quality is of critical importance:

7.2.1 Accuracy of Numerical Weather Prediction (NWP) Models

- **Garbage In, Garbage Out:** NWP models are complex mathematical representations of the atmosphere, designed to simulate its future state. Their performance is highly dependent on the accuracy of the initial conditions provided by observational data. If the input data contains errors, biases, or missing values, the model's initial state will be flawed, leading to **inaccurate forecasts** that diverge from reality.
- **Data Assimilation:** Sophisticated data assimilation techniques integrate observations into model states. High-quality data allows these techniques to optimally adjust the model's representation of the atmosphere, leading to more realistic and skillful predictions. Poor quality data, conversely, can introduce spurious features or degrade the model's performance.

7.2.2 Reliability of Severe Weather Warnings and Public Safety

- **Timely and Accurate Warnings:** Meteorologists rely on real-time observational data (e.g., radar, surface observations) to detect and track severe weather phenomena like thunderstorms, tornadoes, and blizzards. If this data is erroneous or unreliable, it can lead to **missed warnings**, **false alarms**, or **delayed advisories**, directly jeopardizing public safety and property.
- **Decision-Making:** Emergency managers and the public make critical decisions (e.g., evacuations, sheltering) based on meteorological information. Trust in this information is built on its consistent accuracy, which is directly tied to data quality.

7.2.3 Validity of Climate Monitoring and Research

- **Detecting Trends and Variability:** Climate science fundamentally relies on long-term, consistent records of atmospheric parameters. Even small, systematic biases or inhomogeneities in historical data can create **artificial trends** or obscure genuine climate signals, leading to **misleading conclusions** about climate change, variability, or extremes.
- **Model Validation:** Climate models are validated against historical observational data. If the observational data itself is flawed, the validation

process becomes unreliable, potentially leading to overconfidence in an inaccurate model or misdiagnosis of model errors.

- **Policy Implications:** Climate research informs national and international policies. Flawed data quality can lead to incorrect scientific assessments, which in turn can lead to **ineffective or misdirected policy responses** to climate challenges.

7.2.4 Soundness of Atmospheric Research and Scientific Understanding

- **Process Understanding:** Researchers use high-quality observational data to understand fundamental atmospheric processes (e.g., cloud formation, boundary layer dynamics, atmospheric chemistry). If data is compromised, the scientific conclusions drawn from it will be suspect, potentially leading to **flawed theories** or **misinterpretations**of atmospheric behavior.
- **Instrument Development and Calibration:** Data quality control processes often reveal issues with observing systems themselves, leading to improvements in sensor design, calibration procedures, and network deployment. Without rigorous QC, these systemic issues might go undetected.

7.2.5 Efficient Resource Allocation and Economic Impact

- **Optimizing Operations:** Industries like aviation, agriculture, energy, and transportation rely heavily on accurate weather information. Poor data quality can lead to **suboptimal operational decisions**, resulting in economic losses (e.g., flight delays, crop damage, inefficient energy dispatch).
- **Resource Management:** Data on precipitation, temperature, and drought is crucial for water resource management. Inaccurate data can lead to **poor allocation of water resources**, impacting agriculture, industry, and human consumption.

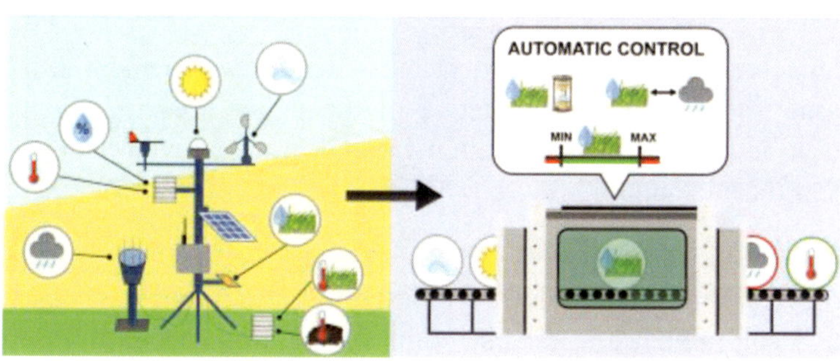

Fig. 7.1: Meteorological data quality control

In conclusion, the importance of data quality in meteorology cannot be overstated. It is the foundation of trustworthy forecasts, credible climate assessments, and robust scientific discovery. Rigorous quality control and preprocessing are therefore not just technical steps but essential safeguards that ensure the integrity and utility of meteorological information, ultimately serving to protect lives, property, and the environment.

7.3 Impact of Errors on Meteorological Analysis and Prediction

In the realm of meteorology, where data is the lifeblood of understanding and forecasting, the presence of errors, no matter how seemingly small, can have profound and cascading impacts on both the interpretation of current atmospheric conditions (analysis) and the projection of future states (prediction). The highly interconnected and chaotic nature of the atmosphere means that even minor inaccuracies can propagate and amplify, leading to significant discrepancies between forecast and reality.

7.3.1 Types of Errors in Meteorological Data

Before discussing the impact, it's useful to briefly categorize common types of errors:

- **Gross Errors (Outliers):** Large, obvious errors that fall far outside expected ranges (e.g., a temperature reading of -999°C or 100°C in a temperate zone). These are often due to sensor malfunction, transmission errors, or human input mistakes.
- **Systematic Errors (Biases):** Consistent deviations from the true value (e.g., a sensor consistently reading 2°C too high, or a radar consistently underestimating precipitation due to calibration issues). These can be subtle and harder to detect.
- **Random Errors:** Unpredictable fluctuations around the true value, often due to instrument noise or environmental variability. While individual random errors might be small, their cumulative effect can degrade data quality.
- **Representativeness Errors:** Occur when a measurement at a specific point is not truly representative of the larger area it is supposed to describe (e.g., a weather station located in a microclimate that differs from the surrounding region).

7.3.2 Impact on Meteorological Analysis

Meteorological analysis involves synthesizing current and past observations to create a comprehensive understanding of the atmosphere's present state. Errors in data directly compromise this understanding:

- **Misleading Interpretation of Current Weather:**
  - **False Features:** A gross error in a pressure reading can incorrectly suggest a high or low-pressure center where none exists, distorting the synoptic pattern.
  - **Incorrect Intensity:** A biased temperature sensor might lead to an overestimation or underestimation of a heatwave or cold snap, misrepresenting the actual conditions.
  - **Distorted Spatial Patterns:** Errors in radar reflectivity can show non-existent precipitation or hide actual precipitation, leading to a false picture of rainfall distribution.
- **Flawed Identification of Weather Features**
  - **Fronts and Boundaries:** Errors in temperature, dew point, or wind data can lead to misplacement or misidentification of fronts, dry lines, or other critical weather boundaries, which are key drivers of local weather.
  - **Storm Structure:** Inaccurate radar velocity data can lead to false indications of rotation (mesocyclones) or wind shear within thunderstorms, or conversely, mask actual dangerous features.
- **Compromised Climate Diagnostics**
  - **Artificial Trends:** Systematic errors or inhomogeneities in long-term observational records (e.g., due to station relocation or instrument changes) can introduce artificial trends that are mistaken for genuine climate change signals. This can lead to incorrect conclusions about warming, cooling, or precipitation shifts.
  - **Misleading Extremes:** Errors can lead to false records for temperature highs/lows or precipitation totals, distorting the assessment of extreme weather events.

7.3.3 Impact on Meteorological Prediction (Numerical Weather Prediction - NWP)

The most significant and far-reaching impact of data errors is on numerical weather prediction. NWP models are highly sensitive to their initial conditions, and errors introduced at this stage can propagate and amplify rapidly.

- **Contaminated Initial Conditions**
  - **Data Assimilation Failure:** Data assimilation systems (which blend observations with short-range model forecasts to create the best estimate of the current atmospheric state) are designed to filter out errors. However, if errors are too large, systematic, or occur in critical

areas, they can be incorrectly assimilated into the model's initial state.

- **Spurious Features:** Erroneous observations can introduce artificial atmospheric features (e.g., a false low-pressure center, an incorrect temperature anomaly) into the model's starting point, which the model will then attempt to evolve.

- **Error Propagation and Amplification (The Butterfly Effect)**
  - **Chaotic Nature:** The atmosphere is a chaotic system. A fundamental principle of chaos theory (often illustrated by the "butterfly effect") states that small differences in initial conditions can lead to vastly different outcomes over time. Even seemingly minor errors in observational data, when assimilated into a model, can grow exponentially as the forecast progresses.
  - **Forecast Divergence:** This leads to forecast "divergence," where the model's prediction deviates significantly from the actual weather that unfolds. The further out the forecast, the more pronounced this divergence becomes.
- **Specific Forecast Impacts**
  - **Incorrect Storm Track and Intensity:** Errors in initial wind fields or moisture profiles can lead to a storm being forecast to move too fast or too slow, or to take an incorrect path. Inaccurate temperature/moisture profiles can lead to over- or under-prediction of storm intensity.
  - **Misplaced or Incorrect Precipitation:** Errors in initial moisture or stability can result in precipitation being forecast in the wrong location, at the wrong time, or with incorrect intensity (e.g., forecasting heavy rain when only light rain occurs, or missing a significant snow event).
  - **Erroneous Temperature and Wind Forecasts:** Biases in initial temperature or pressure can lead to persistent warm or cold biases in subsequent temperature forecasts. Incorrect wind fields can lead to inaccurate wind speed and direction predictions.
- **False Alarms and Missed Warnings for Severe Weather:**
  - **False Alarms:** Errors (e.g., spurious radar echoes, incorrect velocity signatures) can lead a model or a forecaster to predict severe weather that does not materialize, eroding public trust.
  - **Missed Warnings:** Conversely, if critical data is erroneous or missing, the model might fail to predict the development of severe weather, leading to missed warnings and potential loss of life or property.

- **Impact on Aviation and Other Sectors:** Inaccurate forecasts of turbulence, icing, wind shear, or fog due to data errors can have significant safety and economic consequences for aviation, shipping, agriculture, and energy management.

In conclusion, the impact of errors on meteorological analysis and prediction is profound and pervasive. From distorting our understanding of the current weather to fundamentally undermining the accuracy of future forecasts, data quality is not a luxury but a necessity. This underscores the critical importance of robust data quality control and preprocessing procedures, which act as essential gatekeepers, transforming raw, imperfect observations into the reliable information required to navigate and predict the complexities of our atmosphere.

7.4 Common Data Errors and Anomalies in Meteorological Data

Despite significant advancements in sensor technology, communication systems, and automated processing, meteorological data is inherently susceptible to various errors and anomalies. These imperfections can arise at any stage of the data lifecycle: from the initial measurement by a sensor, through transmission, during processing, or due to environmental factors. Identifying and addressing these issues through robust quality control is paramount, as discussed in the previous section.

Here's a detailed look at common data errors and anomalies encountered in meteorological observations:

7.4.1 Sensor-Related Errors

These errors originate directly from the instrument itself or its interaction with the immediate environment.

- **Sensor Drift:** A gradual, systematic change in a sensor's calibration over time, causing it to consistently read higher or lower than the true value. This is a common issue with many electronic sensors due to aging components or environmental exposure.
- *Example:* A humidity sensor that slowly starts reporting relative humidity 5% lower than actual over several months.
- **Stuck/Flat-lined Values:** The sensor outputs the exact same value repeatedly for an extended period, indicating a malfunction where it's no longer responding to changes in the measured parameter.
  - *Example:* A wind speed sensor continuously reporting 0 m/s even during windy conditions, or a temperature sensor stuck at 25°C for hours.

- **Spikes/Outliers:** Sudden, brief, and extreme values that deviate significantly from surrounding valid data points. These are often caused by electrical interference, transient sensor glitches, or communication noise.
    - *Example:* A temperature reading of -50°C for a single minute in the middle of a summer day, surrounded by readings of 25°C.
- **Bias (Systematic Offset):** A consistent difference between the sensor's reading and the true value. Unlike drift, which evolves, a bias is a fixed offset. This can be due to improper calibration, manufacturing defects, or consistent environmental influence.
    - *Example:* A pressure sensor consistently reading 2 hPa higher than the actual atmospheric pressure due to factory calibration error.
- **Lag Errors (Response Time):** Occur when a sensor is slow to respond to rapid changes in the measured parameter. The sensor's output lags behind the actual atmospheric condition. This is particularly noticeable with sensors measuring rapidly fluctuating variables like temperature in turbulent air, or humidity sensors in rapidly changing moisture environments.
    - *Example:* A radiosonde's temperature sensor not fully capturing a sharp temperature inversion because it takes time to equilibrate as it ascends.
- **Icing/Wetting Errors:** Physical interference with the sensor due to precipitation.
    - **Icing:** In freezing conditions, ice accumulation on wind vanes, anemometers, or temperature/humidity probes can cause them to become stuck, provide erroneous readings, or be damaged.
    - **Wetting:** Rain or dew on temperature or humidity sensors can lead to evaporative cooling, causing artificially low temperature readings or incorrect humidity values.
- **Degradation:** General deterioration of sensor performance over time due to exposure to harsh weather elements (UV radiation, dust, salt, corrosion), leading to increased noise, reduced accuracy, or eventual failure.

7.4.2 Communication and Transmission Errors

These issues occur as data is transmitted from the sensor or data logger to the central processing system.

- **Missing Data:** Gaps in the time series of observations. This can be due to power outages at the station, communication link failures, sensor malfunctions, or data logger issues.
    - *Example:* A radar not transmitting data for 15 minutes due to a network outage.
- **Corrupted Data:** Data that is unreadable, garbled, or incorrectly transmitted, often appearing as nonsensical values or characters in the data stream. This can be due to electromagnetic interference, faulty cables, or software bugs in the transmission protocol.
    - *Example:* A digital temperature value appearing as "ABCD" instead of a numerical value.
- **Latency Issues:** Delays in data arrival. While not strictly an "error" in the data itself, significant latency can render real-time data useless for time-sensitive applications like severe weather warnings.
    - *Example:* Radar data arriving 10 minutes late during a rapidly evolving tornado situation.

7.4.3 Environmental and Siting-Related Anomalies (Especially for Radar)

These are not sensor malfunctions but rather phenomena that the sensor correctly detects, but which are not representative of the meteorological event of interest, or are artifacts of the measurement process.

- **Ground Clutter (Radar):** Echoes returned from stationary objects on the ground (buildings, trees, hills, water towers) that are within the radar's beam. These are typically strongest near the radar and at low elevation angles.
    - *Appearance:* Stationary, often very strong echoes that do not move with the wind.
- **Anomalous Propagation (AP) (Radar):** Occurs when the radar beam is bent downwards due to unusual atmospheric refractive index gradients (e.g., strong temperature inversions near the surface). This can cause the beam to hit the ground at unusually long ranges or even "see" beyond the horizon.
    - *Appearance:* Ground clutter appearing far from the radar, or echoes that seem to "pop up" in clear air.
- **Biological Echoes (Radar):** Returns from non-precipitation targets like insects, birds, or bats. These are often present during certain times of day or seasons.

    - *Appearance:* Diffuse, often weak echoes that can move with the wind, sometimes forming "fine lines" (insect boundaries) or "bird rings" (mass bird migrations). Dual-polarization helps differentiate these.

- **Bright Band (Radar):** An artificially enhanced reflectivity signature that occurs at the melting layer (where falling snow melts into rain). Wet snowflakes are highly reflective, leading to a localized increase in reflectivity.
    - *Appearance:* A distinct ring or band of higher reflectivity at a specific altitude, often appearing as a "donut" shape on radar displays. Can lead to overestimation of surface precipitation.
- **Radiation Errors (Temperature):** Even with radiation shields, solar radiation can heat the instrument housing or the sensor itself, causing artificially high temperature readings, especially during calm, sunny conditions. Conversely, longwave radiation from nearby surfaces can affect nighttime readings.
- **Microclimate Effects (Surface Stations):** The immediate environment around a surface weather station (e.g., proximity to a building, a paved surface, a body of water, or a single tree) can create a localized microclimate that is not representative of the broader area.
    - *Example:* A temperature sensor near a blacktop parking lot reading higher than a sensor over grass.
- **Wind Obstructions:** Nearby buildings, trees, or other structures can block or channel wind flow, leading to inaccurate wind speed and direction measurements at a station, especially if the anemometer is not mounted at the standard 10-meter height or if siting guidelines are violated.

7.4.4 Processing and Software Errors

These errors are introduced during the automated or manual processing of raw data.

- **Unit Conversion Errors:** Mistakes in converting raw sensor output (e.g., voltage, resistance) into standard meteorological units (e.g., Celsius, hPa, m/s).
- **Algorithm Bugs:** Flaws in the software algorithms used for quality control, data interpolation, or product generation, leading to incorrect values or displays.
- **Interpolation Artifacts:** Errors introduced when data is interpolated

from irregular observation points to a uniform grid (e.g., for model input). This can lead to smoothing out real features or creating artificial ones.

- **Missing Value Handling:** Inappropriate methods for filling in missing data (e.g., using a simple average when a more sophisticated interpolation is needed) can introduce biases.

7.4.5 Human Errors

Despite automation, human involvement at various stages can introduce errors.

- **Manual Entry Mistakes:** Typographical errors when observers manually input data.
- **Incorrect Siting or Installation:** Improper placement or installation of sensors during setup, violating WMO guidelines.
- **Improper Maintenance or Calibration:** Incorrectly calibrating an instrument or failing to perform necessary maintenance can lead to systematic biases or sensor degradation.

Recognizing these common errors and anomalies is the first crucial step in meteorological data quality control. Robust QC procedures are designed to detect, flag, and mitigate the impact of these issues, ensuring that the data used for analysis, prediction, and research is as accurate and reliable as possible.

7.5 Missing Data in Meteorological Observations

Missing data refers to instances where expected observations or values for a particular meteorological parameter are absent from a dataset. Rather than being an erroneous reading, it's the complete lack of a data point at a specific time or location where one should exist. This is a pervasive and significant challenge in meteorological data management, impacting the completeness and reliability of analyses, forecasts, and research.

7.5.1 Common Causes of Missing Data

Missing data can arise from a multitude of factors, often stemming from the complex nature of observing networks and the environments in which they operate:

- **Sensor Malfunction or Failure:** A sensor might physically break, become disconnected, or experience internal electronic failure, ceasing to provide readings.
- **Power Outages:** Automated weather stations (AWS) rely on a power source (mains, solar, battery). A prolonged power outage will stop data collection and transmission.
- **Communication Link Failures:** Data transmission from remote sensors

to central processing facilities relies on communication networks (cellular, satellite, radio, wired). Outages, interference, or equipment failure in these links can prevent data from reaching its destination.

- **Data Logger Issues:** The on-site data logger, responsible for collecting and temporarily storing sensor readings, can malfunction due to software crashes, memory issues, or full storage capacity.
- **Scheduled Maintenance and Calibration:** Observing stations and instruments require periodic maintenance, calibration, or upgrades, during which data collection may be intentionally paused.
- **Environmental Extremes:** Severe weather conditions can directly impact sensors. For instance, heavy icing can prevent anemometers from rotating or block precipitation gauges, leading to missing wind or rainfall data. Extreme temperatures can also push sensors beyond their operational limits.
- **Human Error:** In the case of manually observed stations, an observer might forget to take a reading, misrecord it, or fail to transmit it.
- **Data Processing/Formatting Errors:** Issues during data conversion, encoding, or decoding can render data unreadable or cause it to be dropped from the dataset.
- **Data Filtering/Quality Control:** Sometimes, data that is deemed erroneous during quality control might be flagged as "missing" rather than being corrected or removed.

7.5.2 Impact of Missing Data on Meteorological Applications

The absence of data points can have significant negative consequences across all facets of meteorology:

- **Reduced Spatial and Temporal Coverage:** Gaps in data mean that the "picture" of the atmosphere is incomplete. This limits the ability to accurately map weather patterns, track storm movements, or understand the full extent of atmospheric phenomena.
- **Compromised Analysis:** When observations are missing, meteorologists must rely on interpolation or extrapolation, which introduces uncertainty. It becomes difficult to draw accurate weather maps, diagnose the precise location of fronts, or identify the true intensity of weather features.
- **Degraded Numerical Weather Prediction (NWP) Model Performance:** Data assimilation systems, which blend observations with short-range model forecasts to create the best estimate of the current atmospheric state, rely heavily on complete and accurate data.

Missing observations mean less information to constrain the model, leading to greater uncertainty in the initial conditions and potentially poorer forecast skill.

- **Challenges for Climate Studies:** Long-term climate records demand continuity and homogeneity. Gaps in historical data can complicate the calculation of accurate averages, trends, and extremes, potentially introducing biases or obscuring genuine climate signals. Researchers may struggle to compare different time periods or regions if data completeness varies.
- **Statistical Issues:** Many statistical analyses and machine learning algorithms require complete datasets. Missing data can complicate these analyses, requiring specialized techniques or leading to biased results if not handled properly.
- **Operational Gaps and Warning Delays:** For real-time operational forecasting, missing data can mean critical information is unavailable when needed most. For example, a missing radar scan during a severe thunderstorm could lead to a delayed or missed tornado warning.

7.5.3 Handling Missing Data (Quality Control and Preprocessing)

Addressing missing data is a crucial part of the quality control and preprocessing pipeline. The approach taken depends on the extent of the missingness, the type of data, and its intended use.

- **Identification and Flagging**
    - Automated systems continuously monitor incoming data for gaps or specific "missing data" codes.
    - Missing values are typically flagged with a specific numerical code (e.g., -9999) or a designated placeholder (e.g., NaN in programming environments) to ensure they are not misinterpreted as valid data points.
- **Reporting and Troubleshooting**
    - Alerts are generated when missing data is detected, prompting operators or engineers to investigate and rectify the cause (e.g., repair a sensor, restore communication).
- **Imputation/Infilling (with caution)**
    - This involves estimating and inserting values for missing data points. However, imputation should always be approached with extreme caution, as it introduces synthetic data that may not reflect the true

atmospheric state. The method chosen depends on the context:

- **Simple Methods**
  - **Last Observation Carried Forward (LOCF):** Using the most recent valid observation. Simple but can be inaccurate if conditions change rapidly.
  - **Mean/Median Imputation:** Replacing missing values with the average or median of the available data. Can smooth out variability and is generally not recommended for dynamic meteorological data.
- **Interpolation**
  - **Temporal Interpolation:** Using values from adjacent time steps (e.g., linear interpolation between the observation before and after the gap). Effective for short, isolated gaps.
  - **Spatial Interpolation:** Using values from nearby valid stations or grid points to estimate the missing value.
  - **Model-Based Imputation:** Using short-range NWP model forecasts or reanalysis data to fill gaps. This leverages the model's physical consistency but introduces model bias.
  - **Statistical Methods:** More advanced techniques like regression-based imputation, Kalman filters, or machine learning algorithms (e.g., K-nearest neighbors, Random Forest) can be used, but require careful validation.
  - **Important Caveat:** For critical applications like severe weather warnings or climate trend analysis, it is often preferable to treat data as truly missing rather than impute, especially if the imputation method introduces significant uncertainty or bias. The decision to impute depends heavily on the acceptable level of uncertainty for the downstream application.
- **Exclusion**
  - For some analyses, particularly in research, it may be more appropriate to simply exclude records or cases that contain missing data, especially if the missingness is extensive or the imputation uncertainty is too high. This reduces the sample size but maintains the integrity of the remaining data.
- **Impact on Downstream Products**
  - Quality control systems must ensure that missing data in base products (e.g., missing reflectivity in radar scans) is appropriately handled and propagated to derived products (e.g., Quantitative Precipitation

Estimates will show "no data" or be significantly underestimated in areas with missing base data).

In essence, while missing data is an unavoidable reality in meteorological observations, robust quality control and preprocessing procedures are designed to identify its presence, understand its cause, and mitigate its impact, ensuring that the meteorological information used for critical decisions is as complete and reliable as possible.

7.6 Outliers in Meteorological Data

Outliers, also known as **gross errors** or **spikes**, are individual data points that deviate significantly from the general pattern or expected range of the surrounding data. In meteorological observations, an outlier is a value that is physically implausible, extremely rare, or inconsistent with observations from nearby sensors or previous time steps. Unlike systematic biases (which are consistent errors) or missing data (which are absent values), outliers are typically isolated, abrupt, and often very large deviations.

7.6.1 Characteristics of Outliers

- **Extreme Deviation:** An outlier's value is far outside the expected or physically possible range for the measured parameter at that time and location.
- **Isolated Nature:** They usually appear as single, sharp "spikes" in a time series, surrounded by otherwise normal and consistent data.
- **Short Duration:** Outliers are typically brief, often lasting only a single observation period before the sensor returns to normal readings.

7.6.2 Common Causes of Outliers

Outliers can arise from various sources, indicating a temporary disruption in the measurement or transmission process:

- **Sensor Malfunction/Glitch:** A momentary electrical fault or physical issue within the sensor itself can cause a brief, erroneous reading.
    - *Example:* A thermistor briefly short-circuiting, causing a momentary reading of -999°C.
- **Electrical Interference/Noise:** Electromagnetic interference from nearby equipment, power surges, or lightning strikes can corrupt the sensor's signal or the data logger's processing.
    - *Example:* A sudden spike in an anemometer reading due to a nearby radio transmission.

- **Transmission Errors:** Errors during data transmission (e.g., bit flips in digital data, packet loss, or corruption over a communication link) can lead to a single, incorrect value being received.
  - *Example:* A single digit in a temperature reading being flipped, changing 25.0°C to 250.0°C.
- **Physical Impact/Interference:** A temporary physical disturbance to the sensor.
  - *Example:* A bird landing on a wind vane, momentarily causing an erratic wind direction reading.
  - *Example:* A piece of debris briefly hitting a rain gauge, causing a false "tip."
- **Software/Firmware Bugs:** Rare instances of bugs in the data logger's firmware or processing software that produce an incorrect value for a single observation.
- **Human Error (for manual observations):** A typo during manual data entry can result in a single, extreme outlier.
  - *Example:* An observer accidentally typing "250" instead of "25.0" for temperature.

7.6.3 Impact of Unaddressed Outliers

If outliers are not detected and handled during quality control, they can significantly degrade the utility of meteorological data:

- **Distorted Statistics:** Outliers can heavily skew statistical calculations such as means, standard deviations, and variances, leading to an inaccurate representation of the true atmospheric conditions.
  - *Example:* A single erroneous temperature reading of 100°C will drastically inflate the average temperature for a day or month.
- **Misleading Analysis:** An outlier on a weather map can create a false high or low pressure center, an erroneous temperature anomaly, or a spurious precipitation area, leading forecasters to misinterpret the current atmospheric state.
- **Degraded Model Performance:** When assimilated into Numerical Weather Prediction (NWP) models, outliers can introduce significant errors into the model's initial conditions. Due to the chaotic nature of the atmosphere, these errors can propagate and amplify rapidly through the forecast, leading to inaccurate predictions of storm tracks, intensity, or precipitation.

- **False Alarms/Warnings:** An outlier in real-time data (e.g., a spurious radar reflectivity spike or a false wind gust) could trigger an automated warning system or lead a forecaster to issue a warning that is not justified by actual conditions, eroding public trust.
  - **Compromised Climate Records:** While less frequent in long-term climate averages, persistent or extreme outliers, if not removed, can contaminate climate records and lead to incorrect assessments of extreme events or long-term trends.
  - **Algorithm Malfunctions:** Downstream algorithms that process meteorological data (e.g., for calculating derived products, identifying weather phenomena, or training machine learning models) can malfunction or produce incorrect results when fed unhandled outliers.

7.6.4 Detection Methods for Outliers (Quality Control)

Robust quality control systems employ various tests to identify outliers:

- **Range Checks (Absolute Limits):** The simplest test, checking if a value falls within physically plausible minimum and maximum limits for that parameter (e.g., temperature between -100°C and +60°C).
- **Rate-of-Change Checks (Temporal Consistency):** Compares a current observation to previous observations from the same sensor to ensure that the change over a short time interval is physically realistic.
  - *Example:* Temperature should not change by 20°C in one minute.
- **Internal Consistency Checks:** Compares different parameters measured by the same station that are physically related.
  - *Example:* If a rain gauge reports heavy rain, but the visibility sensor reports perfectly clear conditions, one of the readings might be suspect.
- **Spatial Consistency Checks:** Compares an observation from one station to observations from nearby stations measuring the same parameter. If a value deviates significantly from its neighbors, it's a potential outlier.
  - *Example:* A station reporting -10°C while all surrounding stations report 20°C.
- **Persistence Checks:** Compares a value to the previous value to ensure it's not "stuck" or unchanging for an unusually long period. (This is more for stuck values, but can catch some forms of outliers if they jump to a value and stay there).
- **Statistical Methods:** More advanced techniques like standard deviation tests (e.g., flagging values more than 3 or 4 standard deviations from the

mean), median absolute deviation (MAD), or Grubbs' test can be used to identify outliers based on statistical properties of the data.

- **Model-Based Checks:** Comparing observations to a short-range NWP model forecast or analysis. Large deviations can flag potential errors.

7.6.5 Handling Strategies for Outliers

Once an outlier is detected, there are several ways to handle it:

- **Flagging:** The most common approach. The outlier is marked with a specific quality control flag indicating that it is suspect or erroneous, but the original value is retained. This allows users to decide whether to use or discard the flagged data based on their application's sensitivity.
- **Removal/Deletion:** The outlier is completely removed from the dataset. This is often done for obvious gross errors that are clearly impossible.
- **Correction/Imputation (with extreme caution):** In some cases, if the nature of the error is well understood (e.g., a known offset), an attempt might be made to correct the value. Alternatively, the outlier might be replaced with an interpolated value from surrounding valid data points. However, this is done with great care, as imputation introduces synthetic data and can mask underlying sensor issues. For critical meteorological data, removal or flagging is generally preferred over correction unless there's high confidence in the correction.

Effective detection and handling of outliers are fundamental to ensuring the high quality and trustworthiness of meteorological data, which in turn underpins reliable weather forecasting, robust climate analysis, and sound scientific research.

7.7 Sensor Drift in Meteorological Data

Sensor drift is a pervasive and challenging type of systematic error in meteorological observations. It refers to the **gradual, unintended, and systematic change in a sensor's calibration or performance over time**, causing its readings to progressively deviate from the true value of the measured parameter. Unlike a sudden malfunction (which would be an outlier) or a fixed bias (which is a constant offset), drift is a continuous, evolving error that can be subtle and difficult to detect without careful monitoring and quality control.

7.7.1 What is Sensor Drift?

Imagine a perfectly calibrated thermometer. Over months or years of continuous operation, due to various factors, its internal components might subtly change, causing it to consistently read, for example, 0.5°C higher than the actual temperature, then 0.6°C higher, then 0.7°C higher, and so on.

This slow, creeping deviation from accuracy is sensor drift. It implies that the relationship between the physical phenomenon being measured and the sensor's output signal is no longer stable.

7.7.2 Common Causes of Sensor Drift

Sensor drift is a complex phenomenon influenced by both the sensor's design and its operating environment:

- **Aging of Components:** Electronic components (resistors, capacitors, integrated circuits) and mechanical parts (springs, bearings) naturally degrade over time, altering their electrical properties or physical characteristics.
- **Environmental Exposure**
  - **Temperature Extremes:** Prolonged exposure to very high or very low temperatures can stress sensor materials and electronics.
  - **Humidity and Moisture:** Water vapor can cause corrosion, alter the dielectric properties of materials, or lead to condensation, particularly affecting humidity and pressure sensors.
  - **UV Radiation:** Ultraviolet light can degrade plastic housings, protective coatings, and even sensing elements directly exposed to the sun.
  - **Contamination:** Dust, dirt, salt spray (in coastal or marine environments), pollutants, and biological growth (e.g., algae on water sensors) can accumulate on sensing surfaces, altering their response. This is a major issue for humidity sensors.
  - **Vibration and Physical Stress:** Continuous vibration (e.g., on a wind turbine or buoy) or physical stress from wind can cause mechanical wear or subtle shifts in sensor alignment.
- **Chemical Reactions:** Exposure to certain atmospheric gases or pollutants can chemically alter the sensing element, particularly in air quality sensors but also potentially affecting humidity or trace gas sensors.
- **Power Supply Fluctuations:** Inconsistent or noisy power supply can sometimes contribute to sensor instability over time.

7.7.3 Characteristics of Sensor Drift

- **Gradual:** The change in accuracy happens slowly over extended periods (weeks, months, years).
- **Systematic:** The deviation is typically in a consistent direction (e.g., always reading higher, or always lower).

- **Often Unidirectional:** While drift can fluctuate, it often shows a net increase or decrease in bias over time.
- **Subtle:** Unlike gross errors, drift can be very small initially, making it difficult to detect without rigorous quality control and calibration.

7.7.4 Impact of Unaddressed Sensor Drift

If sensor drift is not detected and corrected, it introduces a systematic bias into the data, with significant negative consequences:

- **Degraded Forecast Accuracy (NWP):** Biased observations, when assimilated into Numerical Weather Prediction (NWP) models, can systematically pull the model's initial state away from reality. This leads to persistent biases in the model's forecasts (e.g., consistently forecasting temperatures too high or too low), reducing forecast skill and reliability.
- **Compromised Climate Records:** For long-term climate monitoring, sensor drift is a major concern. Even a small, consistent drift over decades can create an **artificial trend** in the climate record, leading to incorrect conclusions about warming, cooling, or changes in precipitation. This is a key reason why **homogenization** (adjusting historical data for non-climatic changes like instrument drift) is crucial for climate studies.
- **Misleading Analysis and Diagnostics:** Meteorologists using real-time data with uncorrected drift might misdiagnose atmospheric conditions. For example, a drifting pressure sensor could lead to misjudgment of a low-pressure system's intensity or location.
- **Flawed Research Findings:** Scientific studies relying on data with uncorrected drift will produce biased results, potentially leading to incorrect conclusions about atmospheric processes or climate variability.
- **Economic Impacts:** For industries sensitive to weather (e.g., agriculture, energy, aviation), biased data from drifting sensors can lead to inefficient resource management or suboptimal operational decisions.

7.7.5 Detection and Mitigation Strategies

Combating sensor drift is a continuous process involving a combination of proactive measures and rigorous quality control:

- **Regular Calibration:** This is the most fundamental and effective method. Sensors are periodically removed from the field and compared against highly accurate reference standards in a controlled laboratory environment. Any detected deviation (drift) is quantified, and the sensor is either recalibrated or replaced.

- **Intercomparison/Collocation:** Placing a new sensor (or a highly accurate reference sensor) side-by-side with an existing operational sensor for a period. Any consistent difference in readings can indicate drift in the operational sensor.
- **Spatial Consistency Checks:** During automated quality control, a sensor's readings are compared against those of neighboring, well-performing sensors in the network. If a sensor consistently deviates from its neighbors in a systematic way, it can be flagged for potential drift.
- **Temporal Trend Analysis:** Monitoring a sensor's long-term performance by plotting its readings over time and looking for a consistent, non-random trend that deviates from expected behavior or from other sensors in the network.
- **Redundancy:** Deploying multiple sensors for the same parameter at a single station. This allows for continuous cross-validation; if one sensor starts to drift, its readings will diverge from the others, flagging the issue.
- **Robust Sensor Design:** Manufacturers continuously work to design more stable sensors that are less susceptible to environmental degradation and aging.
- **Protective Measures:** Using appropriate radiation shields, protective housings, and proper siting to minimize the impact of harsh environmental factors on the sensors.
- **Software Correction/Compensation:** If the drift pattern is well-characterized (e.g., after a calibration), algorithms can be applied to the raw data to compensate for the known drift. However, this relies on accurate and frequent recalibration to determine the drift rate.
- **Scheduled Replacement:** Implementing a proactive schedule for replacing sensors after a certain operational lifespan, even if no obvious drift is immediately apparent, to prevent data degradation.

In essence, sensor drift is a silent threat to data quality. Its insidious, gradual nature demands continuous vigilance through robust quality control, regular calibration, and strategic network management to ensure that meteorological observations remain accurate and reliable over their operational lifetime.

7.8 Instrument Biases in Meteorological Data

Instrument bias refers to a **systematic, consistent deviation** of a sensor's readings from the true value of the measured meteorological parameter. Unlike random errors, which fluctuate unpredictably around the true value, or outliers, which are brief, extreme deviations, a bias represents a persistent offset. If a sensor has a positive bias, it will consistently read higher than the actual value; if it has a negative bias, it will consistently read lower.

While **sensor drift** (discussed previously) is a *type* of bias that evolves over time, "instrument bias" is a broader term that also encompasses **fixed biases** – offsets that remain relatively constant throughout the instrument's operational life or until a specific event (like recalibration or a change in environmental conditions) alters it.

7.8.1 What Defines an Instrument Bias?

- **Systematic:** The error is not random; it consistently affects readings in a predictable way (e.g., always 1°C too high).
- **Consistent:** The offset remains relatively stable over a period, rather than fluctuating wildly.
- **Offset from True Value:** The sensor's output is consistently shifted away from the actual physical quantity it's supposed to measure.

7.8.2 Common Causes of Instrument Biases

Instrument biases can originate from various stages of a sensor's life cycle, from manufacturing to deployment:

- **Manufacturing/Factory Calibration Errors**
  - During production, if the initial calibration of a sensor is slightly off, it will ship with a built-in, fixed bias.
  - *Example:* A pressure transducer calibrated to read 1000 hPa when the true pressure is 1000.5 hPa will always have a -0.5 hPa bias.
- **Improper Field Calibration/Installation**
  - Even if a sensor is perfect from the factory, an incorrect calibration procedure during installation can introduce a bias.
  - *Example:* A rain gauge set to a wrong tipping bucket volume, leading to a consistent over- or underestimation of rainfall.
- **Environmental Influences (Consistent)**
  - **Radiation Bias:** Even with radiation shields, some temperature sensors can consistently experience a slight heating effect from solar radiation, leading to a consistent positive bias, especially during sunny, calm conditions.
  - **Wetting Bias:** Humidity sensors, if consistently exposed to condensation or persistent wetting without proper drying, might develop a consistent negative bias due to saturation effects or altered surface properties.

    - **Placement Bias:** If a sensor is consistently located too close to a heat source (e.g., a building's exhaust vent) or a reflective surface, it can introduce a fixed bias in temperature readings.
    - **Wind Obstruction Bias:** If a wind sensor is consistently obstructed by a nearby structure, it might always read lower wind speeds from certain directions.
- **Sensor Degradation (Leading to a New Fixed Bias)**
    - Unlike drift, which is a continuous change, a sensor might degrade in a way that causes a sudden, permanent shift to a new, biased state.
    - *Example:* A protective coating on a humidity sensor might partially peel off, leading to a new, fixed offset in its readings.
- **Aging of Reference Standards:** If the reference instruments used for calibration are themselves subtly drifting or biased, they can unknowingly transfer that bias to the sensors being calibrated.

7.8.3 Impact of Unaddressed Instrument Biases

The systematic nature of instrument biases makes their impact particularly insidious, as they can lead to consistently flawed interpretations and predictions:

- **Systematic Errors in Analysis**
    - **Distorted Maps:** A network of stations with a consistent positive temperature bias will show an artificially warm region on a weather map, misrepresenting the actual temperature field.
    - **Misleading Diagnostics:** Biased pressure readings can lead to consistent errors in diagnosing the strength of high or low-pressure systems.
- **Biased Numerical Weather Prediction (NWP) Forecasts**
    - **Persistent Forecast Errors:** If biased observations are assimilated into NWP models, they will systematically pull the model's initial state towards the biased values. This can lead to persistent biases in the model's forecasts (e.g., consistently forecasting temperatures that are too warm or too cold across a region), reducing forecast skill and reliability.
    - **Reduced Model Accuracy:** Even if the model's physics are perfect, feeding it biased observations will result in forecasts that are systematically off.
- **Compromised Climate Records and Trends**
    - **Artificial Trends:** A fixed bias introduced or removed at a specific point in a long-term climate record (e.g., due to a sensor replacement

or a change in station location without proper homogenization) can create an artificial jump or trend in the data, leading to incorrect conclusions about climate change or variability.

- **Misleading Averages:** Biased sensors will lead to inaccurate long-term averages for a location, affecting climate normals and comparisons.

- **Flawed Scientific Research**
    - Studies relying on data with uncorrected biases will yield systematically incorrect results, potentially leading to erroneous scientific conclusions about atmospheric processes or climate impacts.
- **Economic Consequences**
    - Industries relying on precise meteorological data (e.g., energy demand forecasting, agricultural irrigation scheduling) can incur significant economic losses due to consistent forecast errors caused by biased observations.

7.8.4 Detection Methods for Instrument Biases (Quality Control)

Detecting instrument biases requires careful monitoring and comparison

- **Absolute Calibration Checks:** The most direct method. Periodically removing the instrument from the field and calibrating it against a highly accurate, traceable reference standard in a controlled laboratory environment. Any consistent difference indicates a bias.
- **Intercomparison/Collocation:** Deploying a new, freshly calibrated sensor or a known "golden standard" reference sensor side-by-side with the operational sensor for a period. Consistent differences in readings between the two indicate a bias in the operational sensor.
- **Spatial Consistency Checks (Buddy Checks):** Comparing an observation from a station to the average or interpolated value from several nearby, well-performing stations. A persistent deviation from the neighbors suggests a bias.
- **Comparison with Independent Data Sources:** Comparing observations to independent data sources that are known to be reliable, such as:
    - **NWP Model Background/Analysis:** Comparing observations to the short-range model forecast or analysis (often called "observation minus background" or "O-B" residuals). Persistent, non-zero O-B values for a specific sensor can indicate a bias.
    - **Satellite Products:** Comparing surface observations to satellite-derived surface temperatures or other parameters.

  - **Reanalysis Data:** For historical data, comparing against high-quality reanalysis products.
- **Time Series Analysis:** Plotting the sensor's readings over time and looking for consistent offsets compared to expected values or other sensors in the network.
- **Maintenance Logs:** Reviewing maintenance records for any changes to the instrument or its environment that might have introduced a bias.

7.8.5 Mitigation and Correction Strategies

Once a bias is detected, the primary goal is to eliminate or compensate for it:

- **Recalibration:** The ideal solution. Adjusting the sensor's output to match the true value, either mechanically or electronically.
- **Bias Correction Algorithms:** If the bias is stable and well-quantified (e.g., from calibration or intercomparison), a correction factor (e.g., adding or subtracting a fixed offset) can be applied to all subsequent readings from that sensor in software.
- **Instrument Replacement:** If a sensor is inherently faulty or cannot be recalibrated effectively, it must be replaced with a new, properly calibrated one.
- **Improved Siting and Installation:** Addressing environmental factors that cause consistent biases (e.g., relocating a sensor away from a heat source, ensuring proper radiation shield ventilation).
- **Homogenization (for Climate Data):** For historical climate records, statistical techniques are used to adjust past data to account for known biases introduced by changes in instruments, observing practices, or station location.

In essence, instrument biases, whether fixed or slowly drifting, pose a continuous threat to the accuracy of meteorological data. Proactive quality control, regular calibration, and sophisticated analytical techniques are indispensable for detecting and mitigating these biases, thereby ensuring the integrity and trustworthiness of the atmospheric information that underpins all of modern meteorology.

7.9 Reporting Errors in Meteorological Data

Reporting errors refer to inaccuracies or inconsistencies that arise specifically during the process of preparing, encoding, transmitting, or decoding meteorological observations. These errors are distinct from issues originating directly from the sensor (like drift or bias) or environmental interference (like ground clutter), although they can be exacerbated by underlying data quality problems. Reporting errors fundamentally compromise the *usability* and *interpretability* of otherwise potentially good data.

7.9.1 What are Reporting Errors?

Reporting errors manifest when the data, as it is presented or transmitted, does not accurately reflect the original measurement or when it violates established communication standards. They can occur at various stages:

- **Encoding:** Converting raw sensor output or human observations into a standardized format (e.g., WMO codes, ASCII, binary).
- **Formatting:** Structuring the data according to specific protocols (e.g., METAR, SYNOP, GRIB, NetCDF).
- **Transmission:** Sending the formatted data over communication networks.
- **Decoding/Parsing:** Interpreting the received data stream back into usable values.

7.9.2 Common Causes of Reporting Errors

Reporting errors can stem from a variety of sources, often a mix of human and automated system failures:

- **Human Input Errors (for manual observations)**
  - **Typographical Mistakes:** Simple typos during manual entry (e.g., "25.0" entered as "2.50" or "250").
  - **Incorrect Coding:** Misapplication of WMO codes (e.g., using the wrong code for present weather or cloud type).
  - **Missing Fields:** Forgetting to enter a required data field.
- **Automated System Software Bugs**
  - **Encoding/Decoding Logic Errors:** Flaws in the software that converts raw sensor data into a transmission format, or that decodes received data.
  - **Unit Conversion Errors:** Incorrect scaling or conversion factors applied during data preparation (e.g., converting temperature from Celsius to Fahrenheit incorrectly).
  - **Truncation/Rounding Errors:** Data being inadvertently truncated or rounded incorrectly during formatting, leading to loss of precision.
  - **Missing Delimiters/Headers:** Errors in the formatting that cause parts of the message to be misinterpreted or ignored by receiving systems.

- **Communication/Transmission Issues**
    - **Bit Errors:** Random errors during digital transmission where a bit (0 or 1) is flipped, leading to corrupted characters or values.
    - **Packet Loss:** In network-based transmission, entire data packets might be lost, resulting in missing sections of a report.
    - **Synchronization Issues:** Problems with timing between transmitting and receiving systems, leading to misaligned data.
- **Outdated Software/Standards**
    - Receiving systems might not be updated to the latest version of a data format standard, leading to misinterpretation of new fields or codes.
- **Hardware Malfunctions**
    - Faulty cabling, network cards, or modems can introduce errors during transmission.

7.9.3 Examples of Reporting Errors

- **Incorrect Units:** A temperature value is reported as "25" but is mistakenly interpreted as Fahrenheit when it should be Celsius, or vice-versa.
- **Invalid Codes:** A METAR report contains a present weather code that does not exist in the WMO standard.
- **Truncated Data:** A wind speed of "15.7 m/s" is reported as "15 m/s" due to incorrect rounding or field length limitations.
- **Missing Fields:** A mandatory field like atmospheric pressure is entirely absent from a surface observation report.
- **Garbled Characters:** Due to bit errors, a numerical value appears as a string of random characters (e.g., "34.5" becomes "3@.F").
- **Incorrect Timestamp:** The reported time of observation is inaccurate, leading to data being associated with the wrong temporal context.
- **Incorrect Location:** A station reports data, but its geographical coordinates are erroneously encoded, causing the data to appear at the wrong location on a map.

7.9.4 Impact of Reporting Errors

The consequences of reporting errors are similar to those of other data quality issues, often leading to misinterpretation and degraded system performance:

- **Data Rejection:** Automated quality control systems at receiving centers are designed to flag or reject reports that do not conform to expected formats or contain invalid codes. This leads to data loss, as the erroneous report is not used.

- **Misleading Analysis:** If an erroneous report slips through QC, it can create false features on weather maps (e.g., a non-existent pressure anomaly from a garbled pressure reading) or misrepresent current conditions.
- **Degraded NWP Model Performance:** Data assimilation systems rely on correctly formatted and valid data. Reporting errors can lead to observations being rejected or, worse, being incorrectly assimilated, introducing spurious information into the model's initial state and degrading forecast accuracy.
- **Operational Delays:** Forecasters may spend valuable time manually reviewing and correcting erroneous reports, diverting attention from critical real-time decision-making.
- **Compromised Data Archives:** If reporting errors are not caught, they can contaminate long-term data archives, making historical analysis and climate studies unreliable until significant effort is spent on post-hoc cleaning.

7.9.5 Detection and Mitigation Strategies

Detecting and preventing reporting errors is a key focus of data preprocessing and quality control:

- **Strict Format Validation:** Automated systems rigorously check incoming data against predefined format specifications (e.g., WMO manuals, XML schemas). Any deviation (missing fields, incorrect character types, invalid lengths) triggers a flag or rejection.
- **Range and Consistency Checks:** Even if formatted correctly, the values within the report are subjected to range checks (e.g., temperature within physical limits) and consistency checks (e.g., dew point cannot be higher than temperature).
- **Checksums and Error Correction Codes:** During transmission, techniques like checksums or more advanced error correction codes can be used to detect and sometimes even correct minor bit errors.
- **Automated Monitoring and Alerting:** Systems continuously monitor the flow of data and generate alerts when a high rate of rejected reports or specific types of errors are detected, prompting human intervention.
- **Regular Software Updates:** Ensuring that all encoding, decoding, and processing software is up-to-date with the latest data format standards and bug fixes.

- **Training and Protocols (for manual input):** For stations with human observers, rigorous training and clear data entry protocols minimize manual errors.
- **Redundancy in Transmission:** Using multiple communication paths or redundant data transmissions can help ensure data integrity.
- **Feedback Loops:** Establishing mechanisms for data users to report detected errors back to the data providers, allowing for investigation and correction at the source.

In summary, reporting errors, though often technical in nature, pose a significant threat to the usability and reliability of meteorological data. Robust quality control, strict adherence to international standards, and continuous monitoring of data streams are essential to minimize their occurrence and impact, ensuring that meteorological information is consistently accurate and actionable.

7.10 Automated Quality Control Algorithms in Meteorology

The sheer volume, velocity, and variety of meteorological data generated by modern observing systems make manual quality control (QC) impractical, if not impossible. This necessitates the widespread use of **Automated Quality Control (AQC) algorithms**. These are sophisticated computational routines designed to rapidly and objectively identify, flag, and sometimes correct errors and anomalies in meteorological data without direct human intervention. AQC is a critical preprocessing step that transforms raw, potentially flawed observations into reliable information suitable for analysis, numerical weather prediction (NWP), and climate studies (Fig. 7.2).

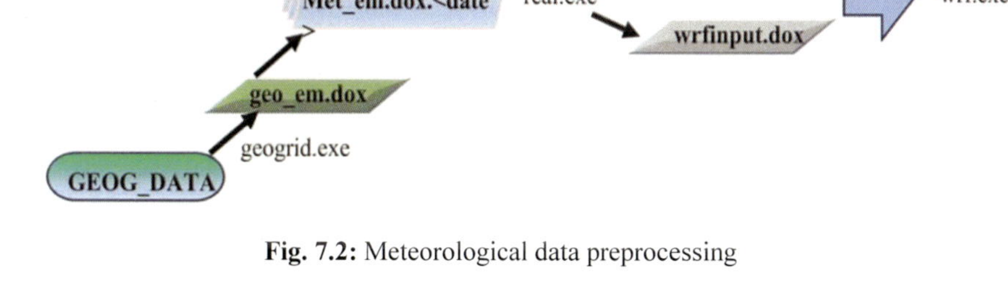

Fig. 7.2: Meteorological data preprocessing

7.10.1 What are Automated Quality Control Algorithms?

AQC algorithms are sets of predefined rules, statistical tests, and computational procedures applied systematically to incoming meteorological data. Their primary goals are to:

- **Detect Errors:** Identify values that are physically impossible, highly improbable, or inconsistent with other related observations.
- **Flag Data:** Assign a "quality flag" to each data point, indicating its perceived reliability (e.g., good, suspect, bad, missing, corrected). This allows downstream users to understand the data's quality and make informed decisions about its use.
- **Correct (Cautiously):** In some cases, for certain types of errors, algorithms may attempt to automatically correct a value, though this is done with extreme caution to avoid introducing new biases.
- **Filter:** Remove or isolate data points identified as erroneous.

7.10.2 Importance of Automated Quality Control

- **Handling Big Data:** AQC is indispensable for processing the massive, continuous streams of data from global observing networks (e.g., thousands of surface stations, hundreds of radar sites, satellite data streams).
- **Speed and Timeliness:** For operational forecasting, AQC provides near real-time quality assurance, ensuring that reliable data is available quickly for forecasters and NWP models.
- **Consistency and Objectivity:** Algorithms apply the same rules uniformly, ensuring consistent quality assessment across all data, unlike potentially subjective manual checks.
- **Efficiency:** Frees up human meteorologists to focus on more complex analysis and forecasting tasks.
- **Foundation for Downstream Applications:** Provides clean, reliable input for data assimilation systems, NWP models, climate archives, and research analyses.

7.10.3 General Principles of AQC

Most AQC algorithms operate based on comparing a data point against:

- **Physical Limits:** Is the value physically possible?
- **Temporal Consistency:** Is the value consistent with previous values from the same sensor?

- **Spatial Consistency:** Is the value consistent with values from nearby sensors?
- **Internal Consistency:** Is the value consistent with other parameters measured by the same sensor or station?
- **Climatological Norms:** Is the value plausible given the typical climate for that time and location?
- **Model Background:** Is the value consistent with a short-range forecast from an NWP model?

7.10.4 Common Automated Quality Control Algorithms/Tests

Here are some of the most widely used AQC algorithms:

- **Range Checks (Absolute Limits)**
  - **Description:** The simplest and first line of defense. Each observed value is checked against predefined, physically plausible minimum and maximum limits for that parameter. If a value falls outside this range, it's flagged as erroneous.
  - **Example:** Air temperature must be between -90°C and +60°C. Relative humidity must be between 0% and 100%. Wind speed cannot be negative.
  - **Purpose:** Catches gross errors and obvious sensor malfunctions.
- **Rate-of-Change Checks (Temporal Consistency)**
  - **Description:** Compares the current observation to one or more previous observations from the *same sensor*to ensure that the change over a short time interval is physically realistic.
  - **Example:** Air temperature should not change by more than 10°C in one minute. Pressure should not drop by more than 5 hPa in 5 minutes.
  - **Purpose:** Detects sudden, unphysical spikes or drops (outliers) that might still be within absolute limits but are impossible given the short time scale.
- **Persistence Checks**
  - **Description:** Checks if a sensor is reporting the exact same value for an unusually long period, indicating a "stuck" sensor or communication failure.
  - **Example:** A wind speed sensor reporting 0.0 m/s for 3 hours straight when conditions are known to be windy.
  - **Purpose:** Identifies sensor malfunctions where the output freezes.

- **Internal Consistency Checks**
  - **Description:** Examines the physical relationships between different parameters measured by the *same station or sensor package.*
- **Example**
  - Dew point temperature must always be less than or equal to air temperature.
  - If a rain gauge reports heavy precipitation, the visibility sensor should not report unlimited visibility.
  - For radiosondes, the pressure, temperature, and humidity profile should be consistent with the hydrostatic equation.
  - **Purpose:** Catches errors where one sensor might be working but another related one is faulty, or where a single reading is inconsistent with physical laws.
- **Spatial Consistency Checks (Buddy Checks)**
  - **Description:** Compares an observation from a particular station to observations from its *neighboring stations*measuring the same parameter. If a value deviates significantly from its neighbors, it's flagged as suspect.
  - **Methodology:** Often involves calculating a weighted average or interpolation from surrounding stations and comparing it to the target station's value.
  - **Example:** A station reporting 30°C while all surrounding stations in a 50 km radius report 10°C.
  - **Purpose:** Identifies localized errors, biases, or representativeness issues that might not be caught by single-station checks. Crucial for detecting errors in sparse networks.
- **Climatological Checks**
  - **Description:** Compares the observed value against historical climatological norms for that specific location and time of year.
  - **Example:** A temperature reading of 40°C in January in Siberia would be flagged, even if it's within absolute physical limits.
  - **Purpose:** Identifies highly improbable but physically possible values that might be errors.
- **Model-Based Checks (Background Checks)**
  - **Description:** Compares the observed value against a short-range forecast (the "background" or "first guess") from a Numerical Weather Prediction (NWP) model.

- **Methodology:** Calculates the "observation minus background" (O-B) residual. If the O-B is excessively large, the observation is flagged.
- **Purpose:** Very powerful for identifying errors, especially in areas with sparse observations, as the model provides a physically consistent estimate. Also helps identify potential biases in the model itself if many observations consistently deviate in one direction.

- **Dual-Polarization Radar QC Algorithms**
  - **Description:** Specialized algorithms leverage dual-polarization products (ZDR, CC, KDP) to identify and filter non-meteorological echoes.
- **Example**
  - **Low CC Filtering:** Automatically removes echoes with very low Correlation Coefficient, which are typically ground clutter, biological targets (insects/birds), or tornado debris (which is a desired signal, so careful handling is needed).
  - **Hydrometeor Classification:** Algorithms classify precipitation types (rain, snow, hail) based on polarimetric signatures, and can flag ambiguous or erroneous classifications.
  - **Purpose:** Significantly improves the quality of radar data by distinguishing precipitation from clutter and identifying specific severe weather signatures.
- **Machine Learning/Statistical Methods**
  - **Description:** Increasingly, more advanced statistical and machine learning techniques (e.g., clustering, anomaly detection, neural networks) are being explored to identify complex error patterns that simpler rule-based systems might miss.
  - **Purpose:** To detect subtle biases, complex error correlations, or to learn from historical error patterns.

7.10.5 Output of AQC: Quality Flags

AQC algorithms typically do not delete data but assign **quality flags** to each data point. These flags are numerical or categorical codes that indicate the perceived quality of the observation. Common flag categories include:

- **(Good):** Data passed all QC checks.
- **(Suspect):** Data is plausible but deviates significantly or triggered a minor flag.
- **(Bad):** Data is highly likely to be erroneous (e.g., failed range check).
- **(Missing):** No data was received.

- **(Corrected):** Data was automatically corrected (e.g., a known bias removed).
- **(Climatologically Improbable):** Data is physically possible but extremely rare for the time/location.

These flags allow downstream systems (e.g., data assimilation) to decide whether to use, weight, or discard the observation based on its quality.

7.10.6 Advantages of AQC

- **Scalability:** Can process enormous volumes of data rapidly.
- **Consistency:** Applies uniform rules, ensuring objective and repeatable quality assessment.
- **Efficiency:** Reduces the need for manual intervention, saving time and resources.
- **Real-time Processing:** Essential for operational forecasting where timely data is critical.

7.10.7 Limitations and Challenges of AQC

- **False Positives/Negatives:** No algorithm is perfect. AQC can sometimes flag good data as bad (false positive) or miss actual errors (false negative), especially for extreme but real weather events.
- **Threshold Dependence:** The performance of many algorithms relies on predefined thresholds, which can be difficult to set optimally across all conditions and regions.
- **Interdependence of Errors:** Sometimes, one error can mask another, or an error in one parameter can cause another parameter to fail a check.
- **Learning from Errors:** Algorithms typically follow predefined rules; they don't "learn" new error patterns unless explicitly programmed or updated.
- **Need for Human Oversight:** AQC is powerful but not infallible. Human meteorologists and data quality experts are still essential for reviewing flagged data, investigating persistent issues, and refining algorithms. AQC is a tool to *assist* human decision-making, not replace it entirely.

In conclusion, Automated Quality Control algorithms are an indispensable component of modern meteorological data processing. By systematically applying a battery of tests, they ensure the integrity and reliability of the vast streams of atmospheric data, transforming raw measurements into trustworthy information that fuels accurate forecasts, robust climate analyses, and insightful scientific research.

7.11 Manual and Semi-Automated Quality Control in Meteorology

While Automated Quality Control (AQC) algorithms are indispensable for handling the sheer volume and velocity of meteorological data, they are not infallible. There are situations where human expertise, judgment, and nuanced understanding are crucial for assessing data quality. This leads to the integration of **manual** and **semi-automated** quality control processes within comprehensive meteorological data management systems. These approaches leverage human cognitive abilities to complement the speed and consistency of algorithms.

7.11.1 Manual Quality Control (MQC)

Manual quality control involves direct human inspection and assessment of meteorological data. While less common for routine, high-volume data streams, it remains vital for specific scenarios.

- **What it involves**
  - **Visual Inspection:** Meteorologists or data quality experts visually examine data plots, time series, and spatial maps to identify patterns, anomalies, or subtle errors that automated algorithms might miss.
  - **Expert Judgment:** Applying deep meteorological knowledge to assess the plausibility of data, especially during extreme or unusual weather events where automated thresholds might be exceeded by valid data.
  - **Cross-Referencing:** Manually comparing data from multiple sources (e.g., a surface observation against a satellite image, a radar product, or a nearby radiosonde sounding) to confirm or deny a suspected error.
  - **Review of Flagged Data:** Detailed investigation of data points that AQC algorithms have flagged as "suspect" or "bad."
- **When it is used**
  - **Post-Analysis/Archival QC:** For historical datasets or climate records, where long-term consistency and accuracy are paramount, and there is less time pressure.
  - **Extreme Weather Events:** During severe weather (e.g., hurricanes, tornadoes, blizzards), where unusual atmospheric conditions can cause AQC algorithms to misflag valid data or miss critical errors. Human review is crucial for warning decisions.
  - **New Observing Systems/Pilot Projects:** When new sensors or platforms are deployed, manual review helps to identify new error

modes or biases that AQC algorithms haven't been trained to detect yet.

- **Algorithm Development/Refinement:** Human-verified "truth" datasets are essential for training and validating AQC algorithms.
- **Complex Cases:** For data points that are persistently flagged by AQC but whose true status is ambiguous.

- **Advantages**
  - **Superior Nuance and Context:** Humans can interpret complex patterns, understand the meteorological context, and apply expert judgment in ways algorithms cannot.
  - **Detection of Subtle Errors:** Can spot subtle biases, gradual drifts, or complex error correlations that AQC might miss.
  - **Handling Extreme Events:** Can correctly identify physically extreme but valid observations that AQC might incorrectly flag as outliers.
  - **Problem Diagnosis:** Humans are better at diagnosing the *cause* of an error (e.g., a specific sensor malfunction, a siting issue) rather than just flagging it.
- **Limitations**
  - **Time-Consuming and Labor-Intensive:** Extremely slow and expensive, making it impractical for large volumes of real-time data.
  - **Subjectivity:** Can be influenced by individual meteorologist's experience, biases, or fatigue, leading to inconsistencies.
  - **Scalability Issues:** Cannot keep up with the data volume generated by modern observing systems.
  - **Limited for Real-time Operations:** Too slow for immediate operational needs.

7.11.2 Semi-Automated Quality Control (SAQC)

Semi-automated quality control represents a hybrid approach, combining the efficiency of automated algorithms with the intelligence and judgment of human meteorologists. It is the most common operational model for real-time meteorological data quality control.

- **What it involves**
  - **Automated Flagging:** AQC algorithms run first, processing the data and assigning initial quality flags (e.g., "good,""suspect,""bad").
  - **Human Review Interface:** Data flagged as "suspect" or "bad" (or data from critical regions/sensors) is presented to a human

meteorologist or data quality analyst via a specialized graphical user interface (GUI).

- **Human Decision:** The human reviews the flagged data point, often alongside surrounding observations, time series plots, and other relevant meteorological information (e.g., satellite imagery, radar data, model analyses). Based on this context, the human makes a final decision:
- Confirm the flag (e.g., confirm it's a bad reading).
- Override the flag (e.g., determine it was a false positive and the data is actually good).
- Adjust the flag (e.g., change from "bad" to "suspect").
- Provide a manual correction if the error is simple and obvious (e.g., a misplaced decimal).
- **Feedback Loop:** Human decisions can be fed back into the AQC system to refine algorithm thresholds, identify new error patterns, or update rules.

- **When it is used**
    - **Operational Real-time Data Streams:** The standard approach for quality control of incoming observations (surface, upper air, radar, satellite) at national meteorological centers.
    - **Critical Data Sources:** For data from key observing sites or systems vital for severe weather warnings.
    - **Complex Weather Situations:** During rapidly evolving or unusual weather events where AQC might struggle.
    - **Data Assimilation Pre-processing:** Ensuring the highest quality data for input into NWP models.
- **Advantages**
    - **Balances Speed and Accuracy:** Leverages the speed of automation for routine checks while reserving human expertise for complex or ambiguous cases.
    - **Improved Accuracy:** Significantly reduces false positives and false negatives compared to purely automated systems.
    - **Efficiency:** Allows meteorologists to focus on the most challenging data quality issues, rather than manually reviewing every data point.
    - **Adaptability:** Humans can quickly adapt to new error modes or unusual meteorological conditions that AQC might not yet handle.

- **Learning and Improvement:** Provides a mechanism for continuous improvement of AQC algorithms through human feedback.
- **Limitations**
    - **Still Resource-Intensive:** Requires trained personnel and specialized software interfaces.
    - **Scalability Bottlenecks:** While more scalable than pure MQC, the human review step can still become a bottleneck during periods of extremely high data volume or widespread errors.
    - **Potential for Human Error:** Though less frequent than in pure MQC, human fatigue or misjudgment can still introduce errors.

7.11.3 The Integrated Quality Control System

In practice, a comprehensive meteorological data quality control system integrates all three approaches:

- **Automated QC (First Pass):** All incoming data undergoes initial, rapid, automated checks to filter out the most obvious errors and assign preliminary flags. This handles the vast majority of data.
- **Semi-Automated QC (Second Pass/Review):** Data flagged as "suspect" or "bad" by AQC, or data from high-priority sources, is routed to human review via interactive interfaces. This is where human judgment refines the quality flags.
- **Manual QC (Deep Dive/Offline):** Used for in-depth investigation of persistent problems, algorithm development, homogenization of historical climate records, or post-event analysis of critical data.

This layered approach ensures that meteorological data is processed efficiently, while maintaining the highest possible level of quality and reliability, which is paramount for the accuracy of forecasts, the integrity of climate records, and the safety of the public.

7.12 Data Imputation and Homogenization in Meteorology

In the continuous quest for complete, accurate, and consistent meteorological datasets, two crucial preprocessing techniques stand out: **data imputation** and **homogenization**. While both aim to improve data utility, they address different types of data quality issues and serve distinct purposes, particularly critical for long-term climate studies and comprehensive analyses.

7.12.1 Data Imputation (Infilling Missing Data)

Data imputation is the process of estimating and filling in missing values within a dataset. It is performed after initial quality control has identified and flagged missing observations. The goal is to create a complete dataset, which is often required by statistical models, numerical weather prediction (NWP) systems, or certain analysis techniques.

- **Purpose**
    - **Completeness:** To provide a continuous time series or spatial grid of data, as many analytical tools and models require complete datasets.
    - **Avoid Data Loss:** To make use of otherwise valuable data points surrounding the missing values, rather than discarding entire records or time periods due to small gaps.
    - **Improved Representativeness:** To provide a more complete picture of atmospheric conditions for analysis and visualization.
- **When is it Applied?**
    - Typically after initial quality control identifies missing data.
    - For short, isolated gaps in time series.
    - For filling in small spatial gaps in gridded datasets.
    - Often used more cautiously for real-time operational data due to the risk of introducing synthetic information.
- **Common Methods of Imputation**
- **Simple Methods (Least Recommended for Meteorology)**
    - **Mean/Median Imputation:** Replacing missing values with the overall mean or median of the observed data. This is generally discouraged for meteorological data as it reduces variability and can bias results.
    - **Last Observation Carried Forward (LOCF):** Using the most recent valid observation to fill the gap. Simple, but can be inaccurate if conditions change rapidly.
- **Interpolation-Based Methods (More Common)**
    - **Temporal Interpolation:** Estimating a missing value based on observations from the same sensor at adjacent time steps.
    - ***Linear Interpolation:*** Drawing a straight line between the valid points before and after the gap.
    - ***Spline Interpolation:*** Using more complex curves to fit the data, often better for non-linear trends.
    - **Spatial Interpolation:** Estimating a missing value based on observations from nearby stations or grid points.

- ***Inverse Distance Weighting (IDW):*** Assigning more weight to closer observations.
- ***Kriging:*** A geostatistical method that considers spatial correlation.
- ***Nearest Neighbor:*** Using the value from the closest valid observation.

- **Model-Based Imputation**
    - **NWP Model Background/Analysis:** Using the short-range forecast (background) or analysis from a numerical weather prediction model to fill the gap. This leverages the model's physical consistency.
    - **Regression-Based Imputation:** Using statistical models to predict missing values based on relationships with other variables or historical data.
- **Advanced Statistical/Machine Learning Methods:**
    - **Kalman Filters:** Can provide optimal estimates for missing values in time series, particularly useful for dynamic systems.
    - **Expectation-Maximization (EM) Algorithm:** Iteratively estimates missing values and model parameters.
    - **K-Nearest Neighbors (KNN):** Finds the 'k' most similar complete data points and uses their values to impute the missing one.
- **Challenges and Considerations**
    - **Introduction of Synthetic Data:** Imputed data is not a true observation. It introduces an element of artificiality and uncertainty into the dataset.
    - **Bias and Variance:** The choice of imputation method can introduce biases or reduce the natural variability of the data, potentially leading to misleading statistical results.
    - **Length of Gap:** Imputation is generally more reliable for short, isolated gaps. Long periods of missing data are very difficult to impute accurately.
    - **Impact on Downstream Applications:** The impact of imputation on subsequent analyses (e.g., trend detection, extreme event analysis, model assimilation) must be carefully considered and ideally quantified. For critical applications, it's often safer to treat data as truly missing rather than impute.
    - **Metadata:** It is crucial to clearly flag imputed values in the metadata so users are aware that these are not original observations.

7.12.2 Homogenization

Homogenization is a specialized quality control process applied primarily to **long-term climate datasets** to identify and adjust for non-climatic changes (inhomogeneities) that can introduce artificial trends or biases into the data. These inhomogeneities are typically caused by changes in the observing system or environment, rather than actual climate variability.

- **Purpose**
  - **Ensure Climatic Representativeness:** To create a consistent, homogeneous time series that accurately reflects only the true climatic variations and trends.
  - **Remove Artificial Signals:** To eliminate biases introduced by non-climatic factors that could otherwise be misinterpreted as climate change or variability.
  - **Enable Reliable Climate Analysis:** To provide trustworthy data for detecting climate trends, assessing extremes, validating climate models, and informing climate policy.
- **When is it Applied?**
  - Almost exclusively for long-term (decades to centuries) observational datasets used in climate research (e.g., daily temperature, precipitation, pressure records from surface stations).
  - It is typically performed offline, after initial QC and sometimes after imputation.
- **Common Causes of Inhomogeneities (Non-Climatic Changes):**
  - **Station Relocation:** Moving a weather station, even a short distance, can change its microclimate (e.g., elevation, exposure to wind, proximity to urban development).
  - **Instrument Changes:** Replacing old instruments with new models (e.g., mercury thermometer to electronic thermistor, manual rain gauge to tipping bucket). Even if both are calibrated, their response characteristics might differ slightly.
  - **Changes in Observing Practices:** Alterations in observation time, reading procedures, or data processing methods.
  - **Urbanization/Land-Use Changes:** Growth of urban areas around a station can create an "urban heat island" effect, causing an artificial warming trend in the station's temperature record.
  - **Changes in Station Environment:** Growth of trees, construction of new buildings, or changes in ground cover near the sensor.

- **Observer Changes:** In manual stations, different observers might have slightly different biases in reading instruments.

- **Homogenization Methods**
  - **Reference Series Method (Relative Homogenization):** The most common approach. It compares the candidate station's time series to one or more "reference series" from nearby stations that are assumed to be homogeneous. Discontinuities or shifts in the difference series between the candidate and reference stations indicate an inhomogeneity in the candidate series.
  - *Statistical Tests:* Various statistical tests (e.g., Standard Normal Homogeneity Test - SNHT, Multiple Analysis of Series for Homogenization - MASH, PRODIGE) are used to detect break points (points in time where a shift occurs).
  - *Adjustment:* Once a break point is detected, the data before or after the break is adjusted by an additive or multiplicative factor to make it consistent with the reference series.
  - **Absolute Homogenization:** Less common, involves comparing observations to a physical model or theory, or to a known absolute standard.
  - **Metadata-Based Homogenization:** Using detailed station metadata (logbooks, site surveys) to identify known change points (e.g., date of relocation, instrument change) and then applying statistical adjustments around those dates.

- **Challenges and Considerations**
  - **Identifying True Break Points:** Distinguishing genuine inhomogeneities from natural climate variability can be challenging, especially in noisy data.
  - **Availability of Reference Series:** Finding truly homogeneous reference stations, particularly in data-sparse regions, can be difficult.
  - **Uncertainty in Adjustments:** The magnitude of the adjustments applied introduces its own uncertainty.
  - **Subjectivity:** While statistical tests are used, human judgment is often involved in confirming break points and selecting appropriate adjustment methods.
  - **Propagating Errors:** Errors in reference series can propagate to the candidate series.
  - **Computational Intensity:** Processing large datasets for homogenization can be computationally demanding.

In essence, data imputation and homogenization are sophisticated, often computationally intensive, processes that are vital for maximizing the utility and trustworthiness of meteorological data. Imputation addresses the problem of missing values to create complete datasets, while homogenization tackles the more subtle and long-term issue of non-climatic biases in historical records, ensuring that our understanding of past climate change is as accurate as possible. Both processes are critical components of a robust meteorological data quality control and preprocessing framework.

7.13 Conclusion: Meteorological Data Quality Control & Preprocessing

This chapter has underscored a fundamental truth in meteorology: the immense value of atmospheric data is not solely in its collection, but in its **quality**. We've delved into the pervasive nature of data errors and anomalies—from sensor drift and instrument biases to missing values and reporting inaccuracies—each posing a unique threat to the integrity of meteorological information. Recognizing that raw observations are inherently imperfect, we explored the indispensable role of **quality control (QC)**, a meticulous process of identifying, flagging, and mitigating these imperfections.

We examined the critical interplay between **automated quality control (AQC) algorithms**, which offer the speed and consistency necessary to handle vast data volumes, and the nuanced judgment provided by **manual** and **semi-automated**approaches. Furthermore, we discussed specialized preprocessing techniques like **data imputation**, which intelligently fills missing gaps, and **homogenization**, a vital process for removing non-climatic biases from long-term records, ensuring the purity of climate signals. Ultimately, these rigorous procedures transform raw, noisy, and potentially incomplete atmospheric signals into trustworthy, actionable information.

In essence, robust data quality control and preprocessing are not merely technical steps; they are the essential guardians of meteorological accuracy. They ensure that the initial conditions fed into complex numerical weather prediction models are as precise as possible, leading to more reliable forecasts that protect lives and property. They guarantee the integrity of climate data records, allowing scientists to confidently assess past climate change and project future scenarios. And they provide the foundation for sound scientific research, enabling a deeper, more accurate understanding of our planet's dynamic atmosphere. As observing systems continue to evolve and data volumes grow, the commitment to rigorous data quality will remain paramount, serving as the bedrock upon which all advancements in meteorological science and its societal benefits are built.

8

Statistical Methods in Meteorological Analysis

8.1 Introduction

Having explored the intricate processes of meteorological data collection, storage, and the crucial steps of quality control and preprocessing, we now turn our attention to how this vast and meticulously prepared information is actually **analyzed and interpreted**. This brings us to the realm of **statistical methods**, which are indispensable tools for making sense of the inherent variability, uncertainty, and sheer volume of meteorological data.

Meteorological phenomena are complex, influenced by numerous interacting factors, and often exhibit chaotic behavior. Statistical methods provide the rigorous framework necessary to quantify these complexities, identify underlying patterns, assess relationships between different atmospheric variables, and ultimately, draw robust conclusions. From summarizing observed climate patterns to evaluating forecast accuracy, and from detecting long-term trends to predicting future weather events, statistics permeate nearly every aspect of modern meteorology.

This chapter will introduce the fundamental statistical concepts and techniques applied in meteorological analysis. We will begin with **descriptive statistics** to summarize and visualize data, then move to **inferential statistics** to draw conclusions about atmospheric processes from samples. We will delve into methods for understanding relationships, such as **correlation and regression analysis**, and explore techniques for analyzing **time series data** to identify trends and periodicities. Furthermore, we will touch upon **spatial statistics** for gridded data and the growing role of **machine learning** approaches. By mastering these statistical tools, meteorologists can transform raw data into meaningful insights, enhance predictive capabilities, and advance our scientific understanding of the Earth's atmosphere.

8.2 Descriptive Statistics in Meteorology

Descriptive statistics play a crucial role in meteorology, helping to summarize and describe large datasets related to weather and climate. Some key applications include:

- **Temperature analysis**: Calculating means, medians, and standard deviations to understand temperature trends and variability.
- **Precipitation patterns**: Using descriptive statistics to characterize rainfall or snowfall distributions, including totals, frequencies, and extremes.
- **Climate trend analysis**: Identifying shifts in weather patterns over time using statistical measures like regression analysis.
- **Weather forecasting:** Descriptive statistics inform forecast models, enabling meteorologists to predict future weather conditions.

Some common descriptive statistical measures used in meteorology include:

- Mean
- Median
- Mode
- Standard deviations
- Variance
- Percentiles

These statistics help meteorologists understand and communicate complex weather data, ultimately informing decision-making in fields like agriculture, urban planning, and emergency management.

can you elaborate and rewrite

Here's a more detailed explanation of descriptive statistics in meteorology:

8.2.1 Understanding Weather Patterns

Descriptive statistics are essential in meteorology for analyzing and interpreting large datasets related to weather and climate. By applying statistical measures, meteorologists can:

- **Summarize temperature data**: Calculate means, medians, and standard deviations to understand temperature trends, variability, and patterns.
- **Analyze precipitation patterns**: Use descriptive statistics to characterize rainfall or snowfall distributions, including totals, frequencies, and extremes.
- **Identify climate trends**: Apply statistical measures like regression analysis to detect shifts in weather patterns over time.

8.2.2 Key Descriptive Statistical Measures

Some common statistical measures used in meteorology include:

- **Mean**: Average value of a dataset (e.g., average temperature)
- **Median**: Middle value of a dataset (e.g., median rainfall)

- **Mode**: Most frequently occurring value in a dataset
- **Standard deviation**: Measure of variability or spread in a dataset
- **Variance**: Measure of the average distance from the mean
- **Percentiles**: Divide a dataset into equal parts (e.g., 25th, 50th, 75th percentiles)

8.2.3 Applications in Meteorology

Descriptive statistics inform various aspects of meteorology, including:

- **Weather forecasting**: Statistical measures help predict future weather conditions.
- **Climate research**: Analyzing trends and patterns in weather data.
- **Agriculture and planning**: Understanding weather patterns informs decision-making.
- **Emergency management**: Descriptive statistics help predict and prepare for extreme weather events.

By applying descriptive statistics, meteorologists can better understand and communicate complex weather data, ultimately supporting informed decision-making in various fields.

8.3 Probability Distributions in Meteorology: Understanding Weather Patterns

Probability distributions are a fundamental tool in meteorology, enabling researchers to model and analyze various weather-related phenomena. By applying these distributions, meteorologists can quantify uncertainty, predict future events, and assess risks associated with weather-related hazards.

8.3.1 Applications of Probability Distributions in Meteorology

- **Precipitation Modeling**: Probability distributions like Gamma, Weibull, or Lognormal are used to model rainfall or snowfall amounts and frequencies. These models help predict precipitation patterns, informing decision-making in agriculture, urban planning, and emergency management.
- **Temperature Forecasting**: Normal or Gaussian distributions often model temperature variations, enabling meteorologists to predict temperature ranges and extremes.
- **Wind Speed Analysis**: Weibull or Rayleigh distributions commonly describe wind speed patterns, helping predict wind-related hazards like storms or gusts.

- **Extreme Weather Events**: Generalized Pareto or Gumbel distributions help model rare events like hurricanes, floods, or droughts, enabling meteorologists to estimate return periods and return levels.

8.3.2 Common Probability Distributions in Meteorology

- **Normal Distribution**: Models temperature, humidity, or other continuous variables with symmetric distributions.
- **Gamma Distribution**: Describes precipitation amounts or durations, often used for modeling rainfall or snowfall.
- **Weibull Distribution**: Models wind speeds, precipitation, or other skewed data, commonly used in wind energy applications.
- **Lognormal Distribution**: Used for variables with skewed, non-negative distributions, such as precipitation amounts or cloud droplet sizes.
- **Poisson Distribution: Modeling Rare Events in Meteorology**

The Poisson distribution is a discrete probability distribution that models the number of events occurring in a fixed interval of time or space, where these events occur with a known constant average rate. In meteorology, the Poisson distribution is used to model rare events, such as:

- **Thunderstorm occurrences**: Modeling the number of thunderstorms in a given area.
- **Tornado frequency**: Estimating the number of tornadoes in a region.
- **Extreme precipitation events**: Modeling the number of heavy rainfall events.

8.3.3 Key Characteristics

- **Discrete events**: The Poisson distribution models count data, such as the number of events.
- **Constant average rate**: The average rate of events is constant over time or space.j
- **Independence**: Events are assumed to be independent of each other.

8.3.4 Applications in Meteorology

- **Risk assessment**: Estimating the probability of rare weather events.
- **Forecasting**: Predicting the number of extreme weather events.
- **Climate research**: Analyzing trends in rare weather events.

8.3.5 Benefits

- **Accurate modeling**: The Poisson distribution provides a realistic model for rare events.

- **Probability estimation**: Enables estimation of probabilities for specific numbers of events.
- **Risk assessment**: Supports decision-making in fields like emergency management and insurance.

The Poisson distribution is a valuable tool in meteorology, helping researchers and forecasters understand and predict rare weather events.

8.3.6 Benefits of Probability Distributions in Meteorology

- **Quantifying Uncertainty**: Estimate probabilities of weather events, enabling meteorologists to communicate uncertainty effectively.
- **Predicting Future Events**: Inform forecasting models, helping predict weather patterns and extremes.
- **Assessing Risks**: Evaluate potential impacts of weather-related hazards, supporting decision-making in various fields.

8.3.7 Real-World Applications

- **Agricultural Planning**: Probability distributions inform crop management decisions, helping farmers mitigate weather-related risks.
- **Emergency Management**: Meteorologists use probability distributions to predict and prepare for extreme weather events, saving lives and reducing damage.
- **Climate Research**: Analyzing probability distributions helps researchers understand shifts in weather patterns, informing climate mitigation and adaptation strategies.

By applying probability distributions, meteorologists can better understand and communicate the complexities of weather patterns, ultimately supporting informed decision-making in various fields.

8.4 Inferential Statistics in Meteorology: Unlocking Insights and Predictions

Inferential statistics are a crucial tool in meteorology, enabling researchers to draw conclusions about large-scale weather patterns, trends, and relationships based on sample data (Fig. 8.1). By applying inferential statistical methods, meteorologists can:

8.4.1 Estimate Population Parameters

- **Mean**: Estimate average temperature, precipitation, or other weather variables, such as average annual rainfall or mean temperature during a specific season.

- **Standard Deviation**: Infer variability in weather patterns, such as temperature fluctuations or precipitation extremes.
- **Proportion**: Estimate the proportion of extreme weather events, such as the percentage of days with heavy rainfall.

8.4.2 Test Hypotheses

- **t-tests**: Compare means of weather variables between different groups or time periods, such as comparing average temperatures between two cities.
- **ANOVA (Analysis of Variance)**: Evaluate differences in weather patterns between multiple groups, such as comparing precipitation patterns in different regions.
- **Regression Analysis**: Model relationships between weather variables, such as the relationship between temperature and precipitation.

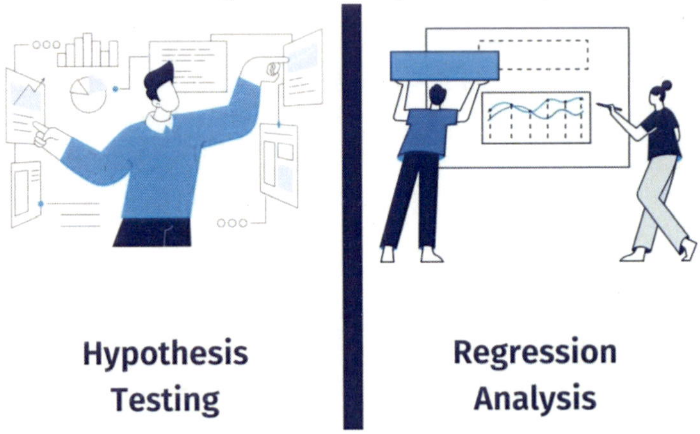

Fig. 8.1: Types of inferential statistics

8.4.3 Predict Future Events

- **Confidence Intervals**: Estimate population parameters with specified confidence levels, such as estimating the average temperature with a 95% confidence interval.
- **Prediction Intervals**: Predict future weather values with a specified level of confidence, such as predicting the temperature range for the next day.
- **Probability Distributions**: Model the probability of extreme weather events, such as the probability of a heatwave or a flood.

8.4.4 Common Inferential Statistical Methods

- **HypothesisTesting**: Evaluate research questions using statistical tests, such as:
  - t-tests for comparing means
  - ANOVA for comparing multiple groups
  - Chi-squared tests for categorical data
- **Regression Analysis**: Model relationships between weather variables, such as:
  - Linear regression for continuous variables
  - Logistic regression for binary variables
- **Time Series Analysis**: Analyze patterns and trends in weather data over time, such as:n
  - ARIMA (AutoRegressive Integrated Moving Average) models
  - Exponential smoothing models
- **Correlation Analysis**: Evaluate relationships between weather variables, such as the correlation between temperature and precipitation.

8.4.5 Statistical Parameters Used

- **p-value**: Determine the significance of statistical tests, with smaller p-values indicating stronger evidence against the null hypothesis.
- **Confidence Level**: Specify the level of confidence for estimates, such as 95% or 99%.
- **Coefficient of Determination (R-squared)**: Evaluate the goodness of fit for regression models, with higher R-squared values indicating better fit.
- **Standard Error**: Estimate the variability of statistical estimates, such as the standard error of the mean.

8.4.6 Applications in Meteorology

- **Climate Research**: Inferential statistics help understand climate variability, trends, and changes, such as analyzing temperature trends over time.
- **Weather Forecasting**: Statistical models inform forecast predictions, enhancing accuracy and reliability, such as predicting precipitation patterns.
- **Hydrology**: Inferential statistics help predict precipitation, runoff, and flood risks, such as estimating the probability of flooding.

- **Agricultural Planning**: Statistical models support decision-making for crop management and yield prediction, such as predicting crop yields based on weather conditions

8.4.7 Benefits in Meteorology

- **Informed Decision-Making**: Inferential statistics support decision-making in various fields, such as agriculture, emergency management, and urban planning.
- **Improved Forecasting**: Statistical models enhance forecast accuracy and reliability, enabling better preparation for weather-related events.
- **Climate Change Understanding**: Inferential statistics help understand climate variability and change, informing policy and decision-making.

By applying inferential statistical methods and understanding various statistical parameters, meteorologists can unlock insights into complex weather phenomena, ultimately informing decision-making and predicting future weather conditions.

8.5 Time Series Analysis (Descriptive)

Meteorological data is inherently dynamic; most atmospheric variables are measured repeatedly over time at fixed locations. This sequential nature makes **time series analysis** a cornerstone of meteorological statistics. A **time series** is simply a sequence of data points measured at successive points in time, typically at uniform intervals (e.g., hourly, daily, monthly, annually). Descriptive time series analysis (Fig. 8.2) focuses on identifying and characterizing patterns, components, and behaviors within such sequences, providing a foundational understanding before any deeper inferential modeling or prediction.

8.5.1 What is Time Series Analysis?

At its core, time series analysis is about understanding how a variable changes over time. For meteorological data, this means examining how temperature, pressure, precipitation, wind speed, or other parameters evolve from one moment to the next, or from one season/year to the next. Descriptive time series analysis helps us answer questions like: “Is it getting warmer over time?”, “Does rainfall follow a regular yearly pattern?”, or “Are there any recurring multi-year cycles in drought conditions?”

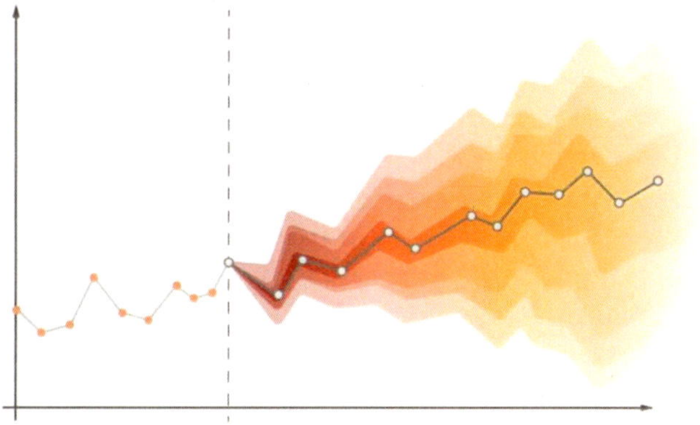

Fig. 8.2: Time series analysis

8.5.2 Components of a Time Series (Descriptive)

A meteorological time series can often be thought of as a combination of several underlying components:

- **Trend:** This is the long-term, underlying direction or movement in the data. It represents a general increase or decrease over an extended period.
- *Meteorological Example:* The gradual increase in global average temperature over the past century, indicating a warming climate trend.
- **Seasonality (or Seasonal Component):** This refers to patterns that repeat over a fixed, known period, typically within a year. These patterns are driven by regular astronomical cycles (Earth's orbit around the sun).
- *Meteorological Example:* The annual cycle of temperature, where summers are consistently warmer than winters, or the predictable pattern of monsoon rainfall in certain regions.
- **Cycles (or Cyclical Component):** These are patterns that repeat over a longer, often irregular, period than seasonality, and their duration is not fixed. They are often associated with natural oscillations in the Earth system.
  - *Meteorological Example:* The El Niño-Southern Oscillation (ENSO) cycle, which typically recurs every 2-7 years, influencing global weather patterns, but without a precise, fixed period. Other examples include the Pacific Decadal Oscillation (PDO) or the North Atlantic Oscillation (NAO).
- **Irregular (or Residual/Noise) Component:** This is the unpredictable, random variation in the data that remains after the trend, seasonal, and

cyclical components have been accounted for. It represents short-term fluctuations or unpredictable events.

- *Meteorological Example:* Day-to-day weather variability, sudden storms, or unexpected cold snaps that deviate from the typical seasonal pattern.

8.5.3 Descriptive Visualizations for Time Series

Visualizing time series data is paramount for descriptive analysis, allowing meteorologists to quickly identify these components:

- **Line Plots (Time Series Plots):** The most fundamental visualization, plotting the variable's value against time. This immediately reveals trends, seasonal cycles, and obvious anomalies.
  - *Example:* A plot of monthly average global temperature over 150 years clearly shows an upward trend and annual seasonality.
- **Seasonal Subseries Plots:** Plots the data for each season (e.g., each January, each February) separately over the years. This helps to highlight seasonal patterns and how they might be changing over time.
  - *Example:* Plotting all January temperatures from 1950-2020 on one graph, then all February temperatures on another, to see if the warming trend is consistent across seasons.
- **Lag Plots:** Plots values of the time series against lagged values (e.g., today's temperature vs. yesterday's temperature). This helps to visualize autocorrelation (how correlated a value is with a previous value).
  - *Example:* A scatter plot of temperature at time 't' versus temperature at time 't-1' to see if there's a strong positive relationship, indicating persistence.
- **Autocorrelation Function (ACF) Plots:** These plots show the correlation of a time series with itself at different time lags. They are crucial for identifying periodicity and the "memory" of the series.
  - *Example:* An ACF plot for daily temperature might show high correlation at a lag of 1 day (today's temperature is like yesterday's), and also at a lag of 365 days (yearly cycle).

8.5.4 Data Requirements for Time Series Analysis

Effective time series analysis requires specific characteristics from the data:

- **Temporal Order:** The data points must be ordered sequentially by time.
- **Regular Intervals:** Ideally, data should be collected at consistent, uniform time intervals (e.g., every hour, every day, every month).

Irregular intervals can complicate analysis, though methods exist to handle them.

- **Sufficient Length:** The time series needs to be long enough to capture the patterns of interest. To identify a seasonal cycle, you need at least a few years of data. To identify a decadal cycle, you need several decades.
- **Data Quality:** As with all analyses, the data must be of high quality. Missing values, outliers, and instrument biases/drift can distort the true patterns and lead to erroneous conclusions. Imputation and homogenization (as discussed in Chapter 7) are often critical preprocessing steps.
- **Stationarity (for advanced inferential methods):** While not strictly a descriptive requirement, many inferential time series models (like ARIMA) assume that the statistical properties of the series (mean, variance) do not change over time. Descriptive analysis helps identify non-stationarity (e.g., a clear trend) that would need to be addressed before applying such models.

8.5.5 Applications of Time Series Analysis in Meteorology

Descriptive time series analysis is widely applied across various meteorological domains:

- **Climate Monitoring**
  - **Trend Detection:** Identifying long-term warming or cooling trends in temperature records, or changes in precipitation patterns.
  - **Seasonality Analysis:** Understanding the typical annual cycle of temperature, rainfall, or sunshine hours for a region.
  - **Extreme Event Frequency:** Analyzing the frequency of heatwaves, cold snaps, or heavy rainfall events over time.
- **Weather Forecasting and Verification**
  - **Forecast Skill Assessment:** Plotting forecast errors over time to see if performance is improving, degrading, or showing seasonal biases.
  - **Identifying Persistence:** Understanding how long a particular weather pattern tends to last (e.g., how long a cold front typically affects a region).
- **Hydrology and Water Resources**
  - **Rainfall Patterns:** Analyzing monthly or annual rainfall totals to understand drought cycles, flood potential, and water availability.
  - **River Flow Analysis:** Studying historical river discharge data to manage water resources and predict flood risks.

- **Air Quality Monitoring**
  - **Pollutant Trends:** Tracking the concentration of pollutants (e.g., ozone, particulate matter) over time to identify trends related to emissions or policy changes.
  - **Diurnal/Seasonal Cycles:** Understanding how pollutant levels vary throughout the day or year.
- **Renewable Energy Assessment**
  - **Wind Resource Assessment:** Analyzing long-term wind speed data to understand typical wind patterns and variability for wind farm siting and energy production forecasts.
  - **Solar Energy Potential:** Studying solar radiation time series to assess solar power generation potential.
- **Agricultural Planning**
  - **Growing Season Analysis:** Understanding the onset and end of growing seasons, frost dates, and accumulated growing degree days.
  - **Pest/Disease Modeling:** Analyzing temperature and humidity time series to predict conditions favorable for agricultural pests or diseases.

8.6 Autocorrelation

Autocorrelation is a fundamental statistical concept in meteorology that describes the **correlation of a variable with itself over different time lags or spatial distances**. In simpler terms, it measures how much a value at a certain point in time or space is related to a value of the same variable at a previous point in time or a nearby point in space. It's about understanding the "memory" or "persistence" within a dataset.

8.6.1 What is Autocorrelation? (Conceptual)

Imagine a sequence of daily temperatures. If today's temperature is warm, it's highly likely that yesterday's temperature was also warm, and tomorrow's temperature will also be relatively warm. This persistence, or the tendency for values to be similar to their immediate predecessors, is a manifestation of **positive autocorrelation**. If, on the other hand, a high temperature was consistently followed by a low temperature, that would suggest **negative autocorrelation**. If there's no relationship between a value and its past values, there's no autocorrelation.

In meteorology, we often deal with two main types:

- **Temporal Autocorrelation:** How a variable at one time point is related to itself at previous time points (e.g., today's temperature vs. yesterday's temperature).

- **Spatial Autocorrelation:** How a variable at one location is related to itself at nearby locations (e.g., temperature at one weather station vs. temperature at a station 50 km away). This was briefly touched upon in the Spatial Statistics section, where "near things are more related than distant things."

8.6.2 Use of Autocorrelation in Meteorology

Understanding autocorrelation is crucial for numerous applications in meteorology:

- **Forecast Persistence**
    - **Short-range Forecasting:** High positive temporal autocorrelation in variables like temperature or pressure means that the current conditions are a good starting point for very short-range forecasts (e.g., "persistence forecasts" where tomorrow's weather is assumed to be like today's). Understanding this persistence helps meteorologists know how long a particular weather pattern is likely to last.
    - **Forecast Verification:** Autocorrelation in forecast errors can reveal systematic biases or patterns in model performance.
- **Climate Variability and Change**
    - **Trend Detection:** When analyzing long-term climate trends, autocorrelation needs to be accounted for. If not, statistical tests might incorrectly suggest a trend is significant when it's just natural persistence.
    - **Climate Oscillations:** Autocorrelation analysis helps identify and characterize natural climate oscillations (like El Niño-Southern Oscillation - ENSO, or the North Atlantic Oscillation - NAO) by revealing recurring patterns at specific time lags (e.g., correlations at lags of 2-7 years for ENSO).
    - **Teleconnections:** It can help identify how weather patterns in one region are correlated with patterns in distant regions over time.
- **Observing Network Design**
    - **Spatial Autocorrelation:** Understanding how spatially correlated meteorological variables are (e.g., temperature, precipitation) helps in designing efficient observing networks. If a variable has very high spatial autocorrelation, stations can be spaced farther apart without losing too much information. If it has low spatial autocorrelation, a denser network is needed.

- **Quality Control:** Spatial autocorrelation is the basis for "buddy checks" in quality control, where an observation is compared to its neighbors. If a station's reading is very different from its highly correlated neighbors, it might be an error.

- **Time Series Modeling**
  - **Model Selection:** Identifying the presence and nature of autocorrelation (e.g., how many past values influence the current one) is a critical first step in building statistical time series models (like AR, MA, ARIMA models) for prediction or simulation. These models explicitly account for the temporal dependence in the data.
  - **Forecasting Future Values:** Models that incorporate autocorrelation can provide more accurate forecasts by leveraging the "memory" of the system.
- **Data Assimilation**
  - **Error Covariance Modeling:** In data assimilation systems for NWP models, understanding the spatial and temporal autocorrelation of observation errors and model errors is crucial for optimally combining observations with model forecasts.
- **Atmospheric Dynamics Research**
  - **Turbulence Analysis:** Autocorrelation can be used to study the characteristics of atmospheric turbulence, where rapid fluctuations are correlated over very short time lags.
  - **Atmospheric Memory:** Researchers use autocorrelation to quantify the "memory" of the atmosphere for various phenomena, which has implications for predictability limits.

8.6.3 Data Used for Autocorrelation Analysis

Autocorrelation analysis can be applied to almost any type of meteorological data that is collected sequentially in time or arranged spatially:

- **Time Series Data**
  - **Surface Observations:** Hourly, daily, monthly, or annual values of temperature, pressure, humidity, wind speed, precipitation totals from individual weather stations.
  - **Upper Air Data:** Time series of parameters at specific pressure levels from radiosonde launches.
  - **Gridded Model Outputs:** Time series of area-averaged values from numerical weather prediction or climate models.

- **Climate Indices:** Time series of indices representing large-scale climate oscillations (e.g., ENSO indices, Arctic Oscillation index).

- **Spatial Data**
  - **Station Networks:** Values of a meteorological variable (e.g., temperature, precipitation) simultaneously observed across a network of weather stations.
  - **Gridded Data:** Values on a regular grid, such as those from radar (e.g., precipitation intensity), satellite (e.g., cloud cover, sea surface temperature), or NWP models (e.g., surface pressure, 500 hPa geopotential height). For spatial autocorrelation, the analysis considers how values vary with distance between grid cells.

In essence, autocorrelation is a powerful concept that acknowledges the interconnectedness of meteorological phenomena across time and space. By quantifying this "memory" or "persistence," meteorologists gain deeper insights into atmospheric processes, improve forecast accuracy, and enhance the reliability of climate assessments.

Descriptive time series analysis provides the essential initial insights into the temporal behavior of meteorological variables. By revealing trends, seasonal cycles, and other patterns, it lays the groundwork for more advanced inferential techniques, allowing meteorologists to build predictive models and draw robust conclusions about the dynamic atmosphere

8.7 Spatial Statistics

Meteorological data is inherently spatial; observations are collected at specific geographical locations, and atmospheric phenomena unfold across vast areas. **Spatial statistics** are a branch of statistics that specifically deals with data collected at different locations, taking into account the geographical relationships and dependencies between those locations. Unlike traditional statistics that assume data points are independent, spatial statistics explicitly acknowledge that observations closer in space are often more similar than those farther apart. This concept is fundamental to understanding and analyzing weather and climate patterns.

8.7.1 Purpose of Spatial Statistics in Meteorology

- **Mapping and Visualization:** Creating continuous maps of meteorological variables (e.g., temperature, precipitation) from discrete station observations.
- **Estimation and Prediction:** Estimating values at unobserved locations (interpolation) or predicting values at future locations.

- **Network Design:** Optimizing the placement of observing stations.
- **Understanding Spatial Patterns:** Identifying clusters, hot spots, or gradients in atmospheric phenomena.
- **Data Assimilation:** Providing spatially consistent data for numerical weather prediction models.

8.7.2 Spatial Autocorrelation

The foundational concept in spatial statistics is **spatial autocorrelation**.

- **Definition:** Spatial autocorrelation describes the degree to which values of a variable at nearby locations are similar (positive spatial autocorrelation) or dissimilar (negative spatial autocorrelation). In meteorology, positive spatial autocorrelation is very common: if it's warm at one weather station, it's likely warm at a nearby station.
- **Tobler's First Law of Geography:** This principle succinctly states: "Everything is related to everything else, but near things are more related than distant things." This law is the philosophical underpinning of spatial autocorrelation.
- **Significance in Meteorology**
  - **Redundancy:** High spatial autocorrelation means nearby stations provide somewhat redundant information, which can be useful for quality control (buddy checks) or for optimizing network density.
  - **Interpolation Basis:** It's the reason why we can reasonably estimate values at unobserved locations based on nearby observations.
  - **Pattern Recognition:** Helps identify whether a meteorological variable is clustered (high values near high values), dispersed (high values near low values), or randomly distributed.
  - **Measurement (Conceptual):** While not using equations here, spatial autocorrelation is often quantified using statistics like Moran's I or by analyzing **variograms** (or semivariograms). A variogram essentially plots how the dissimilarity (or variance) between data points increases as the distance between them increases. It helps characterize the spatial structure of the data.

8.7.3 Inverse Distance Weighting (IDW)

Inverse Distance Weighting (IDW) is a widely used, relatively simple method for spatial interpolation. It's a deterministic interpolation technique, meaning it does not use statistical models of spatial autocorrelation.

- **Principle:** The core idea is that the value at an unobserved location is estimated as a weighted average of the values from nearby observed locations. The "weight" given to each observed point is inversely proportional to its distance from the unobserved location. Closer points receive higher weights, while farther points receive lower weights.

- **How it Works (Conceptual)**
  1. To estimate a value at a specific grid point (where there's no station), the algorithm looks at the values from a predefined number of surrounding weather stations.
  2. For each surrounding station, it calculates a weight based on how close it is to the grid point. Typically, the weight is proportional to 1/dp, where 'd' is the distance and 'p' is a power parameter (commonly 2, for inverse square).
  3. The estimated value at the grid point is the sum of (each station's value multiplied by its weight), divided by the sum of all weights.
- **Advantages in Meteorology**
  - **Simplicity:** Easy to understand and implement.
  - **Intuitive:** Reflects the idea that closer observations are more relevant.
  - **Fast Computation:** Relatively quick for generating maps.
- **Disadvantages**
  - **"Bullseye" Effect:** Can create circular "bullseye" patterns around observation points, especially if the data is sparse or the power parameter is high.
  - **No Extrapolation:** Cannot predict values outside the range of the observed data.
  - **Doesn't Account for Spatial Structure:** Does not consider the actual spatial autocorrelation of the data (e.g., if a mountain range causes values to change rapidly over short distances).
  - **No Error Estimation:** Does not provide a measure of the uncertainty or error associated with the interpolated values.

8.7.4 Kriging

Kriging is a more advanced and powerful geostatistical interpolation technique. Unlike IDW, Kriging uses the statistical properties of the data, specifically its **spatial autocorrelation** (as modeled by a variogram), to determine the optimal weights for interpolation. It's considered a "Best Linear Unbiased Estimator" (BLUE).

- **Principle:** Kriging estimates the value at an unobserved location by weighting the surrounding observed values, but these weights are derived from a statistical model of the spatial correlation structure of the data. It essentially fits a mathematical model (the variogram) to how the data varies with distance and direction.

- **How it Works (Conceptual):**
  - **Variogram Modeling:** First, a variogram is calculated from the observed data. This variogram describes how the dissimilarity (variance) between data points increases with distance. A mathematical model (e.g., spherical, exponential, Gaussian) is then fitted to this experimental variogram. This model captures the spatial structure.
  - **Weight Calculation:** Using the fitted variogram model, Kriging calculates a unique set of weights for each unobserved location. These weights are optimal because they account for:
  - The distance between the observed points and the unobserved point.
  - The spatial autocorrelation between the observed points themselves (i.e., it avoids giving too much weight to a cluster of closely spaced, highly correlated points).
  - **Estimation:** The estimated value at the unobserved location is a weighted average of the observed values, using these optimal weights.
- **Advantages in Meteorology**
  - **Optimal Interpolation:** Provides the "best" linear unbiased estimate, meaning it minimizes the estimation error.
  - **Accounts for Spatial Structure:** Explicitly uses the spatial autocorrelation of the data, leading to more realistic maps.
  - **Provides Error Estimates:** A major advantage is that Kriging not only provides an interpolated value but also an **estimation variance** (or Kriging variance) for each interpolated point. This quantifies the uncertainty or reliability of the estimate, allowing meteorologists to understand where their interpolated maps are more or less trustworthy.
  - **Smoother Maps:** Tends to produce smoother, more realistic maps than IDW.
- **Disadvantages**
  - **More Complex:** Requires a good understanding of geostatistics and variogram modeling.
  - **Computationally Intensive:** Can be slower than IDW, especially for very large datasets.
  - **Sensitivity to Variogram Model:** The quality of the interpolation depends heavily on selecting an appropriate variogram model.

- **Assumptions:** Assumes stationarity (that the statistical properties of the data, like the mean and variance, are consistent across the study area, or that any non-stationarity can be modeled).

8.7.5 Applications of Spatial Statistics in Meteorology

- **Creating Gridded Datasets:** Interpolating point observations (e.g., temperature from weather stations, precipitation from rain gauges) onto a regular grid for use in numerical models, climate analysis, or visualization.
- **Mapping Climate Variables:** Producing continuous maps of average temperature, total precipitation, or extreme event frequency across regions.
- **Rainfall Estimation:** Interpolating rain gauge data to provide more accurate spatial rainfall estimates, especially for hydrological modeling and flood forecasting.
- **Designing Observing Networks:** Using variograms to understand the spatial correlation of variables helps determine the optimal spacing of new stations to maximize information gain while minimizing redundancy.
- **Quality Control:** Spatial interpolation can be used in "buddy checks" where an observed value is compared to an interpolated value from its neighbors to identify potential outliers.
- **Climate Reconstruction:** Estimating past climate conditions in regions with sparse historical data.

Spatial statistics, particularly techniques like Inverse Distance Weighting and Kriging, are indispensable tools for transforming discrete meteorological observations into continuous, meaningful spatial representations, allowing for a deeper understanding and more accurate prediction of atmospheric phenomena across geographical landscapes.

8.8 Conclusion

Concluding our exploration of statistical methods in meteorology, it's clear that these tools are indispensable for extracting meaningful insights from the vast and complex datasets that characterize the atmospheric sciences. From basic descriptive statistics that summarize weather patterns to advanced techniques like time series analysis, spatial correlation analysis, and machine learning, statistical methodologies provide the framework for understanding, modeling, and ultimately predicting atmospheric phenomena. The ability to quantify uncertainty, identify significant relationships, and build predictive models hinges on the rigorous application of these statistical approaches.

As meteorological datasets continue to grow in volume and complexity, driven by advancements in remote sensing, numerical modeling, and in-situ observations, the role of sophisticated statistical methods will only become more critical. Future progress in weather forecasting, climate modeling, and environmental monitoring will increasingly rely on our ability to effectively leverage these statistical tools to uncover hidden patterns, validate model outputs, and ultimately enhance our understanding of the dynamic and ever-changing atmosphere. The journey through statistical meteorology highlights its power not just as a set of analytical techniques, but as a fundamental language for interpreting the world around us, one observation and one analysis at a time.

9

Advanced Analysis Techniques

9.1 Introduction

Building upon our foundational understanding of statistical methods in meteorology, Chapter 9, "Advanced Analytic Techniques," propels us into the forefront of data science applications within the atmospheric sciences. The imperative for these advanced methodologies stems directly from the evolving nature of meteorological data itself. Modern observations, numerical weather prediction (NWP) model outputs, and climate simulations generate an unprecedented **volume** of information, often arriving with incredible **velocity** and exhibiting immense **variety** in format and structure. Traditional statistical approaches, while robust for many tasks, often struggle to efficiently process such vast, high-dimensional datasets, or to fully capture the intricate, often non-linear, relationships inherent in atmospheric processes.

This chapter is dedicated to exploring why and how these sophisticated techniques are not merely supplementary, but increasingly essential. We will delve into how **machine learning and deep learning** algorithms excel at identifying subtle, complex patterns within massive datasets, enabling breakthroughs in areas like severe weather detection, nowcasting, and the bias correction of model outputs. We will also examine **advanced spatial statistics**, which provide powerful frameworks for explicitly modeling the intricate geographical dependencies of meteorological variables, crucial for accurate interpolation, downscaling, and understanding regional climate impacts. Furthermore, the chapter will introduce **functional data analysis**, a method particularly adept at treating continuous meteorological fields (like temperature profiles or pressure maps) as functions, allowing for a more holistic analysis of their evolution and variability. By embracing these advanced analytical techniques, meteorologists and climate scientists can unlock deeper insights, enhance predictive skill, more rigorously quantify uncertainties, and ultimately contribute to a more resilient society in the face of a changing climate. Prepare to expand your analytical toolkit and discover new frontiers in meteorological understanding.

9.2 Principal Component Analysis (PCA) Explained

Principal Component Analysis (PCA) is a fundamental statistical technique that acts like a sophisticated data compressor and pattern finder. In the vast and intricate world of meteorology and climate science, where data pours in from countless sources, PCA is an indispensable tool for **simplifying complex information** and **uncovering hidden, dominant patterns** (Fig. 9.1).

9.2.1 Why Do We Need PCA in Meteorology?

Imagine you're trying to understand the weather across an entire continent. You have measurements for temperature, pressure, humidity, wind speed, and direction, not just at the surface, but at multiple altitudes, and for hundreds or thousands of locations. This creates an enormous dataset, with potentially millions of individual data points for a single time snapshot. This high volume of data presents significant challenges:

- **Information Overload (The "Curse of Dimensionality"):** With so many variables, it becomes incredibly difficult for humans to grasp the overall picture or for computers to process efficiently. It's like trying to understand a complex story by reading every single word in isolation, rather than focusing on the main plot points.
- **Redundancy and Correlation:** Many meteorological variables are not independent. For example, if the temperature rises significantly in one city, it's likely that nearby cities will also see a temperature increase. Similarly, temperature and atmospheric pressure are often related. This means much of our data is redundant, and we're measuring the same underlying atmospheric behavior in slightly different ways.
- **Computational Burden:** Analyzing and building models with hundreds or thousands of interconnected variables can be computationally expensive and slow.
- **Visualization Challenges:** How do you visualize data that exists in 100 or 1000 dimensions? Our brains are limited to three!

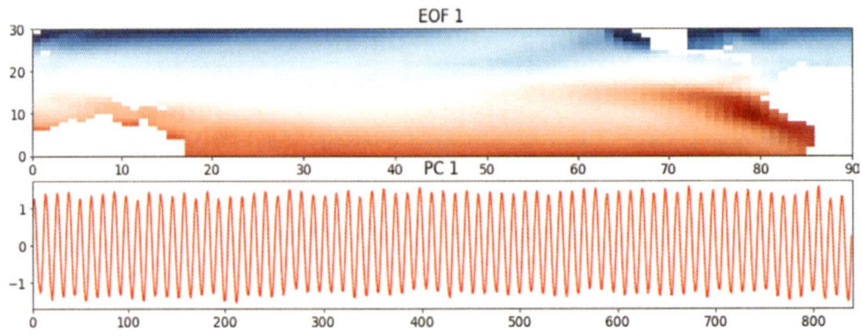

Fig. 9.1: Principal component analysis

PCA steps in to solve these problems by finding a way to represent this high-dimensional data in a much smaller, more manageable set of dimensions, while crucially **preserving as much of the original, important information as possible.**

9.2.2 The Core Idea: Finding the "Main Directions" of Variability

Think of a swarm of bees flying in a three-dimensional room. If you wanted to describe their movement most efficiently, you wouldn't track each bee's individual x, y, and z coordinates. Instead, you might notice that most of the bees are moving primarily along one long diagonal across the room, and then perhaps a little bit up and down, and even less from side to side.

PCA does something similar with data. It finds the "main directions" or "axes" in your multi-dimensional data cloud along which the data points spread out the most. These new axes are called **Principal Components (PCs)**.

- **The First Principal Component (PC1):** This is the single direction that captures the largest amount of variance (spread) in your data. If you were to project all your data points onto this line, they would be as spread out as possible. It represents the most dominant pattern or source of variability.
- **The Second Principal Component (PC2):** This is the next most important direction of variance, but with a crucial condition: it must be **orthogonal (perpendicular)** to the first principal component. This ensures that PC2 captures a *different* aspect of the data's variability that wasn't already explained by PC1.
- **Subsequent Principal Components:** This process continues, with each new PC being orthogonal to all previous ones and explaining progressively less of the total variance.

By focusing on the first few principal components (e.g., the top 2, 3, or 10), which collectively explain a large percentage of the total data variance, we can effectively reduce the dimensionality of the dataset without losing much of its essential character.

9.2.3 The Mathematical Intuition (Without Complex Formulas)

At its heart, PCA relies on understanding how variables move together, which is captured by the **covariance matrix**. This matrix tells us how each pair of variables in our dataset varies in relation to each other.

PCA then performs a special mathematical operation called **eigen-decomposition** on this covariance matrix. This decomposition yields two key things:

- **Eigenvectors:** These are the actual directions of the principal components. Each eigenvector is a set of weights or loadings that tell you how much each original variable contributes to that specific principal component. For example, an eigenvector for PC1 might show that temperature at location A, temperature at location B, and pressure at location A all contribute strongly to this principal component.
- **Eigenvalues:** Each eigenvector has a corresponding eigenvalue. The eigenvalue tells you **how much variance (or "information")** is captured along that eigenvector's direction. A larger eigenvalue means the principal component explains more of the total variability in the data.

By sorting the eigenvalues from largest to smallest, we automatically get the principal components in order of their importance in explaining the data's variance.

9.2.4 Simplified Steps of Performing PCA

- **Standardize Your Data:** Before you start, it's vital to "normalize" your data. This means adjusting each variable so it has a mean of zero and a standard deviation of one. Why? Because if you have temperature measured in Celsius (small numbers) and pressure in Pascals (large numbers), the pressure variable might unfairly dominate the PCA simply because its values are numerically larger. Standardization ensures all variables contribute equally to the variance calculation.
- **Calculate the Covariance Matrix:** This step quantifies how every pair of your standardized variables moves together.
- **Find Eigenvectors and Eigenvalues:** This is the core mathematical step where the "main directions" (eigenvectors) and their corresponding "importance scores" (eigenvalues) are calculated from the covariance matrix.
- **Select the Most Important Principal Components:** You look at the eigenvalues and decide how many principal components you want to keep. Often, you'll look for an "elbow" in a plot of eigenvalues (a "scree plot") or choose enough components to explain a certain percentage of the total variance (e.g., 80%, 90%, or 95%).
- **Transform Your Data:** Finally, you "project" your original (standardized) data onto these selected principal components. This gives you a new dataset where each data point is now described by its scores on these new, uncorrelated principal components, rather than the

original variables. This new dataset has fewer dimensions but retains most of the original data's variability.

9.2.5 Powerful Applications in Meteorology

PCA is a workhorse in meteorological and climate research, providing profound insights:

- **Identifying Dominant Climate Patterns (Empirical Orthogonal Functions - EOFs):** This is perhaps its most famous application. PCA can reveal recurring, large-scale spatial patterns of variability in climate data. For instance, applying PCA to sea surface temperatures in the Pacific can reveal the **El Niño-Southern Oscillation (ENSO)**, showing a distinct pattern of warming and cooling that impacts global weather. Similarly, the **North Atlantic Oscillation (NAO)**, a major driver of European winter weather, can be identified as a leading EOF of sea level pressure. These EOFs are essentially the spatial patterns, and their corresponding principal component time series show how these patterns evolve over time.
- **Weather Pattern Recognition:** Beyond climate, PCA helps identify typical atmospheric circulation patterns associated with specific weather events, such as those leading to prolonged droughts, intense heatwaves, or widespread heavy rainfall.
- **Data Compression and Noise Reduction:** Large datasets from satellites or climate models can be compressed by storing only the scores on the leading principal components, significantly reducing storage needs and transmission times. By focusing on the most dominant patterns, PCA can also help filter out random noise inherent in observations.
- **Forecast Error Analysis:** By applying PCA to the errors of numerical weather prediction models, meteorologists can identify systematic spatial patterns in where and how models tend to go wrong. This helps in understanding model biases and developing strategies for improvement.
- **Optimizing Sensor Networks:** PCA can help determine if certain weather stations or sensors are redundant by showing that their measurements are highly correlated and explained by the same dominant principal components.
- **Teleconnection Studies:** PCA helps in exploring how weather phenomena in one part of the world are statistically linked to conditions in distant regions.

9.2.6 Advantages and Considerations

- **Advantages**
    - **Effective Dimensionality Reduction:** It's highly effective at reducing the number of variables while preserving most of the data's variance.
    - **Noise Filtering:** By focusing on the largest sources of variance, it can inherently reduce the impact of random noise.
    - **Pattern Discovery:** It's excellent for uncovering the underlying structure and dominant modes of variability in complex datasets.
    - **Uncorrelated Components:** The resulting principal components are uncorrelated, which can simplify subsequent statistical modeling.
- **Considerations (Limitations)**
    - **Linearity:** PCA is a linear technique. If the true relationships within your meteorological data are highly non-linear (which is often the case in complex atmospheric systems), PCA might not fully capture them. Other advanced techniques, like non-linear dimensionality reduction methods, might be more appropriate in such scenarios.
    - **Interpretability:** While PCs are mathematically precise, interpreting what a specific principal component *physically represents* can sometimes be challenging. It's a linear combination of many original variables, so its physical meaning might not be immediately obvious.
    - **Sensitivity to Scaling:** As mentioned, the results of PCA are sensitive to the scaling of your original variables. Proper standardization is crucial to ensure that variables with larger numerical ranges don't disproportionately influence the results.
    - **Information Loss:** While PCA aims to minimize information loss, some information is always discarded when you reduce dimensionality. The art lies in choosing enough principal components to capture the essential variability without retaining too much noise or redundancy.

9.2.7 Conclusion

Principal Component Analysis stands as a cornerstone technique in meteorological and climate science. By intelligently reducing the dimensionality of complex datasets and revealing their most significant underlying patterns, PCA empowers researchers to gain profound insights into atmospheric processes, improve the accuracy of weather and climate models, and manage the ever-growing volume of environmental data. It is an indispensable tool for anyone seeking to understand and predict the intricate dynamics of our planet's atmosphere.

9.3 Empirical Orthogonal Functions (EOFs)

Empirical Orthogonal Functions (EOFs) are a widely used statistical technique in meteorology and climate science for **decomposing complex spatio-temporal datasets into a set of orthogonal (uncorrelated) spatial patterns and their corresponding time series**. Essentially, EOF analysis is a specific application of **Principal Component Analysis (PCA)**when applied to gridded or spatially distributed time-series data. It helps identify the dominant modes of variability in a dataset, making it easier to understand and interpret large-scale atmospheric and oceanic phenomena.

9.3.1 The Relationship Between EOFs and PCA

To understand EOFs, it's crucial to recall PCA. PCA finds the directions (principal components) of maximum variance in a multi-dimensional dataset. When this dataset consists of time series at multiple spatial locations (e.g., temperature anomalies at every grid point on a map over several years), PCA's output takes on a specific meaning:

- The **eigenvectors** from PCA become the **Empirical Orthogonal Functions (EOFs)**. These are the **spatial patterns** that explain the most variance in the data. They are fixed in space but represent a recurring pattern of deviation from the mean.
- The **principal component scores** (or time series coefficients) from PCA become the **time series** associated with each EOF. These time series tell us how the amplitude or strength of the corresponding spatial pattern varies over time (Fig. 9.2).

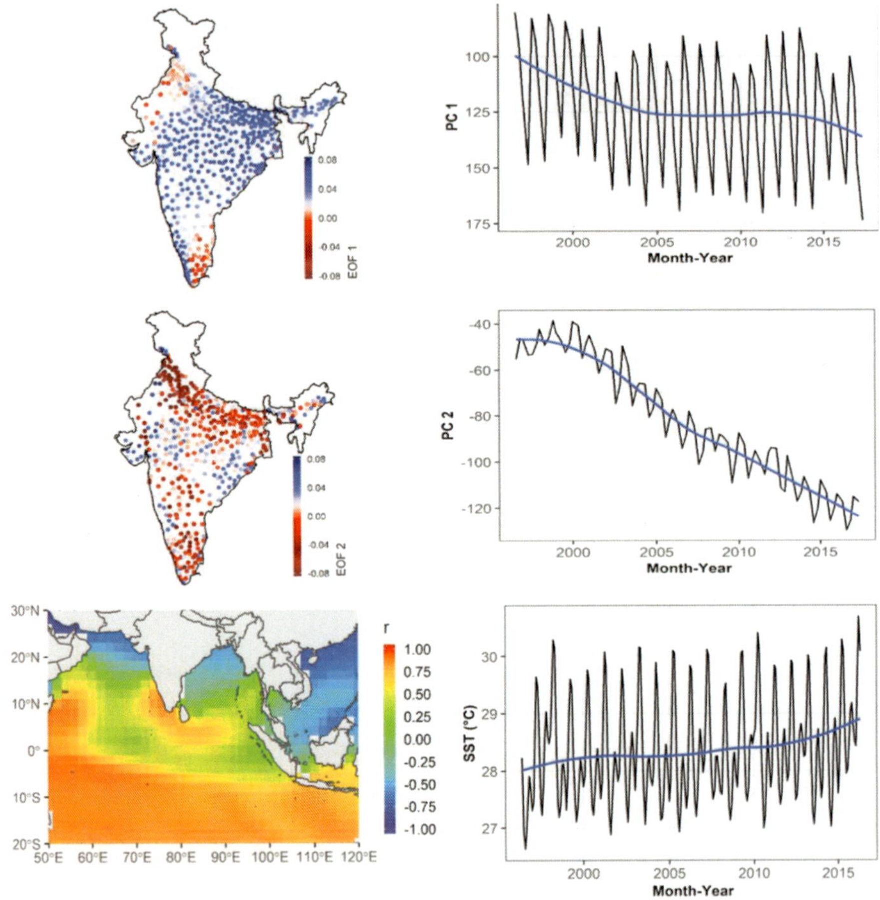

Fig. 9.2: Empirical orthogonal functions analysis

So, in essence, **EOFs are the spatial patterns (eigenvectors) derived from applying PCA to a spatial-temporal dataset.**

9.3.2 How EOF Analysis Works

Imagine you have a dataset of monthly sea surface temperature (SST) anomalies across the Pacific Ocean for 50 years. This data can be thought of as a collection of time series, one for each grid point.

The steps for an EOF analysis are similar to PCA

- **Data Preparation**
  - Organize your data into a matrix where rows represent time steps (e.g., months) and columns represent spatial locations (e.g., individual grid points).

- Often, the data is first **mean-centered** (subtracting the temporal mean at each spatial location) or **standardized** (also dividing by the standard deviation) to focus on anomalies and ensure variables with different magnitudes don't disproportionately influence the results.
- For global or large-scale analyses, **area-weighting** is often applied to account for the varying area represented by grid cells at different latitudes.
- **Compute the Covariance (or Correlation) Matrix:** A covariance matrix is constructed from the prepared data. This matrix describes how the SST anomalies at each pair of grid points vary together over time.
- **Eigen-Decomposition:** The covariance matrix is then decomposed into its eigenvectors and eigenvalues.
- **Eigenvectors:** These are your EOFs. Each EOF is a unique spatial pattern. They are orthogonal to each other, meaning they are statistically independent in terms of the variance they explain.
- **Eigenvalues:** Each eigenvalue corresponds to an EOF and quantifies the amount of variance explained by that specific spatial pattern. The EOFs are typically ordered from the largest eigenvalue to the smallest, meaning the first EOF explains the most variance, the second explains the next most, and so on.
- **Construct Principal Components (Time Series):** Once the EOFs (spatial patterns) are determined, the original (mean-centered) data is projected onto these EOFs to obtain the principal component (PC) time series. Each PC time series represents the temporal evolution of the corresponding EOF pattern. These PC time series are uncorrelated with each other.

9.3.3 Components of an EOF Analysis

An EOF analysis typically yields three main outputs:

- **Empirical Orthogonal Functions (EOFs) / Spatial Modes:** These are the spatial patterns. They are often normalized and presented as maps showing areas of positive and negative anomalies. For example, the first EOF of global SST might show a distinct pattern of warming in the central Pacific and cooling in the western Pacific, characteristic of El Niño.
- **Principal Components (PCs) / Time Series Coefficients:** These are the time series associated with each EOF. They indicate the amplitude and phase of the corresponding spatial pattern over time. A positive value in

the PC time series means the EOF pattern is present with a positive sign, while a negative value means it's present with a negative sign (e.g., La Niña for the El Niño EOF).

- **Eigenvalues / Explained Variance:** These values quantify the amount of total variance in the original dataset that each EOF explains. They are crucial for determining how many EOFs are "significant" or represent substantial variability. Often, the first few EOFs explain a large percentage of the total variance, indicating that the complex spatio-temporal evolution of the system can be largely described by a few dominant patterns.

9.3.4 Why EOFs are Indispensable in Meteorology

EOFs are a cornerstone technique in meteorological and climate research due to their ability to:

- **Identify Dominant Modes of Climate Variability:** This is arguably their most significant application. EOFs have been instrumental in objectively identifying and characterizing major climate phenomena such as:
- **El Niño-Southern Oscillation (ENSO):** The leading EOF of tropical Pacific SST anomalies clearly reveals the ENSO pattern.
- **North Atlantic Oscillation (NAO):** The leading EOF of North Atlantic sea level pressure often corresponds to the NAO.
- **Pacific Decadal Oscillation (PDO):** A long-term pattern of SST variability in the North Pacific.
- These patterns represent the most important "building blocks" of variability in a given climate system.
- **Dimensionality Reduction:** They provide a parsimonious representation of large, high-dimensional datasets. Instead of analyzing thousands of grid points over time, one can focus on the time evolution of a few dominant EOFs. This simplifies analysis and reduces computational burden.
- **Noise Reduction:** By focusing on the leading EOFs that explain the most variance, the analysis effectively filters out random noise or small-scale, incoherent variability, highlighting the more robust and physically meaningful signals.
- **Diagnosis and Attribution:** EOFs help in understanding the physical mechanisms driving observed climate variability. Researchers can correlate EOF time series with other atmospheric or oceanic variables to explore teleconnections and causal relationships.

- **Model Evaluation and Comparison:** EOFs derived from climate models can be compared against observed EOFs to assess how well models simulate realistic patterns of climate variability.
- **Forecasting:** The time series of leading EOFs can sometimes be used as predictors in statistical forecasting models.

9.3.5 Advantages and Limitations

- **Advantages**
  - **Objective Pattern Identification:** EOFs provide an objective, data-driven method for identifying dominant spatial patterns of variability without prior assumptions about their shape.
  - **Data Compression:** They offer a highly efficient way to compress large spatio-temporal datasets.
  - **Orthogonality:** Both the spatial patterns (EOFs) and their associated time series (PCs) are statistically uncorrelated, which simplifies interpretation and subsequent statistical modeling.
  - **Quantification of Variance:** The eigenvalues clearly indicate the proportion of total variance explained by each mode, allowing researchers to focus on the most significant patterns.
- **Limitations**
  - **Linearity Assumption:** EOF analysis is a linear technique. It assumes that the underlying processes can be represented as linear combinations of orthogonal patterns. If the true atmospheric processes are highly non-linear, EOFs may not fully capture their complexity.
  - **Physical Interpretability:** While EOFs are mathematically optimal in terms of variance explained, they don't always correspond directly to a single, easily interpretable physical process, especially for higher-order EOFs. The patterns can sometimes be a mix of different physical phenomena.
  - **Sensitivity to Domain and Weighting:** The resulting EOF patterns can be sensitive to the geographical domain chosen for the analysis and how the data is weighted (e.g., area-weighting for latitude).
  - **Rotation Problem:** While the standard EOFs are orthogonal, sometimes "rotated EOFs" are used (e.g., Varimax rotation) to achieve more localized and potentially more physically interpretable patterns. However, rotated EOFs lose the orthogonality of their corresponding time series.
  - **Sensitivity to Outliers:** Like PCA, EOFs can be sensitive to extreme values or outliers in the data.

9.3.6 Conclusion

Empirical Orthogonal Functions are a cornerstone of modern meteorological and climate analysis. By effectively distilling vast and complex spatio-temporal data into a few dominant, uncorrelated patterns and their temporal evolutions, EOF analysis provides invaluable insights into the fundamental modes of atmospheric and oceanic variability. Despite their limitations, EOFs remain an indispensable tool for understanding climate dynamics, evaluating models, and advancing our predictive capabilities in the atmospheric sciences.

9.4 Dimensionality Reduction Explained

In the realm of data science, particularly in fields like meteorology where datasets are often vast and complex, **dimensionality reduction** is a critical technique. At its core, dimensionality reduction is the process of **reducing the number of random variables under consideration by obtaining a set of principal variables**. In simpler terms, it's about transforming a dataset with many features (or dimensions) into a dataset with fewer features, while striving to **retain as much of the original, important information as possible**.

9.4.1 Why is Dimensionality Reduction Necessary? The "Curse of Dimensionality"

Imagine a dataset where each piece of information (e.g., a weather observation) is described by hundreds or even thousands of different measurements (temperature at various altitudes, wind speed, humidity, pressure, cloud cover, etc.). This scenario is common in meteorology, where data comes from satellites, radars, ground stations, and complex numerical models. While more data might seem better, a high number of dimensions (features) can lead to several significant problems, often referred to as the **"Curse of Dimensionality"**:

- **Computational Burden:** Processing, storing, and analyzing data with many dimensions requires immense computational power and time. Many machine learning algorithms become prohibitively slow or even impossible to run effectively on very high-dimensional data.
- **Data Sparsity:** As the number of dimensions increases, the data points become increasingly sparse in the high-dimensional space. This means that any given data point is very far from almost every other data point. This sparsity makes it difficult for algorithms to find meaningful patterns, relationships, or clusters.
- **Increased Noise and Redundancy:** High-dimensional datasets often contain irrelevant or redundant features. For example, two temperature sensors very close to each other might provide almost identical

information. These redundant features can obscure the true underlying patterns and even introduce noise, making it harder for models to learn effectively.

- **Visualization Challenges:** We can easily visualize data in 2 or 3 dimensions. Beyond that, it becomes impossible for the human brain to directly plot and understand the data's structure. Dimensionality reduction helps project high-dimensional data into a lower-dimensional space that can be visualized.
- **Overfitting:** With too many features, a model might start to learn the noise in the training data rather than the true underlying signal. This leads to a model that performs well on the training data but poorly on new, unseen data.

Dimensionality reduction techniques aim to mitigate these issues by finding a more compact and informative representation of the data.

9.4.2 The Core Principle: Information Preservation

The fundamental goal of dimensionality reduction is to find a lower-dimensional representation that captures the most significant variance or structure in the original data. It's about identifying the "essence" of the data without losing critical information. This can be achieved in two main ways:

- **Feature Selection:** This approach involves selecting a subset of the *original* features that are most relevant to the problem. It's like picking out the most important ingredients from a long recipe. Methods include:
- **Filter Methods:** Using statistical measures (e.g., correlation, chi-squared) to score and rank features.
- **Wrapper Methods:** Using a machine learning model to evaluate subsets of features (e.g., recursive feature elimination).
- **Embedded Methods:** Feature selection is built into the model training process (e.g., Lasso regression).
- **Feature Extraction:** This approach transforms the original features into a new, smaller set of features (components or latent variables). These new features are typically linear or non-linear combinations of the original ones. It's like creating a concentrated extract from a larger substance. This is where techniques like PCA come into play.

9.4.3 Key Techniques for Dimensionality Reduction

- **Principal Component Analysis (PCA):** As discussed previously, PCA is the most common linear dimensionality reduction technique. It finds new, orthogonal axes (principal components) that capture the maximum variance in the data. It's widely used in meteorology to identify dominant modes of climate variability (as Empirical Orthogonal Functions, EOFs).

- **Linear Discriminant Analysis (LDA):** While also a linear transformation, LDA differs from PCA in that it is a *supervised* technique. It aims to find dimensions that best separate different classes of data, making it useful for classification problems.
- **Non-linear Dimensionality Reduction (Manifold Learning):** For data with complex, non-linear structures (often referred to as "manifolds"), linear techniques like PCA might not be sufficient. Non-linear methods aim to preserve the local structure of the data in the lower-dimensional space. Examples include:
- **t-Distributed Stochastic Neighbor Embedding (t-SNE):** Excellent for visualizing high-dimensional data by placing similar points close together and dissimilar points far apart.
- **Uniform Manifold Approximation and Projection (UMAP):** Similar to t-SNE but often faster and better at preserving global data structure.
- **Isomap, Locally Linear Embedding (LLE):** Other manifold learning techniques.

9.4.4 Benefits of Dimensionality Reduction

- **Reduced Computational Cost:** Models train and run faster with fewer features.
- **Less Storage Space:** Smaller datasets require less memory and disk space.
- **Improved Model Performance:** By removing noise and redundancy, models can often generalize better to new data, leading to higher accuracy and reduced overfitting.
- **Enhanced Visualization:** Projecting data into 2 or 3 dimensions allows for visual exploration and understanding of complex datasets.
- **Easier Data Interpretation:** Focusing on the most important underlying components can make the data's structure more comprehensible.

9.4.5 Challenges and Considerations

- **Information Loss:** Some information is inevitably lost during dimensionality reduction. The key is to minimize the loss of *relevant* information.
- **Interpretability of New Features:** While the new features (components) are mathematically derived, their physical meaning can sometimes be less intuitive than the original variables.

- **Choosing the Right Technique:** The best dimensionality reduction technique depends on the nature of the data and the specific problem. Linear methods work well for linearly separable data, while non-linear methods are needed for more complex structures.
- **Parameter Tuning:** Many techniques have parameters (e.g., number of components to keep) that need to be carefully chosen.

9.4.6 Conclusion

Dimensionality reduction is a powerful and essential concept in modern data analysis, particularly for fields like meteorology that deal with massive, intricate datasets. By effectively transforming high-dimensional data into a more concise and meaningful representation, these techniques not only make data more manageable and computationally efficient but also unlock deeper insights, improve the performance of predictive models, and facilitate the visualization of complex atmospheric phenomena. It is a cornerstone for navigating the "big data" challenges in understanding and forecasting our planet's weather and climate.

9.5 Identifying Dominant Modes of Variability

In meteorology and climate science, the atmosphere and oceans are incredibly complex, dynamic systems. They exhibit variability across a vast range of spatial and temporal scales, from localized thunderstorms lasting hours to global climate shifts spanning decades or centuries. To make sense of this complexity, a crucial analytical task is **identifying dominant modes of variability**.

9.5.1 What are Dominant Modes of Variability?

Dominant modes of variability refer to **recurring, large-scale, and coherent patterns of change** in atmospheric or oceanic variables (like temperature, pressure, precipitation, or sea surface temperature) that explain a significant portion of the total variance within a system. These are not random fluctuations but rather organized, often teleconnected, patterns that emerge repeatedly over time. They represent the "principal ways" in which the climate system fluctuates.

Think of it like a symphony orchestra: while many instruments play individual notes, there are underlying themes or melodies that dominate the piece. In the climate system, these dominant modes are the "melodies" that dictate much of the observed weather and climate patterns.

9.5.2 Why is Identifying Them Important?

Understanding and identifying these dominant modes is paramount for several reasons:

- **Simplification and Understanding:** They provide a simplified, low-dimensional representation of a highly complex system. Instead of tracking countless individual measurements, we can focus on the evolution of a few key patterns, making the system more comprehensible.
- **Physical Insights:** Dominant modes often correspond to specific physical processes or interactions within the Earth system (e.g., ocean-atmosphere coupling, large-scale atmospheric waves). Identifying them helps scientists diagnose and understand the underlying mechanisms driving climate variability.
- **Predictability:** Many dominant modes exhibit some degree of predictability, especially on sub-seasonal to interannual timescales. If we can identify and track the phase of these modes (e.g., whether El Niño is strong or weak), we can often make more skillful predictions of regional weather and climate anomalies weeks or months in advance.
- **Attribution:** They help in attributing observed climate anomalies to specific natural drivers. For example, a severe winter might be linked to a particular phase of the North Atlantic Oscillation.
- **Model Evaluation:** Climate models can be evaluated on their ability to accurately simulate the observed dominant modes of variability, which is a strong indicator of their realism.
- **Impact Assessment:** Understanding the typical impacts associated with different phases of these modes helps in assessing risks and planning for future climate scenarios.

9.5.3 How are Dominant Modes Identified?

The primary statistical technique used to identify dominant modes of variability in spatio-temporal datasets is **Empirical Orthogonal Function (EOF) analysis**, which, as discussed, is a direct application of **Principal Component Analysis (PCA)** to gridded data.

Here's how EOFs/PCA help in this context:

- **Data Input:** You start with a dataset of a meteorological variable (e.g., sea surface temperature anomalies, geopotential height, precipitation) measured over a geographical area for a period of time. This forms a large matrix where each column is a spatial location and each row is a time step.

- **Finding Spatial Patterns (EOFs):** EOF analysis mathematically decomposes this matrix into a set of **orthogonal spatial patterns (the EOFs)**. These patterns are fixed in space and represent the most efficient ways to describe the spatial variability in the data. The first EOF explains the most variance, the second the next most, and so on.
- **Tracking Temporal Evolution (Principal Components - PCs):** Associated with each EOF is a **time series (the Principal Component or PC time series)**. This time series tells you how the amplitude or strength of the corresponding spatial pattern varies over time. A high positive value in the PC time series means the EOF pattern is strongly present in its positive phase at that time; a high negative value means it's strongly present in its negative phase.
- **Quantifying Importance (Eigenvalues):** Each EOF also has an associated **eigenvalue**, which quantifies the amount of total variance in the original dataset that the EOF explains. By examining the eigenvalues, scientists can determine which modes are truly "dominant" (explaining a large percentage of the variance) and which are less significant.

By focusing on the first few EOFs and their corresponding PC time series, researchers can effectively identify and characterize the most influential and recurring patterns of variability in the climate system.

9.5.4 Examples of Dominant Modes in Meteorology and Climate

Several well-known climate phenomena have been objectively identified and characterized using EOF analysis:

- **El Niño-Southern Oscillation (ENSO):** The leading EOF of tropical Pacific sea surface temperature anomalies clearly reveals the characteristic El Niño (warm central/eastern Pacific) and La Niña (cold central/eastern Pacific) patterns. The associated PC time series tracks the strength and phase of ENSO over time.
- **North Atlantic Oscillation (NAO):** Often identified as a leading EOF of sea level pressure over the North Atlantic. It describes a seesaw in atmospheric pressure between the Azores high and the Icelandic low, profoundly influencing winter weather in Europe and eastern North America.
- **Pacific Decadal Oscillation (PDO):** A long-lived pattern of Pacific Ocean climate variability characterized by a horseshoe-shaped pattern of sea surface temperature anomalies in the North Pacific.
- **Arctic Oscillation (AO):** A pattern of atmospheric pressure variability over the Arctic that influences weather across the Northern Hemisphere.

- **Madden-Julian Oscillation (MJO):** A major fluctuation in tropical weather on sub-seasonal timescales (30-90 days), often identified through EOFs of outgoing longwave radiation or wind fields.

9.5.6 Conclusion

Identifying dominant modes of variability is a cornerstone of meteorological and climate research. Through techniques like EOF analysis, scientists can distill the immense complexity of atmospheric and oceanic data into a manageable set of key patterns and their temporal evolutions. This not only deepens our fundamental understanding of how the Earth system operates but also provides critical insights for improving weather forecasts, making long-term climate predictions, and assessing the impacts of climate variability on society and ecosystems.

9.6 Cluster Analysis Explained

In the vast and complex world of meteorology and climate science, datasets often contain a multitude of observations, each representing a unique state of the atmosphere or ocean. While techniques like Principal Component Analysis (PCA) help us understand the dominant *patterns of variability* across all data, **Cluster Analysis** offers a different yet equally powerful approach: it focuses on **grouping similar observations together** into distinct clusters.

9.6.1 What is Cluster Analysis?

Cluster analysis is an unsupervised machine learning technique used to **discover natural groupings or structures within a dataset**. Unlike supervised learning, where data is labeled with known categories, cluster analysis works with unlabeled data to find inherent similarities among data points and organize them into clusters such that:

- Data points within the same cluster are highly similar to each other.
- Data points in different clusters are dissimilar from each other.

The "similarity" or "dissimilarity" is typically measured using a **distance metric** (e.g., Euclidean distance, correlation distance) in the multi-dimensional space defined by the variables.

9.6.2 Why is Cluster Analysis Needed in Meteorology?

Meteorological data often represents a continuum of states, but within this continuum, there might be recurring "types" or "regimes" that are not explicitly defined. Cluster analysis helps to objectively identify these types, which is crucial for:

- **Simplifying Complexity:** Grouping millions of individual weather observations into a handful of characteristic "weather types" or "climate regimes" makes the data much more manageable and understandable.
- **Identifying Natural Regimes:** The atmosphere and ocean often settle into certain quasi-stable states or circulation patterns. Cluster analysis can objectively identify these regimes without prior assumptions.
- **Forecasting by Analogy:** If current weather conditions cluster with historical conditions that led to a specific outcome (e.g., a severe storm or a prolonged drought), it can provide valuable insights for future predictions.
- **Understanding Impacts:** Different weather types or climate regimes often have distinct impacts on human activities, ecosystems, or natural hazards. Clustering helps categorize these impacts.
- **Data Reduction for Further Analysis:** Once clusters are identified, subsequent analyses can be performed on the representative characteristics of each cluster rather than on every individual data point.

9.6.3 How Cluster Analysis Works (The Core Idea)

Imagine you have a scatter plot of temperature and humidity readings from many different days. You might visually notice that some days are warm and dry, others are cool and humid, and some are warm and humid. Cluster analysis automates this process of finding these natural groupings.

The general process involves

- **Defining Similarity/Dissimilarity:** Choosing a way to measure how "close" or "far apart" two data points are. Common measures include:
- **Euclidean Distance:** The straight-line distance between two points in a multi-dimensional space.
- **Correlation Distance:** Measures how similar the *shape* of two time series or spatial patterns are.
- **Choosing a Clustering Algorithm:** There are many algorithms, each with its own strengths:
- **K-Means Clustering:** One of the most popular algorithms. You specify the desired number of clusters (K) beforehand. The algorithm iteratively assigns data points to the nearest cluster centroid (the mean of all points in that cluster) and then re-calculates the centroids until the assignments stabilize.
- **Hierarchical Clustering:** Builds a hierarchy of clusters. It can start with each data point as its own cluster and then successively merge

the closest clusters (agglomerative) or start with one large cluster and successively split it (divisive). The result is a dendrogram, which allows you to choose the number of clusters by cutting the tree at a certain level.

- **DBSCAN (Density-Based Spatial Clustering of Applications with Noise):** Identifies clusters based on the density of data points, making it good at finding arbitrarily shaped clusters and identifying outliers (noise).
- **Gaussian Mixture Models (GMM):** Assumes that data points within each cluster are generated from a Gaussian distribution. It's a probabilistic approach that assigns probabilities of belonging to each cluster.
- **Assigning Data Points to Clusters:** Based on the chosen algorithm and distance metric, each data point is assigned to a specific cluster.
- **Interpreting the Clusters:** Once clusters are formed, the next crucial step is to analyze the characteristics of each cluster. What are the typical values of the meteorological variables within each cluster? What weather patterns or climate conditions do they represent?

9.6.4 Applications in Meteorology

Cluster analysis has a wide range of practical applications in atmospheric and climate science:

- **Classification of Weather Regimes:** Identifying recurring large-scale atmospheric circulation patterns (e.g., blocking events, zonal flow, meridional flow) that influence regional weather.
- **Regional Climate Classification:** Grouping geographical areas based on similar climate characteristics (e.g., identifying distinct climate zones based on temperature and precipitation patterns).
- **Synoptic Climatology:** Categorizing daily or weekly weather maps into a finite set of "weather types" that frequently occur, which can then be used to study their frequency, persistence, and impacts.
- **Analysis of Extreme Events:** Identifying specific atmospheric conditions or precursors that tend to cluster together before extreme weather events (e.g., heatwaves, heavy rainfall, droughts).
- **Air Quality Studies:** Grouping days or regions based on similar pollutant concentrations and meteorological conditions to understand pollution sources and dispersion patterns.
- **Cloud Classification:** Categorizing cloud types from satellite imagery based on their radiative properties or structural characteristics.

- **Ensemble Forecast Post-processing:** Grouping similar members of an ensemble forecast to identify distinct possible future scenarios.
- **Downscaling Climate Model Outputs:** Identifying local weather patterns that correspond to large-scale climate model outputs.

9.6.5 Advantages of Cluster Analysis

- **Objective Grouping:** Provides a data-driven way to identify natural groupings without requiring prior knowledge of categories.
- **Pattern Discovery:** Can reveal previously unrecognized structures or relationships within complex meteorological data.
- **Data Summarization:** Reduces the complexity of large datasets by representing them through a smaller number of characteristic clusters.
- **Foundation for Further Analysis:** The identified clusters can serve as categories for subsequent supervised learning, impact studies, or forecasting efforts.
- **Flexibility:** A variety of algorithms and distance metrics allow for adaptation to different data types and research questions.

9.6.6 Limitations and Considerations

- **Defining "K" (for K-Means):** For algorithms like K-Means, determining the optimal number of clusters (K) can be challenging and often requires domain knowledge or heuristic methods (e.g., elbow method, silhouette score).
- **Sensitivity to Outliers:** Some clustering algorithms (like K-Means) can be sensitive to outliers, which might disproportionately influence cluster centroids.
- **Choice of Distance Metric:** The choice of distance metric significantly impacts the clustering results. An inappropriate metric might lead to misleading groupings.
- **Curse of Dimensionality:** In very high-dimensional spaces, distance metrics can become less meaningful, making clustering more challenging. Dimensionality reduction (like PCA) is often performed *before* clustering.
- **Interpretability:** While clusters are formed mathematically, their physical or meteorological interpretation requires careful analysis and domain expertise.
- **Non-Unique Solutions:** Some algorithms can produce different results with different random initializations (e.g., K-Means).

9.6.7 Conclusion

Cluster analysis is an invaluable tool in meteorology and climate science, enabling researchers to uncover hidden structures and natural groupings within complex atmospheric and oceanic datasets. By transforming raw observations into meaningful categories, it simplifies analysis, enhances our understanding of weather and climate regimes, and provides a powerful framework for improving forecasts, assessing impacts, and conducting targeted research. As data continues to proliferate, the ability to objectively identify and interpret these dominant groupings will remain a critical skill for advancing our knowledge of the Earth system.

9.7 Identifying Distinct Weather Patterns or Regimes

In meteorology and climate science, the atmosphere often settles into certain recurring, large-scale configurations that persist for days to weeks. These are known as **distinct weather patterns** or **atmospheric regimes**. Rather than a continuous, chaotic flow, the atmosphere tends to cycle through a finite number of preferred states. Identifying these patterns is a fundamental goal, as it provides a simplified, yet powerful, framework for understanding, predicting, and assessing the impacts of weather and climate.

9.7.1 What are Weather Patterns/Regimes?

A **weather pattern** or **atmospheric regime** refers to a characteristic, quasi-stationary configuration of atmospheric variables (such as pressure, geopotential height, temperature, or wind fields) over a large geographical area. These patterns are often associated with specific weather conditions in different regions. For example, a particular high-pressure pattern over Scandinavia might consistently lead to cold, dry winters in Europe, while a low-pressure pattern in the same region might bring mild, wet conditions.

These regimes are not static; they evolve over time, but they tend to return to similar configurations, making them identifiable and often predictable to some extent.

9.7.2 Why is Identifying Them Crucial?

The ability to identify and categorize these distinct weather patterns is vital for several reasons:

- **Enhanced Understanding:** It simplifies the immense complexity of atmospheric variability by reducing it to a manageable number of characteristic states. This helps scientists understand the underlying physical processes that maintain and transition between these regimes.

- **Improved Sub-seasonal to Seasonal (S2S) Forecasting:** Many weather patterns have predictable influences on regional weather weeks to months in advance. If forecasters can predict the onset or persistence of a particular regime, they can make more skillful probabilistic forecasts for temperature, precipitation, and extreme events.
- **Impact Assessment:** Different regimes are associated with distinct societal and environmental impacts. Identifying them helps in assessing risks related to droughts, floods, heatwaves, cold snaps, and other hazards.
- **Climate Change Studies:** Understanding how the frequency, intensity, and duration of these regimes might change in a warming climate is crucial for projecting future regional climate impacts.
- **Model Evaluation:** Climate and weather models can be evaluated on their ability to accurately simulate the observed weather patterns and their transitions.
- **Communication:** Describing weather in terms of recognizable patterns (e.g., "a strong blocking pattern") can be more intuitive for the public and stakeholders than complex meteorological charts.

9.7.3 How are Distinct Weather Patterns/Regimes Identified?

The identification of these patterns primarily relies on **data-driven, unsupervised learning techniques**, which allow the data itself to reveal its inherent structure. The most common and effective methods include:

- **Empirical Orthogonal Function (EOF) Analysis / Principal Component Analysis (PCA)**
- **How it works:** As discussed previously, EOF analysis decomposes the spatio-temporal variability of a meteorological field (e.g., daily geopotential height anomalies at 500 hPa) into a set of orthogonal spatial patterns (EOFs) and their corresponding time series (PCs).
- **Identifying Regimes:** The leading EOFs often represent the most dominant and recurring large-scale patterns of atmospheric variability. While not always directly interpretable as distinct "regimes" on their own (as they are linear combinations), the *phase* of the associated PC time series can often distinguish between different states. For example, the positive and negative phases of the North Atlantic Oscillation (NAO) are distinct regimes.
- **Strengths:** Objective, identifies patterns that explain maximum variance.

- **Limitations:** Linear method, patterns are orthogonal but may not always correspond to physically distinct regimes.
- **Cluster Analysis**
    - **How it works:** This is a more direct approach to grouping. Meteorological "states" (e.g., daily atmospheric circulation maps, or a vector of key atmospheric variables) are treated as data points in a high-dimensional space. A clustering algorithm (like K-Means, Hierarchical Clustering, or Self-Organizing Maps) then groups these states into a predefined or data-determined number of clusters based on their similarity.
    - **Identifying Regimes:** Each cluster then represents a distinct weather pattern or regime. The mean or centroid of each cluster can be visualized as the representative "map" of that regime.
    - **Strengths:** Directly identifies distinct groups, can capture non-linear relationships depending on the algorithm.
    - **Limitations:** Requires choosing the number of clusters (for some algorithms), sensitivity to distance metrics and outliers.
- **Self-Organizing Maps (SOMs)**
    - **How it works:** SOMs are a type of artificial neural network that performs non-linear dimensionality reduction and clustering. They map high-dimensional data onto a low-dimensional grid (typically 2D) while preserving the topological relationships of the input data. Each node on the grid represents a "prototype" or a characteristic pattern.
    - **Identifying Regimes:** After training, similar input maps will activate nearby nodes on the SOM grid. Each node (or a group of nodes) can then be interpreted as a distinct weather pattern or regime.
    - **Strengths:** Good at visualizing relationships between patterns, can capture non-linearities, and provides a spatial arrangement of regimes.
    - **Limitations:** Requires tuning parameters, interpretation can be complex.

9.7.4 Examples of Well-Known Weather Patterns/Regimes

- **North Atlantic Oscillation (NAO):** A seesaw in atmospheric pressure between the Icelandic Low and the Azores High. Its positive phase typically brings mild, wet winters to Northern Europe and dry conditions to Southern Europe, while the negative phase brings the opposite.

- **Arctic Oscillation (AO):** A pattern of pressure variability over the Arctic, influencing mid-latitude weather.
- **Pacific North American (PNA) Pattern:** A prominent teleconnection pattern in the Northern Hemisphere that affects weather across North America.
- **El Niño-Southern Oscillation (ENSO):** While primarily an ocean-atmosphere phenomenon, its distinct warm (El Niño) and cold (La Niña) phases drive global weather regimes.
- **Blocking Events:** Persistent high-pressure systems that "block" the typical eastward progression of weather systems, leading to prolonged periods of specific weather (e.g., heatwaves, cold snaps, or extended rainfall/drought).
- **Madden-Julian Oscillation (MJO):** An eastward-propagating pulse of clouds and rainfall that circles the globe in the tropics, influencing global weather patterns on sub-seasonal timescales.

9.7.5 Conclusion

Identifying distinct weather patterns or regimes is a cornerstone of modern meteorological and climate research. By employing advanced analytical techniques like EOF analysis, cluster analysis, and Self-Organizing Maps, scientists can objectively distill the immense complexity of atmospheric data into a manageable set of recurring states. This not only deepens our fundamental understanding of atmospheric dynamics but also significantly enhances our ability to predict future weather, assess climate impacts, and communicate complex meteorological information more effectively. It represents a powerful step towards a more comprehensive and actionable understanding of our planet's climate system.

9.8 Machine Learning for Meteorological Data

The field of meteorology and climate science is undergoing a transformative shift with the increasing adoption of **Machine Learning (ML)** techniques. Traditionally, meteorological analysis and forecasting have relied heavily on physical models (e.g., Numerical Weather Prediction - NWP models) based on the fundamental laws of physics. While these models remain indispensable, the sheer volume, velocity, and variety of modern meteorological data, coupled with the computational demands and inherent complexities of atmospheric processes, have created a fertile ground for ML applications.

9.8.1 Why Machine Learning for Meteorological Data?

The need for machine learning in meteorology arises from several key challenges and opportunities:

- **Big Data Volume:** Modern observational platforms (satellites, radar, vast sensor networks) and high-resolution numerical models generate petabytes of data daily. Traditional statistical methods can struggle to process and extract insights from such massive datasets efficiently.
- **Complex Relationships:** Atmospheric processes are often non-linear, chaotic, and involve intricate interactions across scales. ML algorithms, particularly deep learning, are adept at identifying and modeling these complex, non-linear relationships that might be difficult to capture with explicit physical equations or simpler statistical models.
- **Computational Efficiency:** Running high-resolution NWP models is computationally very expensive. ML models, once trained, can often provide predictions or analyses much faster, making them suitable for real-time applications or as computationally cheaper "surrogates" for parts of physical models.
- **Pattern Recognition:** ML excels at identifying subtle, recurring patterns in data that might be missed by human analysts or traditional algorithms. This is crucial for detecting extreme weather events, classifying weather types, or identifying precursors to significant phenomena.
- **Bias Correction and Post-processing:** Physical models, despite their sophistication, often have systematic biases. ML can be used to statistically correct these biases and refine raw model outputs into more accurate and actionable forecasts.
- **Handling Missing or Noisy Data:** ML algorithms can be more robust to missing values and noise in observations compared to some traditional methods.

9.8.2 Types of Machine Learning Applications in Meteorology

Machine learning broadly falls into several categories, each with distinct applications in meteorology:

- **Supervised Learning**
    - **Goal:** Learn a mapping from input features to known output labels.
- **Applications**
    - **Forecasting:** Predicting future values of meteorological variables (e.g., temperature, precipitation, wind speed) based on current and historical data. Techniques include **Regression** (for continuous

values) and **Classification** (for categorical outcomes, like "rain/no rain,""hurricane/no hurricane").

- **Nowcasting:** Short-term, high-resolution forecasting (minutes to a few hours ahead), often using radar and satellite data to predict storm movement or precipitation intensity.
- **Extreme Event Prediction:** Classifying conditions likely to lead to severe thunderstorms, tornadoes, floods, or heatwaves.
- **Bias Correction/Post-processing:** Using observed data to train models that correct systematic errors in raw NWP model outputs.

- **Unsupervised Learning**
  - **Goal:** Discover hidden patterns or structures in unlabeled data.
- **Applications**
  - **Dimensionality Reduction:** Techniques like **Principal Component Analysis (PCA)** (often referred to as Empirical Orthogonal Functions - EOFs in meteorology) to reduce the number of variables while retaining most information, identifying dominant modes of climate variability.
  - **Cluster Analysis:** Grouping similar weather patterns or climate regimes together (e.g., identifying distinct atmospheric circulation types, classifying cloud types from satellite imagery).
  - **Anomaly Detection:** Identifying unusual or novel weather events that deviate significantly from historical patterns.
- **Deep Learning**
  - **Goal:** A subfield of ML using artificial neural networks with multiple layers (deep networks) to learn complex representations from data. Particularly powerful for image, sequence, and large, unstructured datasets.
- **Applications**
  - **Satellite and Radar Image Analysis:** Classifying cloud types, detecting severe storm features (e.g., rotation, hail signatures), tracking storm movement. **Convolutional Neural Networks (CNNs)** are especially suited for this.
  - **Time Series Forecasting:** Predicting the evolution of atmospheric variables using **Recurrent Neural Networks (RNNs)** or **Long Short-Term Memory (LSTM) networks**, which are good at handling sequential data.

- **Generative Models (e.g., GANs):** Generating realistic weather scenarios or downscaling climate model outputs to higher resolutions.
- **Surrogate Models:** Creating computationally cheaper ML models that mimic the behavior of complex physical models, allowing for faster simulations or sensitivity studies.

9.8.3 Specific Machine Learning Techniques and Their Uses

- **Linear Regression / Logistic Regression:** For basic forecasting (continuous values) or binary classification (e.g., presence/absence of an event).
- **Decision Trees / Random Forests / Gradient Boosting Machines (GBMs):** Powerful ensemble methods used for both regression and classification, often providing high accuracy and interpretability. Useful for feature importance analysis.
- **Support Vector Machines (SVMs):** Effective for classification tasks, particularly when data is not linearly separable.
- **K-Nearest Neighbors (KNN):** Simple, non-parametric method for classification or regression, useful for finding similar historical analogs.
- **Neural Networks (ANNs, CNNs, RNNs/LSTMs):** The backbone of deep learning, capable of learning highly complex, non-linear relationships.
- **CNNs:** Excellent for spatial data (images, gridded maps) like satellite imagery or radar scans.
- **RNNs/LSTMs:** Ideal for sequential data like time series or atmospheric profiles.
- **Clustering Algorithms (K-Means, Hierarchical Clustering, DBSCAN, Self-Organizing Maps):** For identifying natural groupings of weather states or climate regimes.

9.8.4 Benefits of Machine Learning in Meteorology

- **Improved Accuracy:** ML models can often outperform traditional statistical methods in complex forecasting tasks, especially for short-term predictions and extreme events.
- **Speed and Efficiency:** Once trained, ML models can provide rapid predictions, crucial for real-time operational meteorology.
- **Automation:** Automates the identification of patterns and relationships that might be too subtle or complex for human detection.

- **Handling Big Data:** Designed to scale with large and high-dimensional datasets.
- **Discovery of New Insights:** Can uncover previously unknown relationships or teleconnections in atmospheric data.
- **Complementary to Physics-Based Models:** ML is increasingly used to augment, correct, and optimize physical models, rather than replacing them entirely.

9.8.5 Challenges and Considerations

Despite the immense potential, applying ML to meteorological data comes with its own set of challenges:

- **Data Quality and Preprocessing:** Meteorological data can be noisy, have missing values, or come in various formats. Extensive preprocessing, quality control, and careful feature engineering are crucial.
- **Interpretability (Black Box Problem):** Deep learning models, in particular, can be "black boxes," making it difficult to understand *why* they make certain predictions. In a field like meteorology, where physical understanding is paramount, this can be a significant hurdle. Explainable AI (XAI) is an active research area addressing this.
- **Generalizability:** Models trained on historical data might not perform well on future, unprecedented weather events (e.g., due to climate change).
- **Causality vs. Correlation:** ML models identify correlations, but these do not necessarily imply causation. Physical understanding is still needed to interpret the relationships learned by ML.
- **Computational Resources:** Training complex deep learning models requires substantial computational power (GPUs).
- **Data Scarcity for Rare Events:** Extreme weather events are by definition rare, leading to limited training data for specific high-impact phenomena.
- **Integration with Physical Models:** Seamlessly integrating ML components into complex NWP systems is an ongoing research challenge.

9.8.6 Conclusion

Machine learning is rapidly becoming an indispensable component of modern meteorological and climate science. From enhancing the accuracy of weather forecasts and identifying subtle climate patterns to optimizing observational networks and improving our understanding of complex atmospheric processes,

ML offers powerful new capabilities. While challenges remain, the synergistic combination of physics-based modeling with advanced machine learning techniques holds the key to unlocking unprecedented insights and delivering more accurate, reliable, and actionable meteorological information for the benefit of society.

9.9 Supervised Learning Explained

In the broad landscape of Machine Learning, **Supervised Learning** is arguably the most common and widely applied paradigm, particularly in fields like meteorology where we often have access to historical data with known outcomes. At its core, supervised learning is about **learning from examples** to make predictions or decisions.

9.9.1 What is Supervised Learning?

Supervised learning is a type of machine learning where an algorithm learns a mapping function from input variables (features) to an output variable (target or label) based on a **labeled dataset**. The term "supervised" comes from the idea that the learning process is "supervised" by the known output labels, much like a teacher supervises a student by providing correct answers during training.

The goal of a supervised learning algorithm is to learn a general rule that can then be used to predict the output for new, unseen input data.

9.9.2 The Core Idea: Learning from Labeled Examples

Imagine you want to predict whether it will rain tomorrow. You have historical data for many past days, and for each day, you know several things (input features): today's temperature, humidity, wind speed, cloud cover, and crucially, whether it *actually rained* tomorrow (the output label).

A supervised learning algorithm would take this historical, labeled data and try to find a relationship between the input features and the "rain/no rain" outcome. It learns patterns like "if temperature is high, humidity is low, and wind is strong, it usually doesn't rain." Once it has learned these patterns from a large number of examples, it can then be given new, unlabeled data (e.g., today's weather conditions) and predict whether it will rain tomorrow.

9.9.3 Key Components of Supervised Learning

- **Labeled Training Data:** This is the foundation. It consists of:
    - **Features (Input Variables):** The measurable characteristics or attributes of each data point (e.g., temperature, pressure, wind speed, satellite image pixels). These are often denoted as X.

- **Labels (Output Variables/Targets):** The known outcome or value associated with each set of features (e.g., "rain,""no rain," actual temperature in Celsius, hurricane category). These are often denoted as Y.
- The algorithm learns from pairs of (X,Y).

- **Learning Algorithm (Model):** This is the mathematical function or process that learns the relationship between Xand Y. Examples include Linear Regression, Decision Trees, Support Vector Machines, Neural Networks, etc.
- **Prediction:** Once the model is trained, it can be used to predict the output (Y′) for new, unseen input data (X′) where the true label is unknown.

9.9.4 Types of Supervised Learning Problems: Regression vs. Classification

Supervised learning problems are generally categorized into two main types based on the nature of the output variable:

- **Regression**
  - **Goal:** To predict a **continuous numerical value**. In regression tasks, the output variable can take on any value within a given range.
  - **How it works:** The algorithm learns to map input features to a specific point on a continuous scale. For example, it might learn that a certain combination of atmospheric conditions corresponds to a temperature of 25.3°C, not just "warm" or "cold."
- **Meteorological Examples**
  - **Temperature Forecasting:** Predicting the exact maximum or minimum temperature for a city tomorrow (e.g., 28.5 °C).
  - **Quantitative Precipitation Forecasting (QPF):** Estimating the precise amount of rainfall (in millimeters or inches) expected over a region in the next 24 hours.
  - **Wind Speed Prediction:** Forecasting the sustained wind speed (e.g., 15 km/h) or gust speed at a specific location.
  - **Sea Level Pressure Forecasting:** Predicting the exact barometric pressure value (e.g., 1012.7 hPa).
  - **Bias Correction of Numerical Weather Prediction (NWP) Models:** Training a regression model to learn the systematic error (bias) of a physical model's temperature output at a given location and then predicting the correction needed for future forecasts.

- **Classification**
  - **Goal:** To predict a **categorical label or class**. In classification tasks, the output variable belongs to a finite set of discrete categories.
  - **How it works:** The algorithm learns to assign an input data point to one of several predefined classes. For example, it might learn that a specific set of radar reflectivity patterns indicates a "hail" storm, while another set indicates "rain."
- **Meteorological Examples**
  - **Rainfall Occurrence:** Predicting whether it will rain tomorrow (Yes/No – a binary classification).
  - **Severe Weather Event Prediction:** Classifying atmospheric conditions as leading to a tornado, severe thunderstorm, or non-severe storm (multi-class classification).
  - **Hurricane Categorization:** Assigning a tropical cyclone to its Saffir-Simpson category (e.g., Category 1, Category 2, etc.) based on satellite imagery and wind speed data.
  - **Cloud Type Identification:** Classifying cloud types (e.g., Cumulus, Stratus, Cirrus, Nimbus) from satellite images or ground-based observations.
  - **Drought/Flood Risk Assessment:** Categorizing regions into different drought severity levels (e.g., "Moderate Drought,""Severe Drought") or flood risk categories (e.g., "Low Risk,""High Risk").
  - **Atmospheric River Detection:** Identifying whether a specific atmospheric pattern constitutes an "atmospheric river" (Yes/No).

9.9.5 Applications of Supervised Learning in Meteorology

Supervised learning is incredibly versatile and has numerous applications in meteorology and climate science:

- **Forecasting:** Predicting future values of meteorological variables (e.g., temperature, precipitation, wind speed) based on current and historical data.
- **Nowcasting:** Short-term, high-resolution forecasting (minutes to a few hours ahead), often using radar and satellite data to predict storm movement or precipitation intensity.
- **Extreme Event Prediction:** Classifying conditions likely to lead to severe thunderstorms, tornadoes, floods, or heatwaves.
- **Bias Correction/Post-processing:** Using observed data to train models that correct systematic errors in raw NWP model outputs.

- **Downscaling Climate Model Outputs:** Using a high-resolution local weather variable (e.g., temperature) as the label and large-scale climate model outputs as features to predict local climate conditions.

9.9.6 Advantages of Supervised Learning

- **Directly Applicable to Prediction:** Its primary strength is its ability to make predictions for new data based on historical examples.
- **High Accuracy:** When trained on sufficient, high-quality labeled data, supervised models can achieve very high accuracy in both regression and classification tasks.
- **Versatility:** A wide range of algorithms are available, suitable for various types of data and problems.
- **Quantifiable Performance:** Model performance can be easily evaluated using metrics like accuracy, precision, recall, F1-score (for classification), or Mean Squared Error, R-squared (for regression).

9.9.7 Limitations and Considerations

- **Requires Labeled Data:** The biggest challenge is the need for large amounts of high-quality, accurately labeled data. Labeling meteorological data can be time-consuming and expensive, especially for rare events.
- **"Garbage In, Garbage Out":** The performance of a supervised model is highly dependent on the quality and representativeness of the training data. Biases or errors in the labels will lead to biased or erroneous predictions.
- **Generalization to Unseen Data:** Models must be able to generalize well to new data that was not part of the training set. Overfitting (where the model learns the training data too well, including its noise) is a common problem that leads to poor generalization.
- **Computational Cost:** Training complex supervised models, especially deep neural networks, can require significant computational resources.
- **Interpretability (Black Box Problem):** Deep learning models, in particular, can be "black boxes," making it difficult to understand *why* they make certain predictions. In a field like meteorology, where physical understanding is paramount, this can be a significant hurdle. Explainable AI (XAI) is an active research area addressing this.
- **Causality vs. Correlation:** ML models identify correlations, but these do not necessarily imply causation. Physical understanding is still needed to interpret the relationships learned by ML.

- **Data Scarcity for Rare Events:** Extreme weather events are by definition rare, leading to limited training data for specific high-impact phenomena.

9.9.8 Conclusion

Supervised learning is a cornerstone of modern meteorological data analysis and forecasting. By enabling algorithms to learn complex relationships from historical observations and model outputs, it empowers meteorologists to make more accurate predictions, classify atmospheric phenomena, and refine the outputs of physical models. As data continues to proliferate and computational power grows, supervised learning will remain an indispensable tool for advancing our understanding and prediction capabilities in the atmospheric sciences.

9.10 Unsupervised Learning Explained

In contrast to supervised learning, where algorithms learn from labeled data, **Unsupervised Learning** deals with **unlabeled data**. The goal is not to predict a specific output, but rather to **discover hidden patterns, structures, or relationships within the data itself**. It's like giving a student a large collection of books without any titles or genre labels and asking them to organize them into meaningful groups based on their content.

9.10.1 What is Unsupervised Learning?

Unsupervised learning algorithms work by analyzing the inherent properties and characteristics of the input data to find underlying structures. There's no "correct answer" provided during training; instead, the algorithm tries to make sense of the data by grouping similar items, reducing complexity, or identifying unusual observations.

9.10.2 The Core Idea: Discovering Hidden Structure

Imagine you have a vast collection of satellite images of clouds, but no one has ever labeled them as "cumulus,""stratus," or "cirrus." An unsupervised learning algorithm could analyze the pixel patterns, textures, and shapes in these images and group them into distinct categories based on their visual similarities. It wouldn't know the names of the cloud types, but it would identify that "these images look alike,""those images look alike, but are different from the first group," and so on.

9.10.3 Key Applications of Unsupervised Learning in Meteorology

Unsupervised learning is particularly powerful in meteorology for tasks where explicit labels are scarce, difficult to obtain, or where the goal is exploratory data analysis. Two prominent applications are:

- **Anomaly Detection (Outlier Detection)**
  - **Goal:** To identify data points, events, or patterns that deviate significantly from the "normal" or expected behavior of the system. Anomalies are often indicative of rare, unusual, or potentially critical events.
  - **How it works:** Anomaly detection algorithms learn the underlying distribution or typical characteristics of the "normal" data. Any new data point that falls outside this learned normal range, or has a very low probability of belonging to the normal distribution, is flagged as an anomaly.
  - **Statistical Methods:** Simple methods might use statistical thresholds (e.g., values more than 3 standard deviations from the mean).
  - **Density-Based Methods:** Algorithms like DBSCAN (Density-Based Spatial Clustering of Applications with Noise) can identify outliers as points that are not part of any dense cluster.
  - **Reconstruction-Based Methods:** Techniques like Autoencoders (a type of neural network) learn to compress and reconstruct normal data. Anomalous data, being different, will have high reconstruction errors.
  - **Proximity-Based Methods:** Algorithms like Isolation Forest work by isolating anomalies in a tree structure, often requiring fewer splits than normal data points.
- **Why it's crucial in Meteorology**
  - **Extreme Weather Events:** Identifying unusually strong storms, unprecedented heatwaves, or extreme cold snaps that fall outside typical historical ranges. This is critical for issuing warnings and preparing for impacts.
  - **Sensor Malfunctions:** Detecting faulty weather sensors that are providing erroneous readings, which could otherwise corrupt datasets or lead to incorrect forecasts.
  - **Climate Change Signals:** Identifying long-term trends or events that are anomalous compared to historical climate variability, potentially signaling the influence of climate change.
  - **Model Performance Monitoring:** Flagging instances where a numerical weather prediction model's output deviates significantly from observations, indicating potential model issues or highly unusual atmospheric conditions.

  - **Data Quality Control:** Automatically flagging suspicious data points in large observational datasets for human review.
- **Meteorological Examples of Anomaly Detection**
  - Flagging a sudden, localized temperature spike that is inconsistent with surrounding station data and historical norms, potentially indicating a sensor error or a microclimate anomaly.
  - Identifying a hurricane that rapidly intensifies at an unprecedented rate, signaling a potentially dangerous and unusual event.
  - Detecting a precipitation event with an intensity far exceeding historical records for a given duration, indicating a flash flood risk.
  - Noting unusual patterns in atmospheric chemistry data that might indicate a novel pollution event or volcanic eruption.
- **Pattern Recognition (Clustering and Dimensionality Reduction)**
- **Goal:** To discover inherent groupings (clusters) or underlying structures (patterns) within unlabeled data. This helps in understanding the natural organization of complex systems.
- **How it works**
- **Clustering:** Algorithms group data points into clusters such that points within a cluster are more similar to each other than to points in other clusters. This allows for the identification of "types" or "regimes." (As elaborated in the previous response on Cluster Analysis).
  - *Example Algorithms:* K-Means, Hierarchical Clustering, DBSCAN, Gaussian Mixture Models (GMMs), Self-Organizing Maps (SOMs).
- **Dimensionality Reduction:** Algorithms transform high-dimensional data into a lower-dimensional representation while preserving essential information. This reveals the most dominant "modes" or "patterns" of variability. (As elaborated in the previous response on Dimensionality Reduction and PCA/EOFs).
  - *Example Algorithms:* Principal Component Analysis (PCA), Independent Component Analysis (ICA), t-Distributed Stochastic Neighbor Embedding (t-SNE), Uniform Manifold Approximation and Projection (UMAP).
- **Why it's crucial in Meteorology**
  - **Identifying Distinct Weather Patterns or Regimes:** As discussed, the atmosphere tends to cycle through preferred states. Unsupervised pattern recognition (especially clustering and EOFs) is the primary method for objectively identifying these recurring large-scale

atmospheric circulation patterns (e.g., blocking patterns, zonal flow, phases of the NAO or PNA).

- **Climate Classification:** Grouping geographical regions based on similar climate characteristics (e.g., identifying distinct climate zones based on temperature, precipitation, and seasonality).
- **Cloud Classification:** Automatically categorizing various cloud types from satellite or ground-based imagery based on their visual features, without needing pre-labeled examples for every type.
- **Atmospheric Feature Extraction:** Identifying coherent structures in complex data, such as fronts, jet streams, or mesoscale convective systems, from raw sensor data.
- **Data Exploration and Hypothesis Generation:** Unsupervised methods can reveal unexpected relationships or patterns in data, leading to new scientific hypotheses about atmospheric processes.
- **Data Compression:** Dimensionality reduction techniques compress large datasets while retaining the most important information, making them more manageable for storage and transmission.

- **Meteorological Examples of Pattern Recognition**
  - Using **Cluster Analysis** on daily sea level pressure maps to identify recurring "weather types" over Europe, such as "Scandinavian Blocking," Zonal Flow," or "Atlantic Ridge."
  - Applying **Empirical Orthogonal Function (EOF) analysis** (a form of PCA) to global sea surface temperature anomalies to objectively discover the **El Niño-Southern Oscillation (ENSO)** as the dominant mode of variability.
  - Using **Self-Organizing Maps (SOMs)** to categorize thousands of daily atmospheric profiles (temperature, humidity, wind at different altitudes) into a few representative "atmospheric states" for a given region.
  - Applying **t-SNE** to high-dimensional atmospheric chemistry data to visualize natural groupings of pollutant mixtures, revealing different pollution sources or transport pathways.

9.10.4 Advantages of Unsupervised Learning

- **No Labeled Data Required:** This is the biggest advantage, making it invaluable when labels are expensive, scarce, or impossible to obtain.
- **Discovery of Hidden Insights:** Can uncover structures, patterns, or anomalies that were previously unknown or not explicitly defined.

- **Data Exploration:** Excellent for initial exploratory data analysis to understand the underlying organization of a dataset.
- **Dimensionality Reduction:** Simplifies complex, high-dimensional data, making it more manageable and interpretable.
- **Foundation for Supervised Learning:** Discovered clusters or reduced dimensions can sometimes be used as features or labels for subsequent supervised learning tasks.

9.10.5 Limitations and Considerations

- **No "Ground Truth":** Without labels, evaluating the "correctness" of the learned patterns or clusters can be subjective and challenging. Interpretation often requires strong domain expertise.
- **Algorithm-Dependent Results:** Different unsupervised algorithms, or different parameter settings for the same algorithm, can yield different results.
- **Scalability:** Some unsupervised algorithms can be computationally intensive for very large datasets.
- **Sensitivity to Noise/Outliers:** Some algorithms can be sensitive to noise or outliers, which might distort the discovered patterns.
- **Interpretability:** While patterns are identified, understanding their physical meaning often requires careful analysis and correlation with other meteorological knowledge.

9.10.6 Conclusion

Unsupervised learning plays a vital and complementary role to supervised learning in meteorology. By enabling the discovery of hidden structures, the identification of anomalies, and the reduction of data complexity without the need for explicit labels, it provides powerful tools for understanding the inherent dynamics of the Earth system. From recognizing extreme events to objectively classifying weather regimes and uncovering dominant climate patterns, unsupervised learning is indispensable for advancing our knowledge and predictive capabilities in atmospheric and climate science.

9.11 Deep Learning for Meteorological Data

Deep Learning (DL) is a specialized subfield of Machine Learning that employs **artificial neural networks** with multiple layers (hence "deep") to learn complex patterns and representations directly from data. Unlike traditional machine learning algorithms that often require manual feature engineering, deep learning models can automatically discover intricate features and hierarchies within raw data, making them exceptionally powerful for handling large, unstructured, and high-dimensional datasets common in meteorology.

The rise of deep learning, fueled by advancements in computational power (especially GPUs) and the availability of massive datasets, has revolutionized many scientific fields, and meteorology is no exception. DL models are proving to be highly effective at tasks that were previously challenging for traditional methods, such as interpreting vast amounts of image data or predicting the evolution of complex time series.

9.11.1 Why Deep Learning for Meteorology?

Deep learning is particularly well-suited for meteorological applications due to:

- **Handling Big Data:** It can process and learn from enormous volumes of data from satellites, radar, and high-resolution models.
- **Non-linear Relationships:** Atmospheric processes are inherently non-linear and chaotic. Deep neural networks are excellent at modeling these complex, non-linear relationships.
- **Automatic Feature Extraction:** DL models can automatically learn relevant features from raw data (e.g., edges, textures, shapes in images; trends, seasonality in time series) without explicit programming.
- **Complex Pattern Recognition:** They excel at identifying subtle and intricate patterns that might be missed by human observers or simpler algorithms.
- **Performance:** Often achieves state-of-the-art performance in tasks like image classification, object detection, and sequence prediction.

9.11.2 Neural Networks for Image Processing (Satellite and Radar Data)

Meteorological data often comes in the form of images or grids, such as satellite imagery (visible, infrared, water vapor), radar reflectivity maps, or gridded model outputs. **Convolutional Neural Networks (CNNs)** are the workhorses of deep learning for image processing and are exceptionally well-suited for these types of meteorological data.

- **Convolutional Neural Networks (CNNs)**
- **Core Idea:** CNNs are designed to automatically and adaptively learn spatial hierarchies of features from input images. They do this by applying small filters (kernels) that "convolve" (slide) across the input data, detecting local patterns like edges, textures, or specific shapes. These detected features are then combined in subsequent layers to form more complex representations.

- **Key Components**
- **Convolutional Layers:** These are the heart of a CNN. Each filter learns to detect a specific feature (e.g., a vertical edge, a cloud texture). Multiple filters are applied, and their outputs form "feature maps."
  - *Analogy:* Imagine small magnifying glasses scanning across a weather map, each looking for a specific pattern (e.g., a strong temperature gradient, a circular radar echo).
- **Pooling Layers (e.g., Max Pooling):** These layers reduce the spatial dimensions of the feature maps, making the model more robust to small shifts or distortions in the input and reducing computational load. They essentially summarize the presence of a feature in a region.
  - *Analogy:* After finding many small features, pooling combines them into a more general summary for a larger area.
- **Activation Functions (e.g., ReLU):** Introduce non-linearity into the network, allowing it to learn more complex relationships.
- **Fully Connected Layers:** After several convolutional and pooling layers have extracted high-level features, these traditional neural network layers take the flattened feature maps and perform the final classification or regression based on the learned features.
- **Why CNNs are good for Satellite/Radar Data**
  - **Spatial Invariance:** CNNs can detect patterns regardless of their exact position in the image (e.g., a storm cell can be detected whether it's in the top-left or bottom-right).
  - **Feature Hierarchy:** They learn hierarchical features, from simple (e.g., cloud edges) to complex (e.g., storm structures, frontal boundaries).
  - **Direct from Raw Data:** They can learn directly from raw pixel data, reducing the need for manual feature engineering.

9.11.3 Meteorological Examples of CNNs for Image Processing

- **Cloud Classification**
  - **Input:** Satellite images (visible, infrared, water vapor channels).
  - **Task:** Classifying cloud types (e.g., Cumulus, Stratus, Cirrus, Thunderstorm clouds).
  - **Benefit:** Automates a tedious task, provides consistent classification, and can identify subtle variations.

- **Severe Weather Detection and Nowcasting**
  - **Input:** Radar reflectivity images, Doppler velocity data, satellite imagery.
  - **Task:** Detecting and predicting the short-term evolution of severe weather phenomena like tornadoes, hail, heavy rainfall, or mesocyclones.
  - **Benefit:** Can identify signatures of severe weather faster and potentially more accurately than human analysis or traditional algorithms, leading to earlier warnings.
- **Hurricane Intensity Estimation**
  - **Input:** Satellite images of tropical cyclones.
  - **Task:** Estimating hurricane intensity (e.g., maximum sustained winds, Saffir-Simpson category) by analyzing cloud patterns, eye structure, and outflow.
  - **Benefit:** Provides rapid, objective intensity estimates, especially in data-sparse ocean regions.
- **Atmospheric River Detection**
  - **Input:** Integrated Water Vapor Transport (IVT) maps from reanalysis or model data.
  - **Task:** Identifying and tracking atmospheric rivers, which are crucial for heavy precipitation events.
  - **Benefit:** Automates detection and can provide early warnings for associated flood risks.
- **Wildfire Smoke Detection**
  - **Input:** Satellite imagery.
  - **Task:** Identifying and tracking smoke plumes from wildfires.
  - **Benefit:** Aids in air quality forecasting and emergency response.

9.11.4 Neural Networks for Time Series Prediction

Meteorological data is inherently sequential; observations and forecasts evolve over time. Predicting future states based on past sequences is a critical task. **Recurrent Neural Networks (RNNs)** and their more advanced variants, **Long Short-Term Memory (LSTM) networks**, are specifically designed for handling sequential data.

- **Recurrent Neural Networks (RNNs)**
- **Core Idea:** Unlike traditional neural networks that process inputs independently, RNNs have “memory.” They process sequences by

passing information from one step in the sequence to the next. This allows them to learn temporal dependencies and context.

- **How it works:** An RNN has a "recurrent" connection that feeds the output (or hidden state) from the previous time step back into the network as an input for the current time step. This internal loop enables it to maintain a "memory" of past information.
    - *Analogy:* Imagine a person reading a sentence. To understand the current word, they need to remember the words that came before it. An RNN does something similar, using its internal state to remember context from previous time steps.
- **Limitations of basic RNNs:** They often struggle with **long-term dependencies**, meaning they might forget information from many steps ago in a long sequence (the "vanishing gradient problem").
- **Long Short-Term Memory (LSTM) Networks**
- **Core Idea:** LSTMs are a special type of RNN designed to overcome the vanishing gradient problem and effectively learn long-term dependencies. They achieve this through a more complex internal structure called a "cell state" and "gates."
- **Key Components (Gates):** LSTMs use three types of "gates" that regulate the flow of information into and out of the cell state:
    - **Forget Gate:** Decides what information to discard from the cell state.
    - **Input Gate:** Decides what new information to add to the cell state.
    - **Output Gate:** Decides what information from the cell state to output at the current time step.
- **Why LSTMs are good for Time Series Data**
    - **Long-Term Memory:** Their gate mechanism allows them to selectively remember or forget information over long sequences, making them ideal for capturing complex temporal patterns in meteorological data.
    - **Handling Variable-Length Sequences:** Can process sequences of different lengths.
    - **Capturing Non-linear Dynamics:** Like other neural networks, they can model non-linear relationships over time.

9.11.5 Meteorological Examples of RNNs/LSTMs for Time Series Prediction

- **Local Weather Forecasting**
    - **Input:** Historical time series of temperature, humidity, pressure, wind speed, etc., from a specific weather station.

- **Task:** Predicting future values of these variables for the next few hours or days.
- **Benefit:** Can capture complex diurnal cycles, seasonal trends, and short-term fluctuations more effectively than traditional statistical time series models.

- **Climate Index Prediction**
  - **Input:** Time series of climate indices (e.g., ENSO indices, NAO index).
  - **Task:** Forecasting the future state of these indices.
  - **Benefit:** Aids in sub-seasonal to seasonal climate outlooks, as these indices strongly influence regional weather patterns.
- **Atmospheric Pollutant Concentration Forecasting**
  - **Input:** Time series of pollutant concentrations (e.g., PM2.5, Ozone), along with meteorological variables (wind, temperature, humidity).
  - **Task:** Predicting future pollutant levels.
  - **Benefit:** Crucial for public health warnings and environmental management.
- **River Flow/Flood Prediction:**
  - **Input:** Time series of upstream rainfall, river levels, and snowmelt.
  - **Task:** Forecasting downstream river discharge and potential flood events.
  - **Benefit:** Supports water resource management and flood preparedness.
- **Wind Power Forecasting**
  - **Input:** Historical wind speed and direction data.
  - **Task:** Predicting future wind speeds at wind farm locations.
  - **Benefit:** Crucial for optimizing renewable energy generation and grid management.

9.11.6 Conclusion

Deep learning, particularly through the application of Convolutional Neural Networks for image processing and Recurrent Neural Networks (especially LSTMs) for time series prediction, represents a paradigm shift in meteorological data analysis. These powerful techniques enable meteorologists to extract unprecedented insights from vast and complex datasets, automate challenging tasks, and ultimately improve the accuracy and timeliness of weather and climate predictions. While challenges related to interpretability and computational resources persist, the synergistic integration of deep learning with physics-based models is paving the way for a new era of understanding and forecasting our dynamic atmosphere.

9.12 Data Assimilation Fundamentals: Integrating Observations into Numerical Models

Data assimilation is a sophisticated and absolutely crucial process in modern meteorology and oceanography. Think of it as the intelligent bridge that connects our understanding of the atmosphere (represented by computer models) with real-world measurements (observations). Its main job is to **combine all available observations of the Earth system with a numerical model's forecast to create the "best possible" picture of the current state of the atmosphere or ocean.**This "best picture," called the **analysis**, then becomes the starting point for the next weather forecast.

9.12.1 Why is Data Assimilation So Important?

We need data assimilation because neither our computer models nor our observations are perfect:

- **Our Numerical Models are Good, but Not Perfect**
    - **Simplifications:** Weather models are based on the laws of physics, but they have to make simplifications and use approximations, especially for complex processes like how clouds form or how tiny air movements behave.
    - **Limited Detail:** Models can only represent the atmosphere down to a certain level of detail (resolution). They can't capture every tiny swirl or gust of wind.
    - **Sensitive Starting Points:** Even a perfect model would give a wrong forecast if it started from an inaccurate description of the current weather.
- **Our Observations are Real, but Imperfect and Incomplete**
    - **Measurement Errors:** Every instrument has some degree of error, whether it's a thermometer on the ground or a satellite orbiting Earth.
    - **Gaps in Coverage:** We don't have sensors everywhere at all times. There are vast areas, like over the oceans or remote polar regions, where observations are sparse. Also, observations are taken at specific times, not continuously.
    - **Indirect Information:** Satellites, for instance, often measure radiation, which then needs to be converted into temperature or humidity. This conversion process can introduce its own uncertainties.

Data assimilation steps in to intelligently merge these two imperfect sources of information. It leverages the strengths of observations (which provide real-world data) and models (which provide a physically consistent and complete picture across space and time). Without it, even the most advanced weather models would quickly drift away from reality because their starting conditions would be wrong.

9.12.2 The Core Idea: Finding the Optimal Balance

At its heart, data assimilation is about finding the **sweet spot** between two pieces of information:

- **The "Background" or "First Guess":** This is the short-range weather forecast from the numerical model. It represents what the model thinks the atmosphere looks like *before* any new observations are considered. It's consistent with the laws of physics that the model follows.
- **The "Observations":** These are the new measurements that have just come in from satellites, weather balloons, radar, and ground stations. They offer real-world updates but can be noisy or incomplete.

The data assimilation system then calculates an **«Analysis»**, which is a carefully weighted average of this "background" and the "observations." The weights are crucial:

- If the model's forecast (background) is believed to be very accurate, and the new observations are thought to be noisy, the system will trust the model more.
- If the observations are considered highly accurate, and the model's forecast is uncertain, the system will lean more heavily on the observations.

This sophisticated weighting is determined by how much uncertainty or "error" we believe exists in both the model's forecast and the observations. The system aims to minimize the overall "misfit" between its final analysis and both the background and the observations, giving more weight to the more reliable information.

9.12.3 How Observations are Integrated into Numerical Models (Key Methods)

The process of integrating observations involves comparing what the model *would* see if it were an observer to what the actual observations show. Then, the model's state is adjusted to reduce this difference.

Here are the main ways this is done:

- **Variational Methods (3D-Var and 4D-Var)**

  These methods treat data assimilation like a puzzle, trying to find the single best atmospheric state that fits both the model's forecast and the observations as closely as possible.

- **3D-Var (Three-Dimensional Variational Assimilation)**
  - **Concept:** This method creates a single "snapshot" of the atmosphere at a specific moment in time. It takes all observations collected within a short window around that time and treats them as if they all occurred at that exact moment.
  - **How it Integrates Observations:** It adjusts the model's initial state (the background) by pulling it towards the observations. The amount of pull is determined by how confident we are in the model versus the observations. This adjustment happens iteratively, meaning the system makes small changes, checks how well it fits, and then makes more changes until it finds the best fit. It uses pre-calculated information about typical model errors.
  - **Limitation:** Because it treats all observations as if they're from one moment, it doesn't fully account for how the atmosphere changes during the assimilation window.

- **4D-Var (Four-Dimensional Variational Assimilation)**
  - **Concept:** This is a more advanced version that considers observations over a **period of time** (e.g., a 6-hour or 12-hour window). It tries to find the best starting point for the model at the *beginning* of this window, such that when the model runs forward, its forecast perfectly matches all the observations at their *actual times*within that window.
  - **How it Integrates Observations:** The clever part here is using something called an **adjoint model**. This is a special mathematical tool that efficiently figures out how sensitive the initial conditions are to changes caused by observations later in the time window. This allows information from observations taken at the end of the window to influence the analysis at the very beginning, creating a much more physically consistent and realistic starting point for the forecast.
  - **Benefit:** Provides a dynamically consistent analysis because it explicitly accounts for the atmosphere's evolution over time. It's considered the most advanced method in many major weather centers.
  - **Limitation:** Extremely complex and computationally intensive, requiring significant computing power.

- **Kalman Filter (KF) and Ensemble Kalman Filter (EnKF)**

  These methods are like continuous updates, adjusting the model's state and its estimated uncertainty as each new observation comes in.

- **Kalman Filter (KF)**
  - **Concept:** This is an ideal method for simpler, linear systems. It keeps track of both the atmosphere's state and how uncertain that state is. When new observations arrive, it uses a formula to update both the state and its uncertainty.
  - **Limitation:** It's too computationally demanding for complex global weather models because it would need to track the uncertainty for every single variable and how they relate to each other, which is an astronomical number. Also, the atmosphere is not perfectly linear.

- **Ensemble Kalman Filter (EnKF)**
  - **Concept:** This is a practical and widely used adaptation of the Kalman Filter. Instead of trying to calculate the exact uncertainty for every variable, it uses a **group (an "ensemble") of slightly different model forecasts**. The spread and relationships among these ensemble members give a good estimate of the current uncertainties in the model's forecast.
  - **How it Integrates Observations:** When new observations arrive, each member of the ensemble is individually adjusted to be closer to the observations. This adjustment is guided by the uncertainties estimated from the ensemble's spread. This process ensures that the ensemble members are nudged towards reality while still maintaining their natural spread, which represents the forecast uncertainty.
  - **Benefit:** Can handle the non-linear nature of the atmosphere and is computationally feasible for large weather models. It also naturally provides an estimate of how uncertain the forecast is.
  - **Limitation:** The quality of the uncertainty estimate depends on how many ensemble members are used, and special techniques are needed to prevent the ensemble members from becoming too similar over time.
  - **Hybrid Methods:** Many modern weather forecasting centers use a combination of these approaches, known as **hybrid methods**. For example, they might use a 4D-Var system but enhance it by incorporating the more realistic, flow-dependent uncertainty information derived from an ensemble (like in EnKF). This allows

them to get the best of both worlds: the dynamic consistency of 4D-Var with the more accurate uncertainty estimates from ensembles.

9.12.4 The Continuous Assimilation Cycle

Data assimilation isn't a one-off task; it's a continuous, repeating cycle that keeps the weather models aligned with reality:

- **Forecast:** The numerical model runs a short forecast (e.g., 6 or 12 hours) starting from the previous analysis. This becomes the "background" for the next step.
- **Observation Collection:** During this forecast period, new observations are continuously gathered from all available sources.
- **Analysis:** The data assimilation system then takes the background forecast and intelligently blends it with these new observations to create the "analysis"– the best possible estimate of the atmosphere's current state.
- **Initialization:** This new analysis then becomes the **starting point** (initial conditions) for the next short-range forecast run by the numerical model.

This cycle repeats constantly, ensuring that weather forecasts are regularly updated and corrected by incorporating the latest real-world information.

9.12.5 Impact and Importance

Data assimilation is truly the unsung hero behind the accuracy of today's weather forecasts. By optimally combining the power of physics-based models with the reality of observations, it provides the crucial starting conditions that allow numerical weather prediction models to produce reliable and skillful predictions. It's a highly active area of research, constantly evolving to integrate new types of observations, improve how errors are understood, and boost computational efficiency, all with the goal of pushing the boundaries of how well we can predict weather and climate.

9.13 Conclusion

Concluding our journey through advanced analytic techniques, it is evident that these sophisticated methodologies are no longer merely supplementary tools but have become **indispensable pillars** in modern meteorology and climate science. We've seen how techniques like Principal Component Analysis (EOFs) and Cluster Analysis enable us to distill vast, high-dimensional datasets into comprehensible patterns and distinct weather regimes, providing unprecedented insights into the dominant modes of atmospheric and oceanic variability. Furthermore, the power of Supervised and Unsupervised Learning, including the transformative capabilities of Deep Learning, has demonstrated how we can extract complex relationships, make highly accurate predictions, and even identify subtle anomalies directly from raw meteorological data, from satellite imagery to time series.

The integration of observations into numerical models through advanced Data Assimilation techniques underscores the critical synergy between physics-based modeling and data-driven approaches. By intelligently blending the strengths of both, we continuously refine our understanding of the Earth system's current state, providing the most accurate possible initial conditions for forecasting. As meteorological data continues to grow in volume, velocity, and variety, and as our computational capabilities expand, these advanced analytic techniques will undoubtedly continue to evolve, pushing the boundaries of scientific discovery, enhancing our predictive capabilities, and ultimately strengthening our resilience in the face of a dynamic and changing climate.

10

Visualization of Meteorological Data

10.1 Introduction

Having explored the advanced analytical techniques that allow us to extract meaningful insights and build predictive models from meteorological data, we now arrive at a crucial stage in the scientific process: **communication and understanding**. The chapter, "Visualization of Meteorological Data," focuses on the art and science of transforming complex numerical datasets into insightful visual representations. In a field as inherently spatial and temporal as meteorology, effective visualization is not just about making pretty pictures; it is a fundamental tool for **exploration, analysis, and the clear conveyance of information** about atmospheric phenomena (Fig. 10.1).

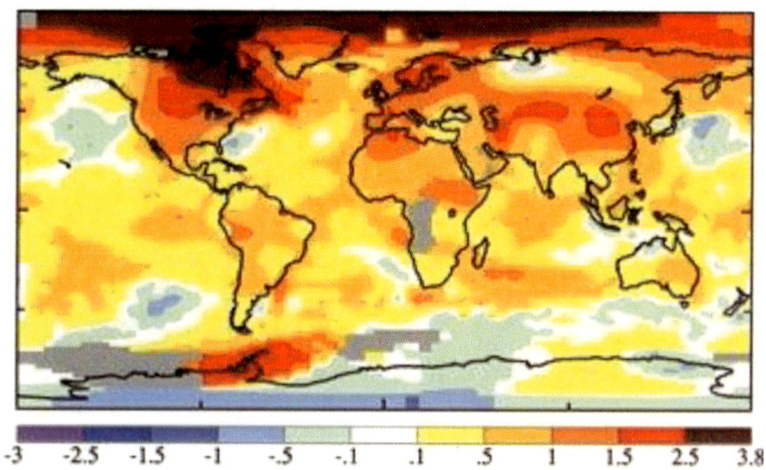

Fig. 10.1: Visualization of weather data

Consider the sheer volume and multifaceted nature of meteorological data. We grapple with variables spanning three spatial dimensions (and often more when considering altitude), evolving dynamically over time, and encompassing a wide array of physical quantities – temperature, pressure, wind vectors, humidity, precipitation rates, radiation fluxes, and more. Raw numbers and statistical summaries, while essential for quantitative analysis, often fail to reveal the intricate spatial patterns, temporal evolutions, and interrelationships that characterize weather and climate. This is where visualization steps in, acting as a bridge between the abstract world of numerical data and our intuitive human capacity for visual pattern recognition.

This chapter will delve into the diverse techniques used to visualize meteorological data, ranging from fundamental methods like contour plots and time series graphs to more advanced approaches such as vector fields, animations, and 3D renderings. We will explore how different visualization methods are suited to specific types of data and the questions we aim to answer. For instance, a contour map might beautifully illustrate the spatial distribution of temperature across a region, while an animation could vividly depict the movement and intensification of a hurricane over time. Vector fields are indispensable for understanding wind patterns, and satellite imagery provides a direct visual snapshot of cloud formations and land surface conditions.

Beyond simply displaying data, effective meteorological visualization aims to **highlight key features, reveal underlying trends, and facilitate the identification of anomalies or significant events**. A well-designed visualization can instantly convey information that might take pages of text or tables to describe. It can aid researchers in their exploratory analysis, allowing them to visually identify patterns that might warrant further statistical investigation. Moreover, clear and compelling visualizations are paramount for communicating weather forecasts, climate projections, and scientific findings to a broader audience, including policymakers and the public, fostering understanding and informed decision-making.

In this chapter, we will not only introduce various visualization methods but also discuss principles of effective design, emphasizing clarity, accuracy, and the importance of tailoring visualizations to their intended purpose and audience. We will touch upon the tools and technologies commonly used in meteorological visualization, from standard plotting libraries to specialized software packages. Ultimately, this chapter aims to equip you with the knowledge and appreciation for how the visual representation of meteorological data transforms abstract numbers into tangible insights, enabling us to better understand and communicate the dynamic beauty and power of our atmosphere. Prepare to see the data in a whole new light.

10.2 Principles of Effective Data Visualization

Effective data visualization is far more than just generating a graph; it's a powerful form of communication, a scientific tool for discovery, and an art form that transforms abstract data into understandable insights. In meteorology, where datasets are inherently complex, multi-dimensional, and often critical for immediate decision-making, adhering to robust visualization principles is not merely good practice—it's essential for accuracy, clarity, and impact. These principles serve as guiding lights, ensuring that the visual stories we tell with meteorological data are truthful, compelling, and actionable.

Here are the key principles for creating truly effective data visualizations in meteorology, with further elaboration:

10.2.1 Clarity and Simplicity (Minimize Clutter and Maximize the Data-Ink Ratio)

The paramount goal of any visualization is to convey information with maximum clarity and minimum effort from the viewer. This means stripping away anything that doesn't directly contribute to the message.

- **Focus on the Signal, Not the Noise:** Every line, label, color, and shape in your visualization should serve a purpose. If an element doesn't help the viewer understand the data or the story you're telling, it's clutter.
- **The Data-Ink Ratio (Tufte's Principle):** Edward Tufte, a pioneer in data visualization, introduced the concept of the "data-ink ratio," which suggests that a large proportion of the ink (or pixels) on a graphic should be used to represent data information, rather than non-data elements.
- **Meteorological Example:** Instead of a radar reflectivity map with a busy background texture, excessive grid lines, and an overly decorative title, a clear map would focus on the radar echoes themselves, using subtle geographical outlines and minimal, well-placed labels. Remove unnecessary borders, shadows, or 3D effects on 2D plots that add no informational value.
- **Legibility:** Ensure all text (titles, labels, legends) is easily readable, with appropriate font sizes and contrasts against the background.

10.2.2 Accuracy and Integrity (Avoid Distortion and Misrepresentation)

A visualization must truthfully represent the underlying data. Any visual distortion, whether intentional or accidental, can lead to incorrect conclusions and erode trust.

- **Proportionality is Key:** The visual magnitude of an element must directly correspond to the numerical value it represents. If a bar representing 10 mm of rain is half the height of a bar representing 20 mm, that's proportional. If it's one-third the height, it's misleading.
- **Appropriate Axis Scaling**
  - **For Bar Charts:** Axes representing quantities should generally start at zero to avoid exaggerating differences. If a non-zero baseline is

used, it must be clearly indicated and justified (e.g., for very small variations around a large mean, but even then, alternative chart types might be better).

- **For Line Charts (Trends):** While starting at zero might not always be necessary for trends, the scale should encompass the full range of relevant data points without overly compressing or stretching the variations.
- **Meteorological Example:** When showing temperature anomalies over decades, starting the y-axis at 0° Cmight obscure subtle but important warming trends. However, for total annual precipitation, starting the y-axis at 0 mm is crucial to accurately represent accumulated amounts.
- **Fair Representation of Spatial Averages:** Be mindful of how spatial averages are displayed, especially on maps. Ensure that the visual representation (e.g., color intensity) accurately reflects the average value over the represented area.
- **Contextualization:** Always provide sufficient context: units of measurement (°C, hPa, m/s), timeframes (daily, monthly, annual), geographical extent, and data sources. This allows viewers to interpret the data correctly.

10.2.3 Purpose-Driven Design (Know Your Audience and Goal)

Before you even begin designing, clearly define *why* you are creating the visualization and *who* your audience is. This dictates every design choice.

- **Exploration vs. Explanation**
    - **Exploratory Visualizations:** For your own analysis, these might be highly interactive, dense with detail, and allow for rapid slicing, dicing, and drilling down into the data. The goal is to discover patterns.
    - **Explanatory Visualizations:** For communicating findings to others, these should be highly curated, focused on a single message, and designed for clarity and impact. The goal is to convey a specific insight.
- **Audience Needs**
    - **Fellow Scientists/Meteorologists:** Can handle technical jargon, complex multi-panel plots, and detailed statistical annotations. They seek precision and depth.

- **Policymakers/Emergency Managers:** Need clear, concise, actionable information. Focus on impacts, probabilities, and decision-relevant thresholds. Simplicity and directness are key.
- **General Public:** Requires highly intuitive, jargon-free visuals. Analogies and relatable scales (e.g., "feels like" temperature) can be helpful. Focus on the "what" and "so what."
- **Meteorological Example:** A research paper on atmospheric blocking might use a complex geopotential height anomaly map with multiple contour levels and statistical significance shading. A public weather forecast, however, would use a simplified temperature map with broad color zones and clear city labels.

10.2.4 Appropriate Chart Type Selection

The choice of chart type is fundamental. Each type is designed to convey specific types of relationships or data structures most effectively.

- **Temporal Data**
    - **Line Charts:** Ideal for showing trends and changes over continuous time (e.g., hourly temperature fluctuations, climate change over decades).
    - **Bar Charts:** Good for comparing discrete values over time (e.g., monthly precipitation totals, annual average temperatures).
    - **Hovmöller Diagrams:** Specialized for showing the evolution of a variable in one spatial dimension over time (e.g., propagation of a wave across longitude).
    - **Spaghetti Plots:** Used in ensemble forecasting to show the spread and possible trajectories of different model runs.
- **Spatial Data**
    - **Contour Maps (Isopleths):** Excellent for visualizing continuous fields like temperature, pressure, or humidity across a geographical area. Lines connect points of equal value.
    - **Choropleth Maps:** Use color shading to represent aggregated data over predefined geographical regions (e.g., average rainfall by county, drought intensity by state).
    - **Vector Plots (Quiver Plots):** Essential for visualizing wind speed and direction, where arrows indicate flow.
    - **Streamlines:** Show the instantaneous direction of flow, often used for wind or ocean currents.

    - **Satellite/Radar Imagery:** Direct visual representation of clouds, precipitation, and land surface features.
- **Relationships & Distributions**
    - **Scatter Plots:** To show the relationship between two continuous variables (e.g., temperature vs. humidity).
    - **Histograms/Density Plots:** To show the distribution of a single variable (e.g., frequency of different wind speeds).
    - **Box Plots:** To compare distributions across different categories (e.g., temperature distribution for different seasons).
    - **Wind Roses:** Specialized plots for showing the distribution of wind speed and direction.

10.2.5 Effective Use of Color

Color is a powerful visual variable, but misuse can lead to confusion or misinterpretation.

- **Purposeful Encoding:** Use color to encode information (e.g., higher values are darker blue), highlight key elements (e.g., a warning threshold in red), or differentiate categories (e.g., different storm types).
- **Perceptual Uniformity:** For continuous data (like temperature, pressure, or precipitation amounts), use **sequential** (e.g., light blue to dark blue for increasing precipitation) or **diverging** (e.g., blue-white-red for temperature anomalies, where white is zero and blue/red indicate negative/positive deviations) color palettes.
- **Avoid "Rainbow" (Jet) Palettes:** These are perceptually non-uniform, meaning equal steps in data values don't correspond to equal perceived changes in color. They can create artificial boundaries and obscure details, making it harder to discern gradients.
- **Accessibility (Colorblindness):** Roughly 8% of males and 0.5% of females have some form of colorblindness. Choose color palettes that are distinguishable by individuals with common forms of colorblindness (e.g., avoid red-green combinations for critical distinctions). Tools exist to check colorblind-friendliness.
- **Consistency:** Use the same color scheme for the same variables across all your visualizations within a report or presentation. This builds familiarity and reduces cognitive load for the viewer.

10.2.6 Clear Labels, Legends, and Annotations

Visualizations rarely stand alone. They need textual context to be fully understood.

- **Descriptive Titles:** A concise title that summarizes the visualization's main message or content (e.g., "Hourly Temperature and Dew Point at Station XYZ,""Projected Global Mean Sea Level Rise by 2100").
- **Axis Labels and Units:** Clearly label both axes with the variable name and its units (e.g., "Temperature (°C)", "Precipitation (mm/day)", "Time (UTC)").
- **Legends:** Essential for explaining what different colors, symbols, line styles, or patterns represent on the plot.
- **Annotations:** Use text annotations to draw attention to specific data points (e.g., "Record High Temperature"), explain anomalies, highlight significant events, or provide brief explanations directly on the plot.
- **Source Information:** Always include the source of the data and any relevant credits.

10.2.7 Interactivity (Where Applicable)

For digital visualizations, interactivity can transform a static image into a dynamic exploration tool.

- **Zoom and Pan:** Allow users to explore specific regions of a map or segments of a time series in detail.
- **Tooltips/Hover Information:** Display precise numerical values or additional context when a user hovers over a data point or region.
- **Filtering and Sorting:** Enable users to customize the view based on specific criteria (e.g., show data only for a certain year, filter by a specific weather phenomenon).
- **Animation/Time Sliders:** For temporal data, animation is incredibly powerful for showing evolution and movement (e.g., storm tracks, atmospheric flow patterns, model forecast progression). A time slider allows users to control the animation speed or jump to specific time steps.
- **Layer Toggling:** On maps, allow users to toggle different data layers (e.g., temperature, pressure, wind vectors) on and off.

10.2.8 Consistency Across Visualizations

Maintaining a consistent visual language across multiple figures or within a larger report is crucial for building viewer trust and reducing cognitive load.

- **Uniformity in Design:** Use consistent font styles and sizes, color palettes, line weights, and symbol choices for similar elements across all your visualizations.

- **Layout and Formatting:** Maintain a consistent layout and formatting for similar types of charts (e.g., all time series plots have the same margin and axis style).
- **Meteorological Example:** If red represents high temperatures in one map, it should represent high temperatures in all related maps. If a dashed line represents a model forecast, it should consistently do so.

10.2.9 Accessibility

Design visualizations that are understandable by the widest possible audience, including individuals with disabilities.

- **High Contrast:** Ensure sufficient contrast between data elements and their backgrounds, and between text and background.
- **Colorblind-Friendly Palettes:** Reiterate the importance of choosing color palettes that are distinguishable by individuals with common forms of colorblindness.
- **Redundant Encoding:** Do not rely solely on color to convey critical information. Use shape, pattern, line style, or direct text labels in addition to color.
- **Descriptive Text:** For web-based or digital visualizations, provide alternative text descriptions for images for screen readers.
- **Clear Language:** Use clear, concise language in titles and labels, avoiding overly technical jargon when the audience is not expert.

10.2.10 Iterative Process and Feedback

Visualization is a design process, not a one-time task. Rarely is the first attempt the best.

- **Draft and Refine:** Create initial drafts, critically evaluate them, and be prepared to make significant revisions.
- **Seek Feedback:** Share your visualizations with colleagues, mentors, or members of your target audience. Ask specific questions about clarity, interpretation, and potential misreadings.
- **Test with Audience:** If possible, observe your target audience interacting with your visualization to identify areas of confusion or difficulty.

By diligently applying these principles, meteorologists can create powerful and insightful visualizations that not only communicate complex atmospheric information effectively but also drive deeper scientific understanding, facilitate collaboration, and support informed decision-making in a world increasingly reliant on accurate environmental intelligence.

10.3 Map Projections in Meteorological Data Visualization

When meteorologists and climate scientists visualize data, they are almost invariably dealing with information that originates from the Earth's curved, three-dimensional surface. Whether it's temperature readings, wind vectors, or precipitation accumulations, these phenomena exist on a sphere (or, more precisely, an oblate spheroid, slightly flattened at the poles). However, our primary tools for display—computer screens, printed maps, and digital displays—are all flat, two-dimensional surfaces (Fig. 10.2). The essential process of transforming this curved geographical information onto a flat plane is known as a **map projection**.

Fig. 10.2: Map projections in weather data visualization

10.3.1 The Fundamental Challenge: Why Distortion is Inevitable

The core problem of map projections is a fundamental mathematical impossibility: **you cannot flatten a curved surface, especially a sphere, onto a flat plane without introducing some form of distortion.** This is a concept often illustrated by imagining peeling an orange and attempting to lay its skin perfectly flat on a table. You would inevitably have to stretch, tear, or compress parts of the peel to achieve a flat surface.

Similarly, every map projection will distort at least one, and typically several, of the following geographical properties:

- **Area (Equivalence):** The true relative size of landmasses or features. If a country is genuinely twice the size of another on the globe, an equal-area projection will show it as twice the size on the map.
- **Shape (Conformality/Orthomorphism):** The true shape of geographical features and the angles between lines (e.g., meridians and

parallels intersecting at right angles). A conformal map preserves local shapes, meaning a small island will look the same on the map as it does on the globe.

- **Distance (Equidistance):** The true distances between points. An equidistant map preserves distances accurately from one or two specific points, or along specific lines (e.g., along all meridians).
- **Direction (Azimuthality):** The true bearings or directions from one point to another. An azimuthal map preserves true directions from a central point to all other points.

The choice of map projection is therefore a deliberate trade-off. It depends entirely on which of these properties is most critical to preserve for the specific meteorological visualization, and which types of distortion are acceptable for the intended analysis or communication.

10.3.2 Key Properties Preserved by Different Projections

Different map projections are mathematically constructed to optimize for the preservation of one or more of these properties, inevitably distorting others:

- **Conformal (or Orthomorphic) Projections**
  - **What they preserve:** These projections preserve local shapes and angles. This means that at any given point on the map, the scale is the same in all directions, and features appear locally as they would on the globe. Meridians (lines of longitude) and parallels (lines of latitude) will intersect at right angles, just as they do on the Earth's surface.
  - **What they distort:** They significantly distort area, especially as you move further away from the projection's standard lines or points (e.g., the equator in a Mercator projection, or the standard parallels in a Lambert Conformal Conic).
  - **Why it's useful in Meteorology:** Conformal projections are **crucial for meteorological analysis** where the true shape of weather features (like fronts, storm systems, or the curvature of jet streams) is important. They are also vital for accurately representing **wind direction** relative to other features, as the angles between wind vectors and geographical lines are preserved. This is essential for understanding atmospheric dynamics and for synoptic (large-scale weather) analysis.
- **Equal-Area (or Equivalent) Projections**
- **What they preserve:** These projections preserve the true relative size of all areas. If a specific landmass or ocean region covers a certain

percentage of the Earth's surface, it will cover the same percentage of the map's surface.

- **What they distort:** They inevitably distort shape, angle, and often distance, particularly at the edges of the map or far from the central meridian.
- **Why it's useful in Meteorology:** Equal-area projections are **essential for visualizing quantitative data where the spatial extent, density, or accumulation matters**. This includes maps of total precipitation accumulation over a season, population exposure to weather hazards, carbon emissions per unit area, or the spatial distribution of energy fluxes. Using an equal-area projection ensures that regions with high values don't appear disproportionately large (or small) due to projection distortion, giving a true visual sense of their actual contribution.

- **Equidistant Projections**
  - **What they preserve:** These projections preserve true distances from one or two specific points, or along specific lines (e.g., along all meridians).
  - **What they distort:** They distort other properties like area, shape, and angles.
  - **Why it's useful in Meteorology:** While less commonly used for general weather maps, equidistant projections can be valuable for **specific applications where true distances from a central point are critical**. For example, visualizing the range of a weather radar, or showing distances from a central forecast office for emergency response planning.
- **True-Direction (or Azimuthal) Projections**
  - **What they preserve:** These projections preserve true directions (bearings) from a central point to any other point on the map. A straight line drawn from the center of the map to any other point will represent the shortest path (great circle route) and its true compass bearing.
  - **What they distort:** They distort other properties like area, shape, and distance as you move away from the central point.
  - **Why it's useful in Meteorology:** Often used for **polar maps** where the central point is a pole, making it easy to visualize circumpolar atmospheric flow. They can also be useful for visualizing phenomena

that radiate outwards from a central point, though this is less common than other uses.

10.3.3 Common Map Projections Used in Meteorology

Meteorologists frequently rely on a select few map projections, each chosen for its specific strengths and suitability for particular geographical regions or global phenomena:

- **Mercator Projection (Cylindrical, Conformal)**
    - **Concept:** Imagine a cylinder wrapped around the globe, typically tangent at the equator. Lines of longitude and latitude are projected onto this cylinder, which is then unrolled into a flat rectangle.
    - **Properties:** It is perfectly conformal, meaning it preserves local shapes and angles. A unique feature is that lines of constant compass bearing (rhumb lines) appear as straight lines, which historically made it invaluable for navigation.
    - **Distortion:** Its most famous distortion is the **extreme exaggeration of area towards the poles**. For example, Greenland appears much larger than South America, even though South America is actually about eight times larger. Antarctica appears as a massive, stretched-out landmass across the bottom of the map.
    - **Use in Meteorology:** While historically popular for navigation, its severe area distortion makes it **less ideal for displaying global meteorological fields** where the relative sizes of features are important (e.g., comparing the size of a storm in the tropics to one in the Arctic). However, it is sometimes used for global patterns where shape preservation near the equator is desired, or for displaying data that is inherently rectangular (like some satellite imagery grids).
- **Lambert Conformal Conic (Conic, Conformal):**
    - **Concept:** Imagine placing a cone over the globe, typically intersecting the globe at two specific "standard parallels" (lines of latitude). Features are projected onto the cone, which is then unrolled into a flat, fan-shaped map.
    - **Properties:** It is conformal, preserving local shapes and angles, particularly well within the region between its two standard parallels. Distances are true along these standard parallels. Meridians are straight lines converging at a pole, and parallels are concentric arcs.
    - **Distortion:** Area distortion increases as you move away from the standard parallels.

- **Use in Meteorology:** This is an **extremely common and preferred projection for regional weather maps in mid-latitude areas** like North America, Europe, or China. Its ability to preserve the shapes of weather systems (such as fronts, cyclones, and anticyclones) and their angles is crucial for understanding their dynamics and for operational forecasting. Its conic shape naturally fits the elongated east-west extent of many mid-latitude weather systems.

- **Polar Stereographic (Azimuthal, Conformal)**
  - **Concept:** This projection is created by projecting points from one pole (e.g., the South Pole) onto a flat plane that is tangent to the globe at the opposite pole (e.g., the North Pole).
  - **Properties:** It is conformal, preserving shapes and angles around the central pole. Meridians appear as straight lines radiating outwards from the pole, and parallels are concentric circles centered on the pole.
  - **Distortion:** Area distortion increases significantly as you move away from the central pole towards the equator.
  - **Use in Meteorology:Widely used for polar and hemispheric weather maps** (e.g., Arctic or Antarctic weather, Northern Hemisphere 500 hPa charts). It's ideal for visualizing circumpolar atmospheric flow patterns (like the polar vortex or jet stream around the pole) and for tracking high-latitude weather systems where shape preservation is important.
  - Plate Carrée (Equirectangular) Projection (Cylindrical, Equidistant along Meridians):
  - **Concept:** This is the simplest projection, where latitude and longitude are directly mapped to x and y coordinates, respectively. It's like simply drawing a grid of squares on a flat piece of paper and then placing the Earth's features onto it.
  - **Properties:** It results in a rectangular grid where meridians are equally spaced vertical lines and parallels are equally spaced horizontal lines. Distances along meridians are true.
  - **Distortion:** It suffers from **significant shape and area distortion**, especially at higher latitudes. Regions near the poles appear stretched horizontally.
  - **Use in Meteorology:** Often used for **global gridded datasets** (e.g., raw model output, reanalysis data, or satellite data that is already processed into a lat/lon grid) because of its computational simplicity

in mapping grid cells directly to pixels. However, it's generally **not recommended for visual analysis of high-latitude features** due to the severe distortion, as it can mislead interpretation of size and shape.

- **Mollweide and Goode Homolosine Projections (Equal-Area):**
  - **Concept:** These are more complex, often "pseudo-cylindrical" or "interrupted" projections specifically designed to preserve area. The Goode Homolosine, for instance, "cuts" the oceans to reduce shape distortion, resulting in a map that looks like an orange peel laid flat in sections.
  - **Properties:** Both are **equal-area**, meaning the relative sizes of all landmasses and ocean regions are accurately represented.
  - **Distortion:** They introduce significant shape distortion, particularly at the edges of the map or where the projection is "interrupted" (in the case of Goode Homolosine).
  - **Use in Meteorology:** These are **ideal for global climate visualizations** where the accurate representation of spatial extent and density is paramount. Examples include maps of global precipitation totals, carbon fluxes, vegetation cover, or the spatial distribution of climate change impacts. They ensure that your visual comparison of quantities across different regions is not skewed by projection artifacts.

10.3.4 Choosing the Right Projection for Meteorological Data

The selection of an appropriate map projection is a critical decision that directly impacts the clarity and accuracy of your meteorological visualization:

- **Geographical Extent**
- **Local/Regional (Mid-Latitudes):** Lambert Conformal Conic is generally the top choice for its balance of shape preservation and manageable distortion.
- **Polar/Hemispheric:** Polar Stereographic is the standard for its conformal properties and natural fit for circumpolar phenomena.
- **Global (Area-focused):** Mollweide or Goode Homolosine are preferred when accurate area comparison is crucial.
- **Global (Simple Grid Display):** Plate Carrée is often used for raw data display due to simplicity, but always be aware of its high-latitude distortion.

- **Variable Type and Purpose of Visualization**
- **Wind Direction/Flow (Shape/Angle Critical):** Conformal projections (Lambert, Stereographic) are best.
- **Quantitative Accumulations/Densities (Area Critical):** Equal-area projections (Mollweide, Goode Homolosine) are mandatory for truthful representation.
- **Temperature/Pressure Fields (General Patterns):** Conformal projections are often acceptable for regional views, but equal-area might be better for global comparisons of total energy.
- **Storm Tracking/Movement:** Conformal projections are usually preferred to maintain the storm's shape.
- **Preserved Property:** What is the single most crucial property for the message you want to convey? If comparing the actual size of two drought-affected areas, choose equal-area. If analyzing the precise rotation within a supercell thunderstorm, choose conformal.
- **Audience:** For general public maps, simpler, more recognizable projections might be preferred, even if they have some distortion, provided the distortion doesn't fundamentally mislead the message. For scientific analysis, the projection must rigorously and accurately represent the relevant physical properties.

10.3.5 Conclusion

Map projections are far more than just a background for meteorological data; they are an integral part of the visualization itself, fundamentally shaping how we perceive and interpret atmospheric information. While often operating silently in the background, the careful choice of projection is paramount for ensuring that our visualizations are not only aesthetically pleasing but, more importantly, scientifically accurate and effective in communicating the complex dynamics of our planet's weather and climate system. A well-chosen projection transforms raw data into a clear, truthful, and actionable visual story.

10.4 Common Meteorological Plots

Meteorological data is inherently complex, spanning multiple dimensions (spatial, temporal, vertical) and encompassing a vast array of atmospheric variables. To effectively analyze, interpret, and communicate this rich information, meteorologists rely on a specialized toolkit of visualizations. These **common meteorological plots** are meticulously designed to highlight specific aspects of the data, reveal intricate patterns, and facilitate a deeper understanding of atmospheric phenomena, from global climate trends to localized severe weather events.

Here's a detailed look at some of the most frequently used and indispensable plots in meteorology:

10.4.1 Spatial Plots (Maps)

These plots are used to visualize the distribution of meteorological variables across a geographical area at a specific point in time. They are fundamental for understanding the horizontal structure and movement of weather systems and climate patterns.

- **Contour Maps (Isopleth Maps)**
  - **Purpose:** To depict the spatial distribution of a continuous variable by drawing lines that connect points of equal value. These lines are called "isopleths" (from Greek "iso" meaning equal, and "plethos" meaning quantity).
- What it shows
- **Isobars:** Lines of equal atmospheric pressure (typically at mean sea level or a constant altitude).
  - *Interpretation:* Closely spaced isobars indicate a strong pressure gradient, which corresponds to strong winds. High-pressure systems (marked 'H') are centers of diverging flow and generally clear, stable weather. Low-pressure systems (marked 'L') are centers of converging flow, often associated with clouds, precipitation, and unsettled weather.
- **Isotherms:** Lines of equal temperature.
  - *Interpretation:* Reveal temperature gradients (areas where temperature changes rapidly over distance), which are key indicators of frontal zones (boundaries between air masses). They also highlight areas experiencing heat waves, cold snaps, or temperature inversions.
- **Isohypses (Contour Lines of Geopotential Height):** Lines of equal geopotential height on a constant pressure surface (e.g., 500 hPa or 850 hPa).
  - *Interpretation:* These upper-air maps are crucial for visualizing the large-scale atmospheric flow, including troughs (elongated areas of low height/pressure, associated with cold air and unsettled weather) and ridges (elongated areas of high height/pressure, associated with warm air and stable weather). The flow tends to be parallel to these lines, and their curvature dictates the steering of surface weather systems.

- **Isotachs:** Lines of equal wind speed.
  - *Interpretation:* Highlight areas of strong winds, such as the core of the jet stream (a narrow band of strong winds in the upper atmosphere) or localized wind maxima (jet streaks). These are important for understanding atmospheric dynamics and severe weather.
- **Isodrosotherms:** Lines of equal dew point temperature.
  - *Interpretation:* Indicate the distribution of moisture in the atmosphere. High dew points suggest abundant moisture and potential for heavy precipitation, while low dew points indicate dry air.
- **Why it's useful:** Contour maps provide a clear, intuitive visual representation of gradients, centers of highs/lows, and the overall spatial structure of atmospheric fields. They are fundamental for **synoptic analysis**, allowing meteorologists to diagnose current weather patterns and infer future developments.
- **Vector Plots (Quiver Plots)**
  - **Purpose:** To visualize vector quantities, primarily wind speed and direction, across a geographical area.
  - **What it shows:** Arrows originating from grid points or observation locations. The **length of the arrow**indicates the wind speed (longer arrow = stronger wind), and the **direction of the arrow** indicates the wind direction (where the wind is blowing *from* or *to*, depending on convention).
  - **Why it's useful:** Directly illustrates atmospheric flow patterns, revealing areas of **convergence** (winds flowing inward, often associated with rising air and precipitation), **divergence** (winds flowing outward, associated with sinking air and clear skies), and **vorticity** (the rotation of the air). These are crucial for understanding the movement of air masses, the development of cyclones and anticyclones, and the steering of weather systems.
- **Streamlines**
  - **Purpose:** To show the instantaneous direction of flow (e.g., wind) at various points in a fluid. Unlike vector plots, they don't explicitly show magnitude but emphasize the overall flow pattern.
  - **What it shows:** Continuous lines that are everywhere tangent to the instantaneous velocity vector. They visually represent the path that an imaginary parcel of air would take at that moment.
  - **Why it's useful:** Provides a smooth, intuitive visualization of large-scale flow, highlighting areas of **cyclonic**(counter-clockwise

in Northern Hemisphere) and **anticyclonic** (clockwise in Northern Hemisphere) circulations, as well as **deformation zones** where flow patterns are stretching or shearing. Often used in conjunction with contour maps to provide a more complete picture of atmospheric dynamics.

- **Satellite Imagery**
- **Purpose:** To provide a visual representation of the Earth's surface and atmosphere from space, capturing broad-scale weather patterns.
- **What it shows**
- **Visible Imagery:** Captures sunlight reflected by clouds and the Earth's surface.
  - *Interpretation:* Shows clouds as they appear to the human eye (white, bright areas). Useful for identifying cloud types, fog, snow cover, and distinguishing between thick, bright clouds (e.g., cumulonimbus) and thinner, darker clouds (e.g., cirrus). Only available during daylight hours.
- **Infrared (IR) Imagery:** Measures the infrared radiation emitted by cloud tops or the Earth's surface, which is directly related to their temperature.
  - *Interpretation:* Colder temperatures (higher clouds, especially active thunderstorms) are typically shown in brighter colors (white, blue, or purple), while warmer temperatures (lower clouds or clear ground) are darker. Useful day or night for tracking storm systems, identifying high-level clouds, and estimating cloud top heights, which correlate with storm intensity.
- **Water Vapor (WV) Imagery:** Measures radiation emitted by water vapor, primarily in the mid to upper troposphere (around 200–500 hPa).
  - *Interpretation:* Reveals the distribution of moisture in the atmosphere. Dry areas appear dark, while moist areas appear bright. Crucial for identifying upper-level atmospheric features like jet streams, troughs, ridges, and areas of deep convection, even in the absence of visible clouds.
- **Why it's useful:** Provides real-time, broad-scale atmospheric context, crucial for **nowcasting** (short-term forecasting), storm tracking, and understanding large-scale weather patterns and their evolution.
- **Radar Imagery**
- **Purpose:** To detect precipitation, measure its intensity, and estimate its motion, particularly at local to regional scales.

- **What it shows**
- **Reflectivity:** Shows the intensity of precipitation (rain, snow, hail) based on the amount of radar energy reflected back by hydrometeors.
  - *Interpretation:* Higher reflectivity values (often depicted in warmer colors like red, orange, purple) indicate stronger precipitation. Meteorologists look for specific reflectivity patterns associated with severe weather, such as hook echoes (indicative of tornadoes) or hail cores.
- **Doppler Velocity:** Measures the speed of precipitation particles towards or away from the radar.
  - *Interpretation:* Red colors typically indicate movement away from the radar, while green/blue colors indicate movement towards the radar. This is crucial for detecting **rotation within thunderstorms** (mesocyclones, which are strong indicators of potential tornadoes) and **wind shear** (changes in wind speed or direction with height).
- **Why it's useful:** Essential for **short-term forecasting (nowcasting)** of precipitation, severe thunderstorm detection, flash flood warnings, and tracking the precise location and movement of storms.

10.4.2 Temporal Plots (Time Series)

These plots illustrate how meteorological variables change over time at a specific location or for a specific region. They are fundamental for understanding trends, cycles, and anomalies.

- **Time Series Plots (Line Graphs)**
- **Purpose:** To display the evolution of one or more variables over a period, typically with time on the horizontal axis and the variable's value on the vertical axis.
- **What it shows:** A line connecting data points ordered by time.
  - *Examples:* Hourly temperature fluctuations at an airport, daily precipitation totals over a month, monthly average global temperature anomalies over decades, or the historical values of climate indices like ENSO (El Niño-Southern Oscillation).
- **Why it's useful:** Reveals **trends** (e.g., global warming), **cycles** (e.g., diurnal temperature cycles, seasonal precipitation patterns), **anomalies** (unusually high or low values), and the overall temporal behavior of a variable. Fundamental for climate analysis, trend detection, and long-term monitoring.

- **Purpose:** To visualize the spatio-temporal evolution of a variable, showing how a feature propagates across a single spatial dimension over time.
- **What it shows:** Typically, time is on the vertical axis, a spatial dimension (e.g., longitude or latitude) is on the horizontal axis, and the variable's value is represented by color shading or contours.
    - *Interpretation:* Diagonal lines or bands indicate movement of features. For example, a diagonal band of high precipitation moving from west to east on a longitude-time Hovmöller diagram indicates an eastward-propagating rainfall system.
- **Why it's useful:** Excellent for tracking the movement of atmospheric waves (e.g., Kelvin waves, Rossby waves), precipitation bands, tropical disturbances, or other features that propagate over time, such as the **Madden-Julian Oscillation (MJO)**, a significant tropical weather pattern.

10.4.3 Vertical Profile Plots

These plots illustrate how meteorological variables change with altitude (or pressure) in the atmosphere. They are crucial for understanding the three-dimensional structure of the atmosphere.

- **Soundings (Skew-T Log-P Diagrams / Tephigrams / Stüve Diagrams):**
- **Purpose:** To graphically represent the vertical profile of temperature, dew point temperature, and wind from a weather balloon (radiosonde) ascent or a numerical model.
- **What it shows**
    - **Axes:** The vertical axis is typically pressure (logarithmic scale, representing altitude), and the horizontal axis is temperature (often "skewed" to make thermodynamic processes appear as straight lines).
    - **Temperature and Dew Point Curves:** Two lines showing how actual air temperature and dew point temperature change with height.
    - **Wind Barbs:** Plotted along the side, indicating wind speed and direction at various pressure levels.
    - **Thermodynamic Lines:** Include dry adiabats (lines of constant potential temperature for dry air), moist adiabats (lines of constant equivalent potential temperature for saturated air), and mixing ratio lines (lines of constant water vapor content).

- **Why it's useful:** Indispensable for assessing **atmospheric stability** (e.g., identifying inversion layers where temperature increases with height, which can trap pollution), forecasting **thunderstorm potential** (e.g., calculating Convective Available Potential Energy - CAPE, and Convective Inhibition - CIN), determining cloud base (LCL) and cloud top levels, and understanding the vertical structure of the atmosphere for aviation.

- **Hodographs**
- **Purpose:** To visualize the vertical profile of **wind shear** (how wind speed and/or direction change with height) in a concise manner.
- **What it shows:** A plot where wind vectors at different altitudes are plotted from a common origin (usually the surface wind), with the end points of the vectors connected sequentially from the surface upwards.
  - *Interpretation:* The shape of the hodograph reveals the type and strength of wind shear. A long, straight hodograph indicates strong unidirectional shear (good for organized linear storms). A curved or "hooked" hodograph indicates significant directional shear (e.g., veering winds with height), which is crucial for the development of rotating thunderstorms (supercells) and tornadoes.
- **Why it's useful:** Crucial for **severe thunderstorm forecasting**, as specific hodograph shapes are strongly associated with different storm modes and their potential for producing tornadoes, large hail, or damaging winds. Often analyzed in conjunction with soundings.
- **Cross-Section Plots**
  - **Purpose:** To show the distribution of one or more meteorological variables along a vertical slice through the atmosphere. This slice can be along a line of latitude, longitude, or a custom path (e.g., across a mountain range).
  - **What it shows:** Altitude (or pressure) on the y-axis, horizontal distance on the x-axis, and the variable's value represented by color shading or contours. Can show temperature, humidity, wind components (e.g., vertical velocity), or potential vorticity.
  - **Why it's useful:** Provides unique insight into the **vertical structure of atmospheric features** that are not apparent on horizontal maps. This includes visualizing the vertical extent of frontal zones, the core of jet streams, the structure of mountain waves, the height of the tropopause, or the vertical distribution of moisture and stability within a storm system.

10.4.4 Multivariate and Diagnostic Plots

These plots combine multiple variables or are designed for specific diagnostic purposes, often summarizing complex relationships.

- **Scatter Plots**
  - **Purpose:** To visualize the relationship between two continuous variables.
  - **What it shows:** Each point on the plot represents an individual observation, with its x-coordinate corresponding to the value of one variable and its y-coordinate to the value of another.
  - *Interpretation:* Helps identify **correlations** (positive, negative, or no correlation), **clusters** (groups of similar data points), and **outliers** (data points that deviate significantly from the main trend).
- **Why it's useful:** Used extensively for model verification (e.g., plotting model forecast temperature vs. observed temperature), exploring relationships between different meteorological parameters (e.g., relative humidity vs. cloud cover), or identifying potential biases in data.
- **Histograms and Box Plots**
  - **Purpose:** To visualize the distribution of a single variable or compare distributions across different categories.
- **What it shows**
  - **Histogram:** A bar chart showing the frequency or count of data points falling into specific ranges (bins).
  - *Interpretation:* Reveals the shape of the data's distribution (e.g., normal, skewed, bimodal), its central tendency, and its spread. Useful for understanding the typical range and variability of a meteorological variable.
  - **Box Plot:** A standardized way of displaying the distribution of data based on five key numbers: the minimum, first quartile (Q1), median (Q2), third quartile (Q3), and maximum. Outliers are often plotted individually.
  - *Interpretation:* Provides a quick visual summary of the data's center, spread, and skewness. Excellent for comparing the distributions of a variable across different categories (e.g., comparing precipitation amounts between different climate models, or temperature distributions for different seasons or regions).

    - **Why it's useful:** Essential for exploratory data analysis, quality control, and comparing statistical properties of meteorological datasets.

- **Wind Roses**
    - **Purpose:** To graphically represent the distribution of wind speed and direction at a specific location over a period (e.g., a month, a season, or a year).
    - **What it shows:** A circular plot with "spokes" radiating outwards, representing different wind directions (e.g., N, NE, E, SE, S, SW, W, NW). The length of each spoke (or segments within it) indicates the frequency of wind blowing *from* that direction, and different colors within the spokes often represent different wind speed ranges.
    - **Why it's useful:** Provides a concise visual summary of local wind climatology. Crucial for applications such as airport operations (runway orientation), air quality dispersion modeling (how pollutants spread), urban planning, and assessing sites for renewable energy (wind farms).

The diverse array of meteorological plots serves as the visual language of atmospheric science. From broad-scale satellite views that capture global weather patterns to intricate vertical soundings that reveal atmospheric stability, each plot type is meticulously designed to extract and convey specific aspects of the atmosphere's behavior. By mastering these visualizations, meteorologists can effectively diagnose current conditions, interpret complex numerical model outputs, communicate critical forecasts to the public and decision-makers, and ultimately deepen our scientific understanding of the dynamic and ever-changing weather and climate systems that shape our world. The ability to translate raw data into compelling and accurate visual narratives is as vital as the data collection and modeling processes themselves.

10.4 GIS for Meteorological Data

Geographic Information Systems (GIS) are powerful digital frameworks designed to capture, store, analyze, manage, and present all types of data that have a spatial or geographic component. At its heart, GIS allows us to link information directly to specific locations on Earth, creating intelligent, interactive maps that go far beyond static images. For meteorology, a field inherently defined by phenomena distributed across space and evolving over time, GIS is not just a useful tool but an **indispensable technology** that has profoundly transformed how atmospheric data is handled, understood, and applied.

10.4.1 Why is GIS Crucial for Meteorological Data? The Power of Spatial Context

Every piece of meteorological information—a temperature reading, a wind measurement, a precipitation amount, or a cloud formation—is tied to a specific location on Earth and a specific moment in time. This fundamental geographic nature makes GIS an ideal platform for managing and analyzing atmospheric information. The true power of GIS for meteorology lies in its ability to provide **spatial context** and, critically, to **overlay meteorological data with other geographic information**.

Consider a weather map: it's not just a collection of numbers. It's a visual representation of how temperature, pressure, or wind interacts with mountains, coastlines, cities, and forests. GIS allows us to formally integrate these diverse layers of information, revealing relationships and impacts that would be impossible to discern from raw data alone.

10.4.2 The Core Concept: Overlaying Meteorological Data with Geographic Information

The term "overlay" in GIS refers to the process of combining two or more distinct spatial data layers to create a new, integrated layer that reveals relationships or generates new insights. For meteorological applications, this means taking dynamic weather or climate data and placing it on top of static geographic features, or even combining multiple meteorological layers.

Imagine your meteorological data (e.g., a map of expected rainfall) as one transparent sheet, and your geographic information (e.g., a map of river networks, or population density) as another transparent sheet. When you lay them on top of each other in a GIS, you can see how the rainfall directly impacts the rivers or where it overlaps with densely populated areas. This simple concept unlocks immense analytical power.

10.4.3 How Overlaying Enhances Understanding and Action

The ability to overlay meteorological data with geographic information provides several profound benefits:

- **Contextualization:** It places abstract meteorological values into a real-world setting. A temperature of 35°C is just a number until you see it overlaid on a map of a city, showing which neighborhoods are experiencing extreme heat, or on a map of agricultural land, indicating potential crop stress.
- **Relationship Discovery:** Overlaying helps uncover how atmospheric phenomena interact with the Earth's surface. For example, how does a

mountain range influence precipitation patterns? How does a large urban area affect local temperatures?

- **Impact Assessment:** This is perhaps the most critical application. By combining meteorological hazard data with maps of human assets, infrastructure, and vulnerable populations, we can assess the potential impacts of weather events.
- **Targeted Decision-Making:** Overlay analysis allows for highly localized and specific decision-making. Instead of broad warnings, we can issue alerts for specific neighborhoods, identify critical infrastructure at risk, or direct resources precisely where they are needed most.
- **Communication Clarity:** Integrated maps are far more intuitive and easier for non-experts to understand. They tell a compelling story of "what is happening where, and what might be affected."

10.4.4 Detailed Examples of Overlaying in Meteorology

Let's explore specific scenarios where overlaying different data layers in a GIS provides invaluable insights:

- **Flood Risk Assessment and Management**
  - **Meteorological Data Layers:** Real-time precipitation radar (raster), quantitative precipitation forecasts (gridded), river gauge levels (point data), historical flood inundation maps (polygon).
  - **Geographic Information Layers:** Digital Elevation Models (DEMs - raster) showing terrain and elevation, river networks (line data), land use/land cover maps (polygon/raster), critical infrastructure (e.g., hospitals, power stations - point/polygon), population density maps (raster/polygon), road networks (line data).
  - **Overlay Analysis:** By overlaying high-intensity rainfall forecasts with river networks and terrain data, GIS can model and visualize potential flood inundation areas. Further overlaying with population density and critical infrastructure layers immediately identifies vulnerable communities and essential services at risk, informing targeted evacuation orders, resource pre-positioning, and emergency response routes.
- **Air Quality Monitoring and Management**
  - **Meteorological Data Layers:** Atmospheric pollutant concentrations (e.g., PM2.5, Ozone - gridded/point), wind speed and direction (vector/gridded), atmospheric stability (gridded).

- **Geographic Information Layers:** Emission sources (e.g., industrial facilities, major roadways - point/line/polygon), population density maps, land use maps (e.g., residential, industrial zones), topographical features (DEM).
- **Overlay Analysis:** Overlaying pollutant concentration maps with wind fields helps visualize pollutant plumes and their dispersion. Combining this with emission source locations can help identify major contributors. Overlaying with population density maps allows for the assessment of human exposure to pollution, guiding public health advisories or policy interventions in affected areas.

- **Agricultural Meteorology (Agrometeorology) and Crop Management**
  - **Meteorological Data Layers:** Daily temperature, precipitation, soil moisture, evapotranspiration (gridded/point).
  - **Geographic Information Layers:** Crop type maps (polygon), soil type maps (polygon), irrigation infrastructure (line/polygon), agricultural field boundaries (polygon).
  - **Overlay Analysis:** Overlaying current temperature and precipitation data with specific crop type maps can identify areas experiencing heat stress or drought conditions for particular crops. Combining this with soil type and irrigation infrastructure maps can help farmers optimize irrigation schedules, apply fertilizers more efficiently, or make informed decisions about planting and harvesting.
- **Severe Weather Impact and Warning Dissemination**
  - **Meteorological Data Layers:** Real-time radar reflectivity (raster), storm tracks (line), tornado/severe thunderstorm warnings (polygon), hail probability (gridded).
  - **Geographic Information Layers:** Population centers (point/polygon), critical infrastructure (e.g., schools, hospitals, emergency services - point/polygon), road networks, power grid lines (line), building footprints (polygon).
  - **Overlay Analysis:** Overlaying a tornado warning polygon with detailed street maps and building footprints allows emergency managers to pinpoint exactly which neighborhoods are under threat. Combining this with critical infrastructure layers ensures that vulnerable facilities are prioritized for protection or evacuation. This enables highly targeted and actionable warnings, moving beyond broad county-level alerts.

- **Climate Change Vulnerability and Adaptation Planning**
  - **Meteorological/Climate Data Layers:** Projected temperature increases, changes in precipitation patterns, sea level rise scenarios, extreme heat/cold event frequency (gridded).
  - **Geographic Information Layers:** Coastal zone maps, floodplains, socio-economic data (income levels, age demographics - polygon), infrastructure (roads, utilities), ecological zones (e.g., wetlands, forests).
  - **Overlay Analysis:** Overlaying sea level rise projections with coastal infrastructure and population data identifies areas at risk of inundation. Combining projected drought frequency with agricultural land use and water resource maps highlights regions facing water scarcity. This comprehensive spatial analysis is crucial for developing targeted adaptation strategies and resilient planning.

GIS has fundamentally transformed the way meteorological data is handled, analyzed, and communicated. Its unparalleled ability to **overlay dynamic meteorological information with static geographic features** is the cornerstone of its power. This integration provides crucial spatial context, reveals intricate relationships, and enables precise impact assessments that are vital for informed decision-making across a vast spectrum of applications—from issuing immediate severe weather warnings to crafting long-term climate adaptation strategies. As the volume and complexity of environmental data continue to proliferate, GIS will remain an indispensable tool, enabling a more comprehensive, actionable, and visually compelling understanding of our dynamic planet

10.5 Interactive Visualization Tools and Libraries for Meteorological Data

In the rapidly evolving landscape of meteorological data analysis, static plots, while foundational, often fall short of fully conveying the dynamic and multi-dimensional nature of atmospheric phenomena. This is where **interactive visualization tools and libraries** become indispensable. Interactive visualizations allow users to explore data dynamically—zooming, panning, filtering, hovering for details, and animating changes over time—thereby fostering deeper understanding, facilitating quicker insights, and enabling more effective communication.

For meteorologists, the ability to interact with data is crucial for:

- **Exploratory Data Analysis:** Rapidly testing hypotheses and discovering hidden patterns.

- **Diagnosis of Weather Systems:** Pinpointing specific features, tracking their evolution, and understanding their interactions.
- **Model Verification:** Comparing model outputs with observations at various scales and resolutions.
- **Communication:** Tailoring presentations to specific audiences by focusing on relevant details or animating complex processes.

This section will delve into popular interactive visualization tools and libraries, categorized by their programming language ecosystems (Python, R) and specialized software, providing detailed explanations of their features and applications in meteorology.

10.5.1 Python Libraries for Interactive Visualization

Python has emerged as a dominant language in scientific computing, and its rich ecosystem of libraries offers powerful capabilities for interactive meteorological data visualization.

- **Matplotlib (with Interactive Backends)**
    - **What it is:** Matplotlib is the foundational plotting library in Python, widely used for creating static, publication-quality figures. While primarily known for static plots, it supports various interactive backends (e.g., TkAgg, Qt5Agg, WebAgg, notebook) that enable basic interactivity.
- **Main Features/Strengths**
    - **Versatility:** Can create a vast array of plot types (line, scatter, contour, vector, histograms, etc.).
    - **Customization:** Offers granular control over every aspect of a plot's appearance.
    - **Interactive Backends:** When used with appropriate backends (e.g., %matplotlib notebook in Jupyter), it allows for zooming, panning, and saving figures directly from the plot window.
    - **Foundation for Others:** Many other Python plotting libraries are built on top of or inspired by Matplotlib.
- **How it's used in Meteorology**
    - Creating standard meteorological plots like contour maps of temperature or pressure, time series of station data, vertical soundings (often with specialized extensions like MetPy).
    - Basic interactive exploration of model outputs or observational data in research and development environments.

    - Often combined with Cartopy for geographical mapping.
    - **Limitations:** Its native interactivity is relatively basic compared to dedicated interactive libraries. Achieving highly interactive web-based dashboards requires additional frameworks.
- **Plotly**
- **What it is:** Plotly is a powerful, open-source graphing library that generates interactive, web-based visualizations. It supports a wide range of chart types and is particularly strong for creating dashboards and web applications. It has APIs for Python, R, and JavaScript.
- **Main Features/Strengths**
    - **Interactive by Default:** Plots are inherently interactive, allowing users to zoom, pan, hover for data details (tooltips), toggle traces, and save images directly from the browser.
    - **Web-Based Output:** Generates plots in HTML, making them easily shareable and embeddable in web pages, Jupyter notebooks, or Dash applications.
    - **plotly.express:** A high-level API that simplifies the creation of complex plots with minimal code.
    - **Wide Range of Plot Types:** Supports 2D and 3D plots, statistical charts, and specialized scientific plots.
- **How it's used in Meteorology**
- Creating interactive time series plots of climate indices, station data, or model forecasts, allowing users to zoom into specific periods.
- Interactive contour maps of weather variables, where hovering over a point reveals its exact value.
- Visualizing ensemble forecasts with interactive spaghetti plots.
- Building interactive web dashboards for real-time weather monitoring or climate data exploration using Dash (see below).
- **Limitations:** Can sometimes be slower for extremely large datasets compared to static plotting or specialized rendering engines.
- **Bokeh**
    - **What it is:** Bokeh is a Python library for creating interactive visualizations that target modern web browsers for presentation. It's designed for building interactive data applications and dashboards, especially with streaming or large datasets.
- **Main Features/Strengths**

- **Interactive Web Output:** Renders plots as HTML and JavaScript, enabling highly interactive features like zooming, panning, selection tools, and custom callbacks.
- **Streaming Data:** Optimized for visualizing streaming or large datasets efficiently.
- **Server-Side Capabilities:** Comes with a Bokeh server that allows for complex, interactive web applications with Python callbacks.
- **Customization:** Offers flexibility in styling and layout.

- How it's used in Meteorology:
  - Building interactive web interfaces for real-time weather data streams (e.g., displaying live sensor readings).
  - Creating custom interactive tools for forecasters to explore model outputs or observational networks.
  - Visualizing large climate model ensembles where data might be too extensive for static plots.
  - **Limitations:** Can have a steeper learning curve than plotly.express for simple plots.
- **Folium**
  - **What it is:** Folium is a Python wrapper for the popular JavaScript library Leaflet.js, designed specifically for creating interactive maps.
- **Main Features/Strengths**
  - **Interactive Basemaps:** Easily adds interactive basemaps (e.g., OpenStreetMap, Esri, Stamen) to plots.
  - **Geospatial Overlays:** Allows overlaying various geospatial data types (markers, lines, polygons, heatmaps, GeoJSON) on interactive maps.
  - **Pop-ups and Tooltips:** Supports interactive pop-ups that display information when clicking or hovering over map features.
  - **How it's used in Meteorology**
  - Visualizing the locations of weather stations, radar sites, or buoy networks.
  - Displaying storm tracks, frontal boundaries, or warning areas as interactive overlays on a map.
  - Plotting gridded meteorological data (e.g., temperature, precipitation) as interactive raster overlays, allowing users to zoom and pan.

   - Creating interactive maps for public outreach or educational purposes, showing weather impacts in specific geographical contexts.
   - **Limitations:** Primarily focused on mapping; not a general-purpose plotting library for charts like time series or histograms.
- **Cartopy (for Geospatial Context, often combined with Matplotlib)**
   - **What it is:** Cartopy is a Python library designed for geospatial data processing and visualization, providing cartographic projections, transforming points, lines, vectors, and polygons, and drawing coastlines and political boundaries. While its output is primarily static, it's crucial for providing the correct geographical context for meteorological plots.
- **Main Features/Strengths**
   - **Map Projections:** Supports a wide range of map projections (e.g., Lambert Conformal, Stereographic, Plate Carrée), essential for accurate meteorological mapping.
   - **Geospatial Features:** Easily adds coastlines, country borders, state boundaries, rivers, and grid lines.
   - **Integration with Matplotlib:** Works seamlessly with Matplotlib, allowing meteorologists to create scientifically accurate and aesthetically pleasing maps.
- **How it's used in Meteorology**
   - Creating the geographical base for almost all spatial meteorological plots (contour maps, vector plots, radar/satellite imagery).
   - Ensuring that the underlying map projection accurately represents the meteorological phenomena being visualized (e.g., using Lambert Conformal for mid-latitude storm systems).
   - **Limitations:** Primarily a static mapping library. Interactivity typically comes from combining it with Matplotlib's interactive backends or converting its output to formats compatible with Plotly/Bokeh.
- **Dash (Plotly's Framework for Web Apps)**
   - **What it is:** Dash is a Python framework for building analytical web applications. It's built on top of Plotly.js, React.js, and Flask, allowing Python developers to create interactive dashboards entirely in Python, without needing to write JavaScript.
- **Main Features/Strengths**
   - **Full Web Applications:** Enables the creation of complex, multi-page, interactive web dashboards.

- **Pure Python:** All development is done in Python, leveraging familiar data science libraries.
- **Interactive Components:** Provides a wide range of interactive components (sliders, dropdowns, buttons) that can trigger updates to Plotly graphs.
- **Scalability:** Can be deployed to serve many users.

- **How it's used in Meteorology**
    - Building operational dashboards for real-time weather monitoring, allowing forecasters to select regions, variables, and time steps.
    - Creating interactive climate data portals for researchers or the public to explore climate change projections.
    - Developing custom tools for visualizing ensemble forecasts, allowing users to filter ensemble members or analyze specific scenarios.
    - **Limitations:** Can have a steeper learning curve for complex layouts and callbacks.

10.5.2 R Packages for Interactive Visualization

R is another powerful language for statistical computing and graphics, with a robust set of packages for interactive data visualization, particularly strong in statistical modeling and specialized graphics.

- **ggplot2 (with Interactive Extensions)**
    - **What it is:** ggplot2 is R's most popular and powerful package for creating static, elegant, and highly customizable statistical graphics, based on the "Grammar of Graphics" concept.
- **Main Features/Strengths**
    - **Grammar of Graphics:** Allows users to build plots layer by layer (data, aesthetics, geoms, facets, scales, etc.), providing immense flexibility.
    - **High-Quality Graphics:** Produces aesthetically pleasing and publication-ready plots.
    - **Extensibility:** A vast ecosystem of extensions builds upon ggplot2.
- **How it's used in Meteorology**
    - Creating static versions of almost all common meteorological plots (time series, scatter plots, histograms, box plots for climate data).
    - Visualizing statistical relationships between meteorological variables.

- **Interactive Extensions**
  - **plotly (R version):** The R version of Plotly can convert ggplot2 objects into interactive web-based plots, adding zoom, pan, hover, and toggle features. This is a very common way to add interactivity to ggplot2 plots.
  - **ggiraph:** Another package that adds interactive capabilities (tooltips, hover effects, clickable elements) to ggplot2 plots, making them suitable for web display.
  - **Limitations:** Native ggplot2 is static. Interactivity requires additional packages.
- **Leaflet**
  - **What it is:** Leaflet is a leading open-source JavaScript library for interactive mobile-friendly maps. The R leaflet package provides a convenient interface to create and customize these maps directly from R.
- **Main Features/Strengths:**
  - **Interactive Basemaps:** Easily integrates with various online basemaps (e.g., OpenStreetMap, Esri, CartoDB).
  - **Geospatial Data Overlays:** Supports adding markers, lines, polygons, circles, and raster image overlays.
  - **Pop-ups and Labels:** Enables interactive pop-ups and labels for map features.
  - **Clustering:** Can cluster large numbers of markers for better visualization.
- **How it's used in Meteorology**
  - Displaying weather station locations, buoy positions, or radar sites on interactive maps.
  - Visualizing storm tracks, warning areas, or areas of interest for field campaigns.
  - Overlaying gridded meteorological data (e.g., temperature forecasts, precipitation amounts) as interactive raster layers.
  - Creating interactive maps for public weather advisories or climate impact visualizations.
  - **Limitations:** Primarily a mapping library; not for general-purpose data plotting.

- **Shiny**
  - **What it is:** Shiny is an R package that makes it incredibly easy to build interactive web applications (dashboards) directly from R, without needing to know HTML, CSS, or JavaScript.
- **Main Features/Strengths**
  - **Pure R Development:** Develop full web applications using only R code.
  - **Interactive Widgets:** Provides a wide range of interactive input widgets (sliders, dropdowns, text inputs) that dynamically update plots and tables.
  - **Reactive Programming:** Built on a reactive programming model, where changes in inputs automatically trigger updates in outputs.
  - **Deployment:** Applications can be deployed easily on Shiny Server or RStudio Connect.
- **How it's used in Meteorology**
  - Building custom interactive dashboards for forecasters to explore real-time weather observations, NWP model outputs, and ensemble forecasts.
  - Creating interactive climate data portals for researchers to analyze climate change scenarios or for the public to explore local climate trends.
  - Developing tools for visualizing and comparing different meteorological datasets or model runs.
  - **Limitations:** Can become computationally intensive for very large datasets if not optimized. Requires a Shiny Server for public deployment.
- **dygraphs**
  - **What it is:** dygraphs is an R package that provides an interface to the dygraphs JavaScript library, specifically designed for creating interactive time series plots.
- **Main Features/Strengths**
  - **Interactive Time Series:** Excellent for exploring long time series data with features like zooming, panning, range selectors, and interactive legends.
  - **Handles Large Data:** Optimized for displaying large datasets efficiently.
  - **Date/Time Axis:** Robust handling of date and time axes.

- **How it's used in Meteorology**
    - Visualizing long-term climate records (e.g., temperature, CO2 concentration).
    - Exploring hourly or daily meteorological station data.
    - Displaying model forecast time series with uncertainty bands.
    - **Limitations:** Specialized for time series; not a general-purpose plotting tool.

10.5.3 Specialized Software for Meteorological Data Visualization

Beyond programming libraries, several dedicated software packages are designed specifically for meteorological and climate data visualization and analysis. These often offer highly optimized performance for specific data formats and complex scientific rendering capabilities.

- **GrADS (Grid Analysis and Display System)**
    - **What it is:** GrADS is a powerful, venerable, and widely used interactive desktop tool for easy access, manipulation, and visualization of earth science data. It's command-line driven but highly versatile.
- **Main Features/Strengths**
    - **Handles Gridded Data:** Optimized for working with gridded binary data, NetCDF, HDF, and GRIB formats.
    - **Scripting:** Supports a powerful scripting language for complex analyses and visualizations.
    - **Interactive Exploration:** Allows for zooming, panning, slicing (cross-sections), and animating data.
    - **Diagnostic Capabilities:** Built-in functions for calculating meteorological diagnostics (e.g., vorticity, divergence).
- **How it's used in Meteorology**
    - Rapidly visualizing and analyzing numerical model outputs (e.g., GFS, ECMWF) and reanalysis datasets.
    - Creating custom plots for research and operational forecasting.
    - Generating animations of atmospheric fields.
    - **Limitations:** Command-line interface can be less intuitive for new users. Graphics are functional but not as aesthetically polished as modern web-based tools.

- **Panoply**
  - **What it is:** Developed by NASA's Goddard Institute for Space Studies (GISS), Panoply is a cross-platform application for plotting geospatial data from NetCDF, HDF, GRIB, and other formats.
- **Main Features/Strengths:**
  - **User-Friendly GUI:** Very easy to use with a graphical user interface (GUI), making it accessible for quick data exploration.
  - **Direct File Support:** Directly opens and displays data from common meteorological data formats.
  - **Basic Plotting:** Supports creating 2D contour plots, color-filled plots, and animations.
  - **Cross-Platform:** Available for Windows, macOS, and Linux.
- **How it's used in Meteorology**
  - Quickly viewing and verifying satellite data, climate model outputs, or reanalysis products stored in NetCDF/HDF files.
  - Educational purposes and initial data exploration for researchers.
  - **Limitations:** Less powerful for complex analysis or highly customized plots compared to scripting environments.
- **IDV (Integrated Data Viewer)**
  - **What it is:** Developed by Unidata (UCAR), the IDV is a powerful, highly interactive, and extensible Java-based application for visualizing and analyzing a wide variety of meteorological and other geoscience data.
- **Main Features/Strengths**
  - **Multi-Dimensional Visualization:** Excellent for visualizing 2D, 3D, and 4D (time-varying 3D) data.
  - **Diverse Data Formats:** Supports a vast array of real-time and archived data formats (NetCDF, GRIB, BUFR, McIDAS, NEXRAD Level II/III, etc.).
  - **Interactive Exploration:** Allows for sophisticated interactive manipulation of displays, including zooming, panning, slicing, rotating 3D views, and animating data.
  - **Analysis Capabilities:** Includes tools for calculating meteorological diagnostics, cross-sections, and soundings.

- **How it's used in Meteorology**
- Operational forecasting workstations for real-time weather monitoring and severe weather analysis.
- Research and education, allowing detailed exploration of model outputs, radar data, and observational networks.
- Visualizing complex atmospheric phenomena in 3D.
- **Limitations:** Can be resource-intensive. The Java-based interface might feel dated to some users.

- **GEMPAK / NAWIPS (National Advanced Weather Interactive Processing System)**
  - **What it is:** GEMPAK is a classic, command-line driven meteorological analysis and display package that has been a staple in the US National Weather Service (NWS) and academia for decades. NAWIPS is the NWS's operational workstation built upon GEMPAK.
- **Main Features/Strengths**
  - **Operational Focus:** Designed for the demanding environment of operational weather forecasting.
  - **Comprehensive Tools:** Provides a wide range of plotting, analysis, and diagnostic capabilities for various meteorological variables and levels.
  - **Real-time Data Ingest:** Optimized for ingesting and displaying real-time observational and model data.
- **How it's used in Meteorology**
  - Primary workstation for NWS forecasters for many years.
  - Used in university meteorology departments for teaching and research.
  - **Limitations:** Steep learning curve due to command-line interface. Graphics are functional but not modern. Being phased out in some operational contexts by newer systems.
- **ArcGIS / QGIS (Geographic Information Systems Software)**
  - **What it is:** These are full-fledged GIS software platforms. ArcGIS (Esri) is a commercial suite, while QGIS is a powerful open-source alternative. They are designed for comprehensive geospatial data management, analysis, and mapping.

- **Main Features/Strengths**
  - **Robust Spatial Analysis:** Offers extensive tools for spatial analysis, including overlay, buffering, network analysis, terrain analysis, and geoprocessing.
  - **Data Integration:** Excellent for integrating diverse geospatial datasets (vector, raster, point) from various sources.
  - **Professional Mapping:** Produces high-quality, cartographically sound maps.
  - **Interactive Exploration:** Provides interactive tools for zooming, panning, querying features, and managing layers.
- **How it's used in Meteorology**
  - **Impact Assessment:** Overlaying meteorological hazard maps (e.g., flood zones, storm surge areas) with population data, infrastructure, and land use to assess vulnerability and risk.
  - **Climatology:** Mapping climate zones, analyzing spatial patterns of climate change impacts, and visualizing long-term trends.
  - **Environmental Monitoring:** Integrating meteorological data with environmental datasets (e.g., air quality, pollution sources, ecological zones).
  - **Emergency Management:** Supporting disaster response and planning by providing a common operating picture that combines weather information with critical assets.
  - **Limitations:** While they can display meteorological data, they are not specifically designed for atmospheric science diagnostics or complex 3D atmospheric visualizations out-of-the-box like IDV or GrADS.
- **ParaView / VisIt (Scientific Visualization Software)**
  - **What it is:** ParaView (Kitware) and VisIt (LLNL) are powerful, open-source, multi-platform data analysis and visualization applications designed for large, complex scientific datasets, particularly those from simulations (e.g., climate models, computational fluid dynamics).
- **Main Features/Strengths**
  - **High-Performance Visualization:** Optimized for handling extremely large datasets (terabytes) and parallel processing.
  - **3D Visualization:** Excellent for visualizing complex 3D structures, volumetric data, and vector fields.

- **Advanced Filtering and Analysis:** Offers a wide range of filters for data manipulation, extraction, and quantitative analysis.
- **Scripting:** Both support scripting (Python for ParaView) for automation and customization.

- **How it's used in Meteorology**
  - Visualizing outputs from high-resolution climate models or atmospheric simulations (e.g., cloud resolving models, LES models).
  - Exploring complex 3D atmospheric structures like turbulent eddies, convective updrafts, or pollutant plumes.
  - Analyzing and visualizing ensemble forecast data in a comprehensive 3D context.
  - **Limitations:** Can have a steep learning curve. Overkill for simple 2D plots.
- **Metview**
  - **What it is:** Developed by the European Centre for Medium-Range Weather Forecasts (ECMWF), Metview is a powerful, interactive workstation application and a Python scripting interface for meteorological data analysis and visualization.
- **Main Features/Strengths**
  - ECMWF Data Integration: Optimized for working with ECMWF's vast archives of operational and reanalysis data (e.g., ERA5).
  - GRIB Support: Excellent support for GRIB format data, common in NWP.
  - Comprehensive Plotting: Offers a wide range of meteorological plot types, including cross-sections, vertical profiles, and advanced contouring.
  - Python Interface: Provides a robust Python API for scripting and automation.
- **How it's used in Meteorology**
  - Primary tool for ECMWF users to access, analyze, and visualize their model outputs and reanalysis data.
  - Research and operational forecasting, particularly in Europe and for global-scale analysis.
  - **Limitations:** Primarily focused on ECMWF data and conventions, though it can read other formats.

The evolution of interactive visualization tools and libraries has fundamentally transformed how meteorologists interact with and understand atmospheric data. From the flexibility of Python and R libraries that enable custom web applications and dynamic maps, to specialized software designed for high-performance 3D rendering and real-time operational analysis, these tools empower scientists to move beyond static images. They facilitate deeper exploration, more nuanced interpretation, and ultimately, more effective communication of the complex and ever-changing dynamics of our planet's weather and climate system. The ability to dynamically explore and present meteorological data is now an essential skill for anyone working in the atmospheric sciences.

10.6 Conclusion

Concluding our exploration of the visualization of meteorological data, it's clear that this discipline is far more than just an aesthetic exercise; it is an **indispensable cornerstone** of modern atmospheric science. We've traversed the landscape of common meteorological plots, from the foundational contour maps that reveal the spatial structure of weather systems to the intricate soundings that unlock the vertical secrets of the atmosphere, and the dynamic time series that chronicle atmospheric evolution. Each visualization technique, whether depicting spatial distributions, temporal trends, or vertical profiles, serves a unique purpose in translating raw numerical data into comprehensible and actionable insights. The adherence to principles of clarity, accuracy, and purpose-driven design ensures that these visual narratives are not only compelling but also scientifically sound.

Furthermore, the advent of powerful interactive visualization tools and libraries, spanning versatile programming ecosystems like Python and R, alongside highly specialized meteorological software, has revolutionized how we interact with atmospheric information. These tools empower meteorologists to dynamically explore vast datasets, pinpoint critical features, diagnose complex atmospheric phenomena, and communicate findings with unprecedented clarity to diverse audiences. As the volume, velocity, and complexity of meteorological data continue to escalate, driven by advanced sensors and ever-more sophisticated numerical models, the role of cutting-edge visualization will only grow in importance, serving as the essential bridge between abstract data and intuitive understanding, ultimately enhancing our ability to predict, prepare for, and respond to the challenges of a dynamic Earth system.

11

Numerical Weather Prediction

11.1 Introduction

Having explored the sophisticated statistical methods for analyzing meteorological data and the powerful visualization techniques for interpreting it, we now turn our attention to the very engine that drives modern weather forecasting: **Numerical Weather Prediction (NWP)**. This chapter delves into the intricate science and technology behind predicting the future state of the atmosphere using mathematical models and supercomputers.

The challenge of weather forecasting is immense. The Earth's atmosphere is a chaotic, non-linear fluid system, constantly in motion, with interactions occurring across vast scales, from global circulation patterns to microscopic cloud droplets. For centuries, weather prediction relied heavily on empirical rules, human experience, and the interpretation of synoptic charts. While these methods achieved some success, they were inherently limited in their ability to accurately forecast beyond a day or two, especially for complex or rapidly evolving phenomena.

The visionary idea of predicting weather mathematically was first conceived by Lewis Fry Richardson in the early 20th century. He envisioned solving the fundamental equations governing atmospheric behavior using numerical methods. However, the sheer volume of calculations required made his manual attempts impractical, famously taking him weeks to produce a 6-hour forecast that had already passed. It was only with the advent of the electronic computer in the mid-20th century that Richardson's dream began to materialize, marking the birth of modern NWP (Fig. 11.1).

Fig. 11.1: Three dimensional grids for numerical weather prediction

At its core, Numerical Weather Prediction is the process of **solving a complex set of mathematical equations** that describe the physical laws governing the atmosphere (e.g., conservation of mass, momentum, energy, and water vapor). These equations are too complex to solve analytically, so they are approximated and solved numerically on a three-dimensional grid that spans the globe or a specific region. The process can be broadly understood as a continuous cycle:

- **Observations:** Gathering vast amounts of real-time data from satellites, radar, weather balloons, surface stations, and aircraft to describe the current state of the atmosphere.
- **Data Assimilation:** As we discussed in a previous chapter, this crucial step intelligently combines these imperfect and incomplete observations with a previous short-range model forecast to produce the most accurate possible estimate of the atmosphere's current state – the "initial conditions" for the new forecast.
- **Model Integration:** Using these initial conditions, powerful supercomputers then numerically integrate the atmospheric equations forward in time, step by step, to predict how the atmosphere will evolve over minutes, hours, and days.
- **Post-processing and Verification:** The raw model output is then processed and interpreted into user-friendly forecasts, and its accuracy is continuously verified against new observations.

This chapter will explore the fundamental principles underlying NWP, including the types of atmospheric models used, the role of physical parameterizations, and the computational challenges involved. We will delve into the concept of ensemble forecasting, which addresses the inherent chaotic nature of the atmosphere by running multiple model simulations. Finally, we will discuss the continuous efforts in model verification and improvement that drive the relentless progress in forecast skill. Numerical Weather Prediction is a testament to scientific ingenuity and technological advancement, underpinning everything from our daily weather reports to critical warnings for severe storms, and playing an increasingly vital role in climate research and environmental decision-making.

11.2 Global Versus Regional Numerical Weather Prediction Models

In the realm of Numerical Weather Prediction (NWP), atmospheric models are broadly categorized into two main types based on their geographical coverage: **global models** and **regional models** (also known as Limited Area Models or LAMs). Both play crucial, yet complementary, roles in generating the comprehensive forecasts we rely on daily. The fundamental difference lies in their **spatial domain** and, consequently, their **resolution** and **forecast range**.

11.2.1 Global Models

Global models, as their name suggests, cover the **entire Earth's atmosphere**. They simulate atmospheric processes from pole to pole and from the Earth's surface up to the stratosphere or even higher. To manage the immense computational task of simulating the entire globe, these models typically operate at a **coarser spatial resolution** compared to regional models. This means their grid cells (the discrete boxes into which the atmosphere is divided for calculations) are larger, representing average conditions over broader areas.

- **Advantages**
    - **Complete Global Picture:** They provide a seamless, physically consistent forecast for the entire planet, capturing large-scale atmospheric circulation patterns and their interactions across continents and oceans.
    - **No Lateral Boundary Issues:** Because they cover the entire globe, global models do not suffer from "lateral boundary conditions" problems, which are a significant challenge for regional models. They are self-contained in their domain.
    - **Capture Teleconnections:** They are essential for understanding and forecasting **teleconnections** – atmospheric and oceanic phenomena in one part of the world that influence weather patterns in distant regions (e.g., El Niño's impact on global weather).
    - **Longer Forecast Ranges:** Due to their ability to capture global-scale dynamics, global models are the foundation for medium-range (3-10 days) and long-range (10+ days) forecasts, and are also used for climate modeling.

- **Disadvantages**
  - **Lower Resolution:** Their coarser resolution means they cannot resolve small-scale atmospheric features like individual thunderstorms, local terrain effects (e.g., mountain-valley breezes), or fine details of frontal systems. These sub-grid scale processes must be represented through **parameterizations** (simplified mathematical formulas), which introduce uncertainties.
  - **Computational Cost:** Simulating the entire global atmosphere is incredibly computationally expensive, requiring supercomputers with massive processing power and memory.
  - **Less Detail for Local Forecasts:** While providing a good large-scale picture, their output needs further refinement for highly localized forecasts.
- **Typical Applications**
  - Medium-range and long-range weather forecasting (e.g., 5-day, 7-day, 14-day outlooks).
  - Providing the **boundary conditions** for higher-resolution regional models.
  - Climate modeling and climate change projections.
  - Global atmospheric research.
  - Forecasting large-scale phenomena like tropical cyclones (hurricanes/typhoons) and global circulation patterns.
- **Examples**
  - **GFS (Global Forecast System):** Operated by the U.S. National Oceanic and Atmospheric Administration (NOAA).
  - **ECMWF (European Centre for Medium-Range Weather Forecasts) Integrated Forecasting System (IFS):** Widely regarded as one of the leading global models.
  - **UK Met Office Unified Model:** Used for both weather and climate prediction.
  - **FV3 (Finite-Volume Cubed-Sphere Dynamical Core):** A new generation global model used by NOAA as the basis for the GFS.

11.2.2 Regional Models (Limited Area Models - LAMs)

Regional models focus on a **specific, smaller geographical area** (e.g., a continent, a country, or even a smaller metropolitan region). Their defining characteristic is their **significantly higher spatial resolution** within this limited domain. This means their grid cells are much smaller, allowing them to explicitly resolve finer-scale atmospheric processes and terrain features.

- **Advantages**
  - **High Resolution and Detail:** The primary advantage is their ability to represent small-scale atmospheric phenomena (e.g., individual thunderstorms, sea breezes, lake-effect snow, local mountain-induced flows) and the influence of complex topography (mountains, coastlines, urban areas) with much greater detail.
  - **Improved Local Forecasts:** This higher resolution leads to more accurate and specific forecasts for local weather conditions, which is crucial for severe weather warnings, aviation, and hydrological forecasting.
  - **Better Representation of Convection:** Many regional models are "convection-permitting," meaning their resolution is fine enough (typically 1–4 km) to explicitly simulate the formation and evolution of individual convective storms (thunderstorms) rather than relying solely on parameterizations.
  - **Computational Efficiency (for their domain):** While still computationally intensive, they are less demanding than global models for their specific domain, allowing for more frequent updates or larger ensembles over a smaller area.
- **Disadvantages**
  - **Lateral Boundary Conditions (LBCs):** This is the main challenge. Since regional models only cover a limited area, they need information from a global model at their boundaries to know what's coming into or leaving their domain. Errors or inconsistencies in these **Lateral Boundary Conditions** can propagate into the regional model's forecast, limiting its useful forecast range.
  - **Shorter Forecast Ranges:** Due to the influence of boundary errors and the rapid evolution of small-scale features, regional models typically have shorter useful forecast ranges (e.g., 0-48 hours, sometimes up to 72 hours).
  - **Lack of Global Context:** They cannot capture global teleconnections on their own and rely entirely on the global model for the larger-scale atmospheric steering.
- **Typical Applications**
  - Short-range (0-48 hours) high-resolution weather forecasting.
  - Severe weather forecasting (tornadoes, hail, flash floods).

    - Air quality modeling and dispersion forecasts.
    - Hydrological forecasting (e.g., river basin precipitation, flood prediction).
    - Urban meteorology and microclimate studies.
    - Specialized forecasts for aviation, energy (wind/solar), and agriculture.
- **Examples**
    - **WRF (Weather Research and Forecasting Model):** A widely used, open-source, community model.
    - **HRRR (High-Resolution Rapid Refresh):** A convection-permitting model operated by NOAA, covering the contiguous U.S.
    - **HARMONIE-AROME:** A regional model used by many European meteorological services.
    - **COSMO (Consortium for Small-scale Modeling):** Another prominent European regional model.

11.2.3 Interaction and Nesting: A Symbiotic Relationship

Global and regional models are not isolated but work in a **symbiotic relationship** through a process called **nesting**. Regional models typically get their initial conditions and, crucially, their **lateral boundary conditions (LBCs)** from a global model. This means that the global model provides the large-scale atmospheric context, dictating what flows into and out of the regional model's domain.

This hierarchical approach allows forecasters to leverage the strengths of both: the global model provides the broad, long-range picture, while the regional model "zooms in" to provide high-resolution detail for specific areas of interest.

11.2.4 Evolution and Future Trends

The distinction between global and regional models is becoming somewhat blurred as computational power continues to increase. Global models are constantly increasing their resolution, with some now operating at resolutions approaching those of older regional models. The concept of "global convection-permitting" models is a long-term goal for some centers, aiming to run a single, very high-resolution model for the entire globe. However, even with these advancements, regional models will likely continue to play a vital role for specialized, ultra-high-resolution applications and rapid update cycles over specific domains.

Global and regional numerical weather prediction models are two sides of the same coin, each essential for modern forecasting. Global models provide the necessary large-scale context and longer-range outlook, while regional models zoom in to deliver the fine-scale detail crucial for local impacts and severe weather warnings. Their complementary nature, facilitated by nesting, ensures that meteorologists have a comprehensive and increasingly accurate picture of the atmosphere, from global teleconnections to localized thunderstorms, enabling better preparedness and decision-making worldwide.

11.3 Ensemble Forecasting Explained

In the complex and inherently chaotic world of weather prediction, a single, deterministic forecast (a "single run" of a numerical weather prediction model) can only provide one possible outcome (Fig. 11.2). However, the atmosphere is a chaotic system, meaning that tiny, unobservable differences in the initial conditions or small imperfections in the model itself can lead to vastly different forecast outcomes over time. This fundamental limitation led to the development of **Ensemble Forecasting**.

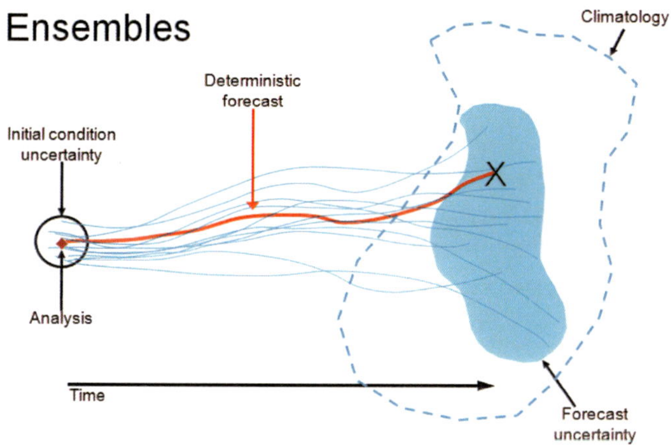

Fig. 11.2: Ensemble forecasting

11.3.1 What is Ensemble Forecasting?

Ensemble forecasting is a technique that addresses the chaotic nature of the atmosphere by running a **collection (an "ensemble") of multiple forecasts** from slightly different initial conditions and/or using slightly different versions of the same numerical weather prediction (NWP) model. Instead of providing a single "best guess," an ensemble forecast provides a **range of possible future scenarios** and, crucially, an estimate of the **probability or confidence** associated with those scenarios.

Think of it like this: If you ask a single expert to predict the stock market, you get one prediction. If you ask 50 experts, you get 50 different predictions, and the spread of those predictions gives you a sense of how certain or uncertain the future is. Ensemble forecasting applies this same principle to the atmosphere.

11.3.2 Why is Ensemble Forecasting Necessary? The Problem of Chaos

The need for ensemble forecasting stems directly from the concept of **deterministic chaos** in the atmosphere:

- **Imperfect Initial Conditions:** We can never perfectly observe the entire atmosphere at any given moment. There are always small errors or gaps in our initial observations (e.g., a tiny unobserved temperature fluctuation over the ocean, or a slight error in a satellite measurement).
- **Model Imperfections:** NWP models are approximations of reality. They contain simplifications, parameterizations for processes they can't explicitly resolve, and inherent numerical errors.
- **Sensitive Dependence on Initial Conditions (The "Butterfly Effect"):** Even if initial conditions and models were almost perfect, a tiny, unobservable difference (like a butterfly flapping its wings in Brazil) can, over time, amplify and lead to vastly different weather outcomes in a distant location (like a tornado in Texas). This divergence happens rapidly in chaotic systems.

A single model run (deterministic forecast) cannot account for these uncertainties. Ensemble forecasting explicitly attempts to capture the range of possible atmospheric evolutions given these inherent uncertainties.

11.3.3 How Ensemble Forecasting Works (The General Process)

The core idea is to create a set of slightly different "starting points" for the NWP model and then run the model multiple times from each of these perturbed starting points.

- **Perturbing Initial Conditions**
  - The most common method is to create an ensemble of initial conditions, each slightly different from the "best guess" analysis. These perturbations are designed to represent the uncertainty in the initial observations and the data assimilation process.
  - Techniques like **Singular Vectors** (used by ECMWF) or **Ensemble Transform Kalman Filter (ETKF)**(used by NCEP) are used to generate these perturbations in meteorologically relevant ways, focusing on areas where small uncertainties are most likely to grow rapidly.
  - Each perturbed initial condition is then used as the starting point for one "ensemble member" forecast.

- **Perturbing Model Physics (Optional but Common)**
    - Some ensemble systems also introduce small perturbations or variations in the physical parameterizations within the model itself (e.g., slightly different ways of calculating cloud formation or radiation). This accounts for the uncertainty in our understanding of atmospheric physics.
    - This can involve running different versions of the model or introducing stochastic (random) perturbations to the parameterization schemes.
- **Running the Ensemble**
    - Each of these slightly different initial conditions and/or model physics configurations is run forward in time by the NWP model. This creates a collection of, say, 20, 50, or even hundreds of individual forecasts.
    - These individual forecasts are called **ensemble members**.
- **Analyzing the Ensemble Output**
    - Instead of looking at just one forecast, meteorologists analyze the **spread** and **clustering** of all the ensemble members.
    - **Spread:** The degree to which the ensemble members differ from each other. A large spread indicates high uncertainty in the forecast, while a small spread suggests higher confidence.
    - **Mean/Median:** The average or median of the ensemble members often provides a more robust and skillful forecast than any single deterministic run, as it smooths out some of the chaotic noise.
    - **Probabilities:** The ensemble allows for the calculation of probabilities for specific events (e.g., "There is a 70% chance of precipitation exceeding 10mm," or "There is a 40% chance of temperature dropping below freezing"). This is done by counting how many ensemble members predict that event.

11.3.4 Key Benefits of Ensemble Forecasting

- **Quantifies Forecast Uncertainty:** This is the most significant advantage. Ensemble forecasts provide a direct measure of how confident meteorologists should be in a given forecast. If the ensemble members are tightly clustered, confidence is high. If they are widely spread, confidence is low, indicating a less predictable situation.
- **Provides Probabilistic Forecasts:** Instead of "It will rain," an ensemble can say "There is an 80% chance of rain." This probabilistic information is invaluable for decision-makers (e.g., farmers, emergency managers, event planners) who need to weigh risks.

- **Identifies Multiple Possible Scenarios:** When the atmosphere is highly uncertain, the ensemble can reveal distinct "clusters" of forecasts, indicating that there are several plausible future weather scenarios. This allows for contingency planning.
- **Improved Forecast Skill:** The ensemble mean (average of all members) often outperforms any single deterministic forecast, particularly for medium-range forecasts, because it averages out the random noise and biases of individual runs.
- **Early Warning of Predictability Changes:** A rapidly increasing ensemble spread can serve as an early warning that a previously predictable forecast is becoming highly uncertain, prompting forecasters to communicate this uncertainty.
- **Guidance for Data Assimilation:** Ensemble systems are often used to estimate the "flow-dependent" background error covariance in advanced data assimilation systems (like Ensemble Kalman Filters), leading to better initial conditions.

11.3.5 Visualizing Ensemble Forecasts

Specialized plots are used to visualize ensemble output:

- **Spaghetti Plots:** Show the trajectories or contours of a single variable (e.g., a geopotential height contour or a storm track) from each ensemble member overlaid on a map. This vividly illustrates the spread and agreement among forecasts.
- **Plume Diagrams:** For a single location, these show the time series of a variable (e.g., temperature) from all ensemble members, often with the ensemble mean and percentile ranges (e.g., 10th-90th percentile) highlighted.
- **Probability Maps:** Show the probability of a specific event occurring (e.g., probability of precipitation exceeding a threshold, or probability of temperature falling below freezing) by counting the number of ensemble members that predict that event.
- **Stamp Plots:** Display small maps for each individual ensemble member, allowing for a quick visual scan of the range of scenarios.

11.3.6 Types of Ensemble Systems

- **Global Ensembles:** Run by major operational centers (e.g., ECMWF, NCEP, UK Met Office) to provide medium-range and long-range probabilistic forecasts for the entire globe.

- **Regional Ensembles (Convection-Allowing Ensembles - CAE):** Higher-resolution ensembles focused on limited areas, designed to forecast small-scale phenomena like individual thunderstorms and their associated hazards (e.g., HRRR Ensemble in the U.S.).
- **Multi-Model Ensembles:** Combine forecasts from different NWP models (e.g., combining forecasts from GFS, ECMWF, and UKMO) to leverage the strengths of each and potentially reduce systematic errors.

11.3.7 Challenges and Future Directions

- **Computational Cost:** Running multiple high-resolution models is extremely computationally expensive, limiting ensemble size.
- **Under-Dispersion:** Ensembles sometimes underestimate the true forecast uncertainty (i.e., the observed weather falls outside the ensemble spread). Research focuses on improving initial condition perturbations and model physics variations to address this.
- **Interpretation:** Communicating probabilistic forecasts and the concept of uncertainty to the public and decision-makers effectively remains a challenge.
- **Post-processing:** Raw ensemble output often needs statistical post-processing (e.g., calibration, bias correction) to improve its reliability and skill.

Ensemble forecasting represents a profound shift in meteorological prediction, moving beyond a single deterministic "best guess" to provide a more realistic and comprehensive picture of future weather. By explicitly accounting for the inherent uncertainties in the atmosphere and our models, ensembles deliver invaluable information on forecast confidence, probabilities of specific events, and the range of plausible future scenarios. This probabilistic approach is fundamental for risk assessment, contingency planning, and ultimately, for making more informed decisions in a world increasingly impacted by weather and climate extremes. It is the gold standard for modern weather prediction.

11.4 NWP Model Output: Gridded Forecasts and Reanalysis Products

Numerical Weather Prediction (NWP) models are the backbone of modern weather forecasting, producing vast amounts of data that describe the future (and sometimes past) state of the atmosphere. This output comes in various forms, but two of the most fundamental and widely used are **gridded forecasts** and **reanalysis products**. While both are derived from NWP models, they serve distinct purposes and are generated through different processes.

11.4.1 Gridded Forecasts (NWP Model Output)

Gridded forecasts are the primary output of numerical weather prediction models. Instead of providing a single value for a specific location, these forecasts provide predictions for a multitude of atmospheric variables (e.g., temperature, pressure, humidity, wind components, precipitation) at discrete points arranged on a **regular grid** in three dimensions (latitude, longitude, and altitude/pressure level) and extending forward in time.

- **How they are Generated**
    - **Initial Conditions:** The process begins with the "analysis"– the best estimate of the current atmospheric state, derived by intelligently combining observations with a previous model forecast through data assimilation.
    - **Model Integration:** Using these initial conditions, the NWP model (a complex set of mathematical equations representing atmospheric physics) is numerically integrated forward in time on a supercomputer.
    - **Output Generation:** At predefined time steps (e.g., every 1 hour, 3 hours, 6 hours), the model's computed atmospheric state is saved onto the specified grid. This generates a series of "forecast hours" or "forecast steps."
- **Characteristics**
    - **Spatial Resolution:** Defined by the spacing between grid points (e.g., 0.25×0.25°latitude/longitude, or 13 kmhorizontal resolution). Higher resolution means finer detail but greater computational cost.
    - **Vertical Resolution:** Defined by the number of vertical levels (e.g., 64 pressure levels from the surface to the stratosphere).
    - **Temporal Resolution:** Defined by the frequency of output (e.g., hourly, 3-hourly, 6-hourly forecasts).
    - **Variables:** Typically include standard meteorological variables like temperature, geopotential height, wind components (u, v), specific humidity, sea level pressure, precipitation, cloud cover, etc., at various levels and the surface.
    - **Forecast Range:** Can range from short-range (0-48 hours) from high-resolution regional models to medium-range (3-10 days) and long-range (10+ days) from global models.
- **Purpose and Applications**
    - **Primary Source for Weather Forecasts:** Gridded forecasts are the foundation for almost all modern weather forecasts, from daily local predictions to severe weather warnings.

  - **Input for Downstream Models:** They provide boundary conditions for higher-resolution regional models, and initial conditions for impact models (e.g., hydrological models, air quality models).
  - **Research and Development:** Used extensively by researchers to study atmospheric processes, evaluate model performance, and develop new forecasting techniques.
  - **Specific Sector Applications:** Used in aviation, agriculture, energy (wind/solar power), transportation, and disaster management.
- **Advantages**
  - **Physically Consistent:** The forecasts are generated by a model based on physical laws, ensuring a degree of internal consistency.
  - **Spatially and Temporally Complete:** Provides data for every grid point and every output time step within the forecast range, filling in gaps where observations are sparse.
  - **Predictive Capability:** Their core function is to predict future atmospheric states.
- **Disadvantages**
  - **Model Biases and Errors:** All models have inherent biases and errors due to simplifications, parameterizations, and initial condition uncertainties. These errors grow with forecast lead time.
  - **Resolution Limitations:** Even high-resolution models cannot perfectly resolve all sub-grid scale processes, leading to some inaccuracies.
  - **Computational Cost:** Running and storing these vast datasets is extremely expensive.
- **Examples**
  - Output from the GFS (Global Forecast System), ECMWF IFS, HRRR (High-Resolution Rapid Refresh), WRF (Weather Research and Forecasting Model).

11.4.2 Reanalysis Products

Reanalysis products are a special type of gridded dataset that provides a **retrospective, physically consistent, and spatially/temporally complete record of the past state of the atmosphere and/or ocean**. Unlike forecasts, which predict the future, reanalysis aims to provide the "best estimate" of what actually happened in the past, often spanning several decades or even over a century.

- **How they are Generated:** Reanalysis is essentially a continuous, highly sophisticated **data assimilation process applied to historical observations**.
- **Fixed NWP Model and Data Assimilation System:** A single, state-of-the-art NWP model and data assimilation system (often a 4D-Var or Ensemble Kalman Filter based system) is chosen and **fixed in time**. This means the model's physics, resolution, and the assimilation methodology do not change over the entire reanalysis period.
- **Assimilation of Historical Observations:** This fixed system then systematically processes *all available historical observations* (surface, upper-air, satellite, aircraft, buoy data) from the beginning of the reanalysis period to the end. These observations are assimilated into the model, just like in real-time forecasting.
- **Backward and Forward Passes (often):** To ensure maximum consistency, some reanalysis projects might even perform multiple "passes" over the data, going both forward and backward in time to optimize the fit between the model and observations.
- **Characteristics**
    - **Consistency:** The most defining characteristic is its **temporal consistency**. Because the model and assimilation system are fixed, any changes or trends observed in the reanalysis product are due to actual changes in the atmosphere/ocean, not due to changes or improvements in the forecasting system itself.
    - **Completeness:** Provides a complete, gridded dataset across all three spatial dimensions and over long historical periods, filling in observational gaps using the model physics.
    - **Resolution:** Typically global, with resolutions varying depending on the reanalysis project (e.g., ERA5 is $0.25^{o}\times0.25^{o}$).
    - **Variables:** Similar to forecast variables, but also often includes derived diagnostic variables.
    - **Period:** Can span from decades (e.g., ERA5 from 1950 to present) to over a century (e.g., 20th Century Reanalysis).
- **Purpose and Applications**
    - **Climate Monitoring and Research:** The primary use. Reanalysis products are invaluable for studying climate variability, trends, and change over long periods, as they provide a consistent view of the atmosphere.

- **Model Validation:** Used as a reference dataset to validate and improve climate models and future NWP systems.
- **Driving Impact Models:** Provides consistent meteorological inputs for hydrological models, agricultural models, and air quality models for historical periods.
- **Extreme Event Analysis:** Studying the atmospheric conditions associated with past extreme weather events (e.g., major hurricanes, droughts, heatwaves) to understand their mechanisms and frequency.
- **Energy Sector Applications:** Assessing historical wind and solar resources for renewable energy planning.

- **Advantages**
  - **Temporal Consistency:** Crucial for climate studies, as observed trends are real and not artifacts of changing model systems.
  - **Spatially and Temporally Complete:** Provides a continuous, gap-free record of the atmosphere, even in data-sparse regions or periods.
  - **Physically Consistent:** The underlying NWP model ensures that the derived atmospheric variables adhere to physical laws.
  - **Integrates All Available Observations:** Leverages the full historical record of observations to create the most accurate possible representation of the past.
- **Disadvantages**
  - **„Frozen" System:** While consistency is an advantage, it also means that improvements in model physics or assimilation techniques developed *after* the reanalysis system was frozen are not incorporated.
  - **Resolution Limitations:** Like forecasts, they are limited by the resolution of the underlying model, meaning they might not fully resolve very small-scale historical events.
  - **Computational Cost:** Generating reanalysis products is a massive computational undertaking.
  - **Observational Changes:** While the system is fixed, changes in the observational network over time (e.g., introduction of satellites) can still introduce subtle "regime shifts" in the reanalysis, which researchers must be aware of.
- **Examples**
  - **ERA5:** The fifth generation ECMWF reanalysis for the global climate and weather, covering 1950 to present.

- **NCEP/NCAR Reanalysis (R1 and R2):** Older but widely used reanalysis products from the National Centers for Environmental Prediction.
- **JRA-55:** The Japanese 55-year Reanalysis.
- **20th Century Reanalysis (20CR):** Extends back to the late 19th century, assimilating only surface pressure observations.

11.4.3 Complementary Tools for Understanding the Atmosphere

Gridded forecasts and reanalysis products are both indispensable outputs of the NWP modeling framework, but they serve distinct and complementary roles. Gridded forecasts provide our best estimate of the **future** atmosphere, guiding daily decisions and warnings. Reanalysis products, on the other hand, offer the most consistent and comprehensive view of the **past** atmosphere, serving as a critical dataset for climate research, historical event analysis, and understanding long-term environmental change. Together, they form the bedrock of modern atmospheric science, enabling us to both predict the future and understand the history of our dynamic planet.

11.5 Verification and Evaluation of NWP Forecasts and Metrics for Assessing Forecast Skill

Numerical Weather Prediction (NWP) models are the sophisticated engines driving modern weather forecasts, producing detailed predictions of the atmosphere's future state. However, no forecast is perfect. To ensure that these predictions are reliable, useful, and continuously improving, a rigorous and systematic process of **verification and evaluation** is absolutely essential. This involves carefully comparing what the model predicted against what actually happened in the real world.

11.5.1 Why is Verification and Evaluation So Crucial?

The process of checking and assessing forecasts is vital for several reasons:

- **Assessing Forecast Quality:** It provides an objective way to measure how "good" a forecast truly is. This helps meteorologists and the public understand how much confidence to place in a prediction.
- **Identifying Strengths and Weaknesses:** By systematically comparing forecasts to observations, we can pinpoint exactly where a model performs well (e.g., predicting temperature on a clear day) and where it consistently struggles (e.g., forecasting heavy rain in mountainous areas). This feedback is invaluable.
- **Guiding Model Development:** Knowing a model's weaknesses allows scientists and engineers to focus their efforts on improving specific

aspects of the model's physics, its resolution, or how it incorporates new observations. It's a continuous cycle of improvement.

- **Informing Decision-Makers:** People who use forecasts to make critical decisions – like emergency managers planning for a storm, farmers deciding when to plant, or energy companies managing power grids – need to understand not just the forecast itself, but also its inherent uncertainty and reliability.
- **Tracking Progress:** Over time, consistent verification allows us to see if forecast skill is genuinely improving, staying the same, or even getting worse. This helps justify investments in new observing systems or supercomputing power.
- **Comparing Systems:** It provides a fair and objective way to compare the performance of different NWP models or different versions of the same model.

11.5.2 What is Verification?

Verification is the direct act of comparing a forecast to what was actually observed. The observation is often referred to as the "ground truth."**Evaluation** goes a step further, interpreting these comparisons to understand the broader implications of the forecast's performance.

When we verify, we look at several aspects of forecast quality:

- **Accuracy:** How close was the prediction to the actual outcome?
- **Bias:** Did the forecast consistently predict too much or too little, or too warm or too cold?
- **Reliability:** For forecasts that give probabilities (e.g., "70% chance of rain"), did the event actually happen 70% of the time when that probability was given?
- **Resolution:** Can the forecast distinguish between different outcomes (e.g., can it tell the difference between heavy rain and light rain)?
- **Skill:** How much better is the forecast than a very simple, basic prediction (like just guessing, or assuming tomorrow's weather will be exactly like today's)?
- **Value:** How useful is the forecast for a particular user's specific decision? (This is often harder to measure purely with meteorological metrics).

11.5.3 Metrics for Assessing Forecast Skill

The way we measure forecast skill depends on what kind of forecast we're looking at (a single prediction or a probability) and what kind of weather variable it is (a number like temperature, or a category like "rain/no rain").

- **Metrics for Deterministic Forecasts**

These are forecasts that give a single, definite answer (e.g., "The temperature will be 20°C," or "It *will* rain»).

For Continuous Variables (e.g., Temperature, Pressure, Wind Speed)

These metrics quantify how far off the forecast was from the actual observation.

- **Mean Absolute Error (MAE)**
  - **What it measures:** This tells you, on average, how big the errors were, regardless of whether the forecast was too high or too low. You simply take the difference between each forecast and observation, ignore the plus or minus sign, and then average all those differences.
  - **Interpretation:** A smaller MAE means the forecasts were, on average, closer to the actual observations. It's a very intuitive measure of typical error size.
- **Root Mean Squared Error (RMSE)**
  - **What it measures:** Similar to MAE, this also tells you the average magnitude of the error. However, because it squares the errors before averaging, it gives more weight to larger forecast mistakes.
  - **Interpretation:** A smaller RMSE indicates better accuracy. If an RMSE is much higher than an MAE for the same forecast, it suggests there were some particularly large errors that the RMSE is highlighting.
- **Bias (Mean Error)**
  - **What it measures:** This reveals if the forecast has a consistent tendency to be too high or too low. You average the differences between forecasts and observations, keeping the plus or minus signs.
  - **Interpretation:** If the bias is a positive number, the model tends to over-predict (e.g., consistently forecasts temperatures that are too warm). If it's negative, it tends to under-predict (e.g., consistently forecasts temperatures that are too cold). A bias close to zero is ideal, but remember that a low bias doesn't automatically mean high accuracy (large positive and negative errors could cancel each other out).

- **Correlation Coefficient (e.g., Pearson Correlation Coefficient)**
  - **What it measures:** This tells you how well the forecast and the observation move together in a linear fashion. For example, if the observed temperature goes up, does the forecast temperature also tend to go up?
  - **Interpretation:** A value close to positive one means they track each other very well. A value close to zero means there's no clear linear relationship. While a high correlation is good, it doesn't mean the forecast is perfectly accurate or unbiased; it just means the *pattern* of change is similar.

11.5.4 For Categorical Variables (e.g., Rain/No Rain, Severe Storm/No Severe Storm)

These metrics are calculated by counting how many times the forecast was right or wrong for specific categories, often organized in a table.

Imagine a table with rows for "Forecasted Rain" and "Forecasted No Rain," and columns for "Observed Rain" and "Observed No Rain."

- **Hits (H):** Forecasted Rain, and it *did* rain.
- **False Alarms (FA):** Forecasted Rain, but it *did not* rain.
- **Misses (M):** Forecasted No Rain, but it *did* rain.
- **Correct Negatives (CN):** Forecasted No Rain, and it *did not* rain.
- **Probability of Detection (POD) / Hit Rate:**
  - **What it measures:** The proportion of actual events (e.g., observed rain) that the forecast correctly predicted.
  - **Interpretation:** Ranges from zero to one. A POD of one means the forecast caught every single instance of rain. It's good for knowing if you're catching the events, but it doesn't tell you if you're also issuing a lot of false alarms.
- **False Alarm Ratio (FAR)**
  - **What it measures:** The proportion of "yes" forecasts (e.g., forecasted rain) that turned out to be wrong (it didn't rain).
  - **Interpretation:** Ranges from zero to one. A FAR of zero means the forecast never cried wolf. It's good for knowing how often your "yes" forecasts are reliable, but it doesn't tell you if you're missing a lot of actual events.

- **Critical Success Index (CSI) / Threat Score**
  - **What it measures:** A balanced measure that considers hits, false alarms, and misses. It's particularly useful for rare but important events (like severe thunderstorms).
  - **Interpretation:** Ranges from zero to one. A CSI of one means a perfect forecast. It's a good overall score because it penalizes both false alarms and misses.
- **Bias Score (BIAS)**
- **What it measures:** Compares the number of times the forecast predicted an event to the number of times the event actually occurred.
- **Interpretation:** A BIAS of one means the forecast system predicted the correct *number* of events (even if they weren't the exact right ones). A BIAS greater than one means it over-forecasted the event (predicted it too often); a BIAS less than one means it under-forecasted.
- **Accuracy (ACC)**
  - **What it measures:** The overall proportion of all forecasts that were correct (both hits and correct negatives).
  - **Interpretation:** A high accuracy seems good, but it can be misleading for rare events. For example, if it only rains 5% of the time, a forecast that *always* predicts "no rain" would have 95% accuracy, but it would be completely useless when it actually does rain.
  - **Metrics for Probabilistic Forecasts (e.g., Probability of Rain):** These forecasts provide a likelihood (a percentage or probability) of an event happening (e.g., "There's a 70% chance of rain").
- **Brier Score (BS)**
  - **What it measures:** This is a widely used score that measures the "goodness" of probability forecasts. It essentially calculates the average squared difference between the forecast probability and the actual outcome (where the outcome is 1 if the event happened, 0 if it didn't).
  - **Interpretation:** Ranges from zero to one. A lower Brier Score means better accuracy. A perfect forecast would have a Brier Score of zero. It encourages honest and well-calibrated probability forecasts.
- **Reliability Diagram**
- **Purpose:** This is a visual tool to check the **reliability** of probabilistic forecasts. It asks: "When the forecast said there was a 30% chance of rain, did it actually rain about 30% of the time?"

- **What it shows:** It plots the actual observed frequency of an event (on the vertical axis) against the forecast probability (on the horizontal axis).
- **Interpretation:** If the forecast system is perfectly reliable, all the points on the diagram will fall along a 45-degree diagonal line. If points are above the line, the forecast system is "over-forecasting" (e.g., it said 30% chance, but it rained 50% of the time). If points are below the line, it's "under-forecasting." The diagram also often shows how often each probability was issued.
- **ROC Curve (Receiver Operating Characteristic Curve)**
  - **Purpose:** This visual tool assesses the **resolution** or **discriminating ability** of probabilistic forecasts – how well the forecast can separate instances where an event occurred from instances where it didn't. It's independent of any overall bias in the forecast.
  - **What it shows:** It plots the "Probability of Detection" (how many events were caught) against the "False Alarm Rate" (how many non-events were incorrectly predicted as events) for various probability thresholds.
  - **Interpretation:** A perfect forecast would have a curve that shoots straight up to the top-left corner and then across. A diagonal line means the forecast has no skill (it's no better than random guessing). The **Area Under the Curve (AUC)** is a single number summarizing this skill: an AUC of one is perfect, an AUC of 0.5 is no skill.
- **Skill Scores**
  - Many of the metrics above give an absolute measure of performance. **Skill scores** take it a step further by comparing the performance of a forecast system to that of a **simple reference forecast** (a very basic, often non-scientific prediction). This tells you the *true skill* of the forecast beyond what you'd get by just guessing or assuming average conditions.
  - **How it works:** A skill score essentially calculates how much better your forecast is compared to a baseline forecast. If your forecast is perfect, the skill score is typically one. If it's no better than the simple reference forecast, the skill score is zero. If it's worse, it's a negative number.

Common simple reference forecasts include

- **Climatology:** Always predicting the long-term average weather for that time of year and location.

- **Persistence:** Always predicting that tomorrow's weather will be exactly the same as today's.
- **Random Chance:** A forecast based on pure randomness.

11.5.5 Challenges in Verification

Despite the sophisticated metrics, verifying NWP forecasts comes with its own set of practical difficulties:

- **Observational Errors:** The "ground truth" observations themselves are not perfect; they contain their own measurement errors and uncertainties.
- **Representativeness:** A single point observation from a weather station might not perfectly represent the average conditions over a model's much larger grid cell. This is especially true for highly localized phenomena like thunderstorms.
- **Scale Dependency:** A model might be very accurate at predicting large-scale weather patterns but struggle with small-scale features. Verification needs to be done at appropriate scales to be meaningful.
- **Rare Events:** It's very challenging to verify forecasts of rare but high-impact events (like tornadoes or extreme floods) because there are simply not many historical examples to compare against.
- **«Double Penalty»:** For phenomena like precipitation, a forecast might predict rain, but if it falls just a few kilometers away or an hour too late, the verification system might penalize the forecast twice: once for missing the observed rain at the correct spot/time, and once for forecasting rain where none occurred.
- **Human Perception vs. Metrics:** Sometimes, a forecast might "look" good to a human meteorologist but score poorly on a specific metric, or vice versa. This highlights the need for a suite of metrics and expert judgment.

Verification and evaluation are not just academic exercises; they are indispensable components of Numerical Weather Prediction. By systematically comparing forecasts to observations using a comprehensive suite of carefully chosen metrics, meteorologists can objectively quantify forecast quality, pinpoint areas for improvement, and track the relentless progress in our ability to predict the atmosphere. This continuous feedback loop is vital for advancing NWP models, enhancing forecast skill, and ultimately providing more reliable and valuable weather information for a wide range of societal and economic applications, helping us better prepare for and respond to the challenges of a dynamic atmosphere.

11.6 Post-Processing of NWP Data: Bias Correction and Downscaling

Numerical Weather Prediction (NWP) models are incredibly powerful tools that provide the foundation for modern weather forecasts. However, the raw output directly from these complex models is rarely used without further refinement. **Post-processing** refers to a suite of techniques applied to NWP model output after the model simulation is complete. The primary goal of post-processing is to enhance the accuracy, utility, and interpretability of forecasts, making them more relevant and reliable for specific applications and users.

The need for post-processing stems from inherent limitations in NWP models

- **Systematic Biases:** Models often have consistent errors (biases) due to simplifications in their physics, approximations in their numerical schemes, or limitations in how they represent sub-grid scale processes.
- **Resolution Mismatch:** Global and even regional models may not have a fine enough resolution to capture local weather phenomena influenced by complex terrain, urban environments, or small-scale atmospheric processes.
- **Direct Output vs. User Needs:** Raw model output (e.g., temperature at 2 meters above ground) might need to be converted into more user-friendly or specific products (e.g., "feels like" temperature, probability of frost, energy consumption forecasts).

Two of the most critical and widely used post-processing techniques are **Bias Correction** and **Downscaling**.

11.6.1 Bias Correction

Bias correction is the process of identifying and removing or reducing systematic errors (biases) in NWP model forecasts. A bias occurs when a model consistently over-predicts or under-predicts a particular variable (e.g., always forecasting temperatures that are too warm, or precipitation amounts that are too low).

Why NWP Models Have Biases: Despite their sophistication, NWP models are not perfect representations of the atmosphere. Biases can arise from:

- **Parameterization Schemes:** Sub-grid scale processes (like cloud formation, convection, radiation, land-surface interactions) cannot be explicitly resolved by the model's grid. They are represented by simplified mathematical formulas called parameterizations, which are imperfect.

- **Initial Conditions:** While data assimilation aims for the best initial state, some residual errors or imbalances can persist.
- **Numerical Approximations:** The mathematical equations are solved numerically, introducing small errors.
- **Boundary Conditions:** For regional models, biases can be inherited from the global model providing their lateral boundary conditions.
- **Representativeness:** The model's grid cell average might not perfectly match a point observation from a weather station.

How Bias Correction Works: Bias correction methods typically learn the systematic relationship between past model forecasts and corresponding observations. Once this relationship is learned, it is applied to future raw model forecasts to "correct" them.

- **Simple Mean Bias Correction**
  - **Concept:** The most basic form. It calculates the average difference between past forecasts and observations for a specific variable at a specific location. This average bias is then simply subtracted from (or added to) future forecasts.
  - **Example:** If a model consistently forecasts temperatures 2°C too high at a particular airport, 2°C is subtracted from all future temperature forecasts for that airport.
  - **Limitations:** Assumes the bias is constant over time and doesn't depend on other factors (like atmospheric conditions or forecast lead time).
- **Model Output Statistics (MOS)**
  - **Concept:** A more sophisticated statistical technique. MOS develops regression equations (or other statistical relationships) that relate raw NWP model output variables (predictors) to actual observed weather conditions (predictands). These equations are derived from a historical dataset of paired forecasts and observations.
  - **How it works:** For example, an MOS equation for maximum temperature might use the model's forecast of 2 m temperature, 850 hPa temperature, cloud cover, and wind speed as predictors to statistically correct the forecast.
- **Advantages**
  - **Accounts for Model Bias:** Directly corrects for systematic model errors.

- **Accounts for Local Effects:** Can implicitly incorporate local effects (e.g., terrain, urban heat island) that the model's resolution might miss, as these are reflected in the historical observations.
- **Tailored to Specific Variables:** Can be developed for various variables (temperature, precipitation probability, wind gusts, ceiling, visibility).

- **Limitations**
  - **Stationary Assumption:** Assumes that the relationship between model output and observations remains constant over time. If the model is updated, MOS equations need to be re-derived.
  - **Data Intensive:** Requires a long historical record of both model forecasts and observations.
  - **Does not improve "Resolution":** MOS can make a forecast more accurate and reliable, but it cannot add detail or information that wasn't already in the raw model output.
- **Machine Learning Approaches**
  - **Concept:** Modern bias correction increasingly uses advanced machine learning algorithms (e.g., Random Forests, Gradient Boosting Machines, Neural Networks) to learn the complex, non-linear relationships between raw model output and observations.
  - **Advantages:** Can capture more intricate and non-linear biases than traditional MOS, potentially leading to higher accuracy. More adaptable to complex interactions.
  - **Limitations:** Can be more computationally intensive to train and may require larger datasets. Interpretability can be challenging.
- **Benefits of Bias Correction**
  - **Improved Accuracy:** Reduces systematic errors, making forecasts closer to reality.
  - **Increased Reliability:** For probabilistic forecasts, bias correction ensures that the stated probabilities are more consistent with observed frequencies.
  - **Enhanced Trust:** More accurate and reliable forecasts build greater confidence among users.
  - **Better Decision Support:** Provides more precise information for critical decisions.

11.6.2 Downscaling

Downscaling is the process of transforming meteorological information from a coarse spatial resolution (e.g., a global climate model grid cell of 100 km) to a finer spatial resolution (e.g., a local forecast for a specific valley, or a climate projection for a 1 km grid). This is crucial because many local impacts and decisions require highly detailed weather or climate information that coarse-resolution models cannot provide.

- **Why Downscaling is Necessary (The Resolution Mismatch)**
  - **Model Limitations:** Global and even regional NWP models, while powerful, operate at resolutions that are too coarse to explicitly resolve local-scale phenomena influenced by complex topography (mountains, valleys), coastlines, urban heat islands, or small-scale convective systems.
  - **Local Impacts:** Many critical applications (e.g., flash flood warnings, agricultural planning, urban air quality, renewable energy site assessment) require weather information at scales of kilometers or even meters, not tens or hundreds of kilometers.

There are two main categories of downscaling

- **Dynamical Downscaling**
- **Concept:** Involves running a higher-resolution **Regional Climate Model (RCM)** or **Limited Area Model (LAM)** over a smaller geographical domain, using the output from a coarser-resolution global model as its **lateral boundary conditions**. The RCM/LAM contains more detailed physics and topography, allowing it to simulate fine-scale atmospheric processes.
- **How it works**
  - A global model (e.g., a Global Climate Model for climate projections, or a Global NWP model for weather forecasts) provides the large-scale atmospheric context (temperature, pressure, winds) at its boundaries.
  - The higher-resolution RCM/LAM is "nested" within this global model's domain.
  - The RCM/LAM then uses its finer grid and more detailed physical parameterizations (or even explicitly resolves processes like convection if the resolution is high enough, e.g., 1–4 km) to simulate the weather within its smaller domain. It effectively "translates" the large-scale global forcing into local weather.

- **Advantages**
  - **Physically Consistent:** The downscaled output is generated by a physical model, ensuring dynamic and thermodynamic consistency.
  - **Resolves Local Processes:** Can explicitly simulate small-scale atmospheric phenomena and the influence of complex terrain.
  - **Provides New Information:** Can generate information about processes that were not explicitly resolved by the coarser global model.
- **Disadvantages**
  - **Computationally Expensive:** Running high-resolution regional models is very demanding, especially for long climate simulations.
  - **Boundary Condition Dependence:** The quality of the downscaled output is highly dependent on the accuracy of the global model's boundary conditions. Errors from the global model can propagate into the regional model.
  - **Model Biases:** The regional model itself can have its own biases.
- **Statistical Downscaling**
  - **Concept:** Establishes statistical relationships between large-scale atmospheric variables (from a coarse-resolution model or reanalysis) and local-scale surface variables (from observations). Once these relationships are learned from historical data, they are applied to future coarse-resolution model outputs to predict local conditions.
- **How it works**
  - **Training Phase:** A statistical model (e.g., regression, machine learning algorithm) is trained using a historical dataset that includes both large-scale atmospheric variables (predictors) and corresponding local observations (predictands).
  - **Application Phase:** Once the statistical relationship is learned, it is applied to the output of a coarse-resolution NWP or climate model. The model's large-scale forecast is fed into the statistical model, which then predicts the local-scale weather.
- **Types of Statistical Downscaling Methods**
  - **Regression-based:** Simple or multiple linear regression, or more advanced non-linear regression techniques.
  - **Weather Generators:** Statistical models that generate synthetic local weather sequences based on large-scale atmospheric states.

- **Machine Learning:** Increasingly, advanced ML algorithms (e.g., Artificial Neural Networks, Random Forests, Support Vector Machines) are used to capture complex, non-linear relationships.
- **Advantages**
  - **Computationally Inexpensive:** Much less demanding than dynamical downscaling, making it feasible for long climate projections or large ensembles.
  - **Directly Corrects Bias:** Can implicitly correct for systematic biases in the coarse-resolution model, as it's trained on observed local data.
  - **Handles Local Effects:** Can capture local effects that are present in the historical observations, even if the model doesn't explicitly resolve them.
- **Disadvantages**
  - **Stationarity Assumption:** Assumes that the statistical relationships learned from the past will hold true in the future. This assumption can be problematic, especially under significant climate change.
  - **Cannot Create New Information:** It cannot generate information about physical processes not present in the coarse-resolution model or the historical observations. It essentially redistributes existing information.
  - **Limited Physical Consistency:** The downscaled output may not be as physically consistent as dynamically downscaled output.

11.6.3 Enhancing the Utility of NWP Data

Post-processing techniques like bias correction and downscaling are indispensable for maximizing the utility and accuracy of NWP model output. Bias correction systematically refines forecasts by removing consistent errors, making them more reliable. Downscaling, whether dynamical or statistical, bridges the critical gap between the coarse resolution of global models and the fine-scale detail required for local impact assessment and decision-making. Together, these methods transform raw model data into actionable, high-quality meteorological information, enabling more precise forecasts, better risk management, and a deeper understanding of our dynamic atmosphere at all scales.

11.7 Conclusion

Concluding our in-depth exploration of Numerical Weather Prediction, it is abundantly clear that NWP stands as the **cornerstone of modern meteorology**, underpinning virtually every aspect of our daily weather forecasts and climate understanding. We've delved into the intricate process of transforming the fundamental laws of atmospheric physics into solvable mathematical equations, integrated forward in time by powerful supercomputers. From the crucial role of data assimilation in providing the most accurate initial conditions, to the strategic interplay between global and regional models, and the vital insights gained from ensemble forecasting's probabilistic approach, NWP is a testament to scientific ingenuity and relentless technological advancement. The continuous cycle of verification, evaluation, and post-processing, including essential techniques like bias correction and downscaling, ensures that these complex model outputs are constantly refined, made more accurate, and tailored to meet the diverse needs of users, from emergency managers to agricultural planners.

The journey of NWP, from Richardson's manual calculations to today's petascale supercomputing, highlights a relentless pursuit of greater accuracy and longer lead times. While inherent atmospheric chaos and model imperfections will always present challenges, the ongoing advancements in observational technologies, data assimilation schemes, model physics, and computational power promise even more skillful and reliable forecasts in the future. Numerical Weather Prediction is not just a scientific endeavor; it is a critical societal service, empowering communities to better prepare for and respond to the impacts of weather and climate, ultimately enhancing safety, economic efficiency, and environmental resilience across the globe.

12

Climate Data and Analysis

12.1 Introduction

Having explored the dynamic realm of Numerical Weather Prediction, which focuses on forecasting the atmosphere's state over days to weeks, we now shift our focus to a broader, longer-term perspective: **Climate Data and Analysis**. While weather describes the atmospheric conditions at a specific place and time, **climate** refers to the long-term average pattern of weather in a region, typically spanning decades or longer. Understanding and analyzing climate data is paramount for comprehending our planet's past climate, identifying ongoing changes, and projecting future scenarios in a warming world.

The study of climate is fundamentally data-driven. Unlike the instantaneous snapshots used in weather forecasting, climate analysis requires vast historical records, often extending back centuries or even millennia, to discern natural variability from human-induced changes. This data comes from a diverse array of sources, including instrumental records (thermometers, rain gauges, barometers), satellite observations, and proxy records (tree rings, ice cores, sediment layers) that provide indirect evidence of past climate conditions. The sheer volume and heterogeneous nature of this data present unique challenges and opportunities for analysis.

This chapter will delve into the essential aspects of climate data and the methodologies used to analyze it. We will begin by exploring the various **types of climate data** available, understanding their strengths, limitations, and how they are collected and curated. Crucially, we will examine how this data is used to characterize **climate variability** – the natural fluctuations within the climate system – and to detect **climate trends**, particularly those associated with global warming. We will introduce key **climate indices** that summarize large-scale atmospheric and oceanic patterns, providing simplified metrics for understanding complex interactions.

Furthermore, the chapter will cover the critical role of **climate models** in projecting future climate scenarios, distinguishing them from weather models by their longer time scales and broader scope. We will discuss how observed climate data is used to validate and improve these models. Finally, we will touch upon the vital application of climate data analysis in **impact assessment**, helping us understand how changes in climate affect various sectors, from agriculture and water resources to human health and ecosystems.

Ultimately, this chapter aims to equip you with the knowledge and tools necessary to navigate the complex world of climate data, enabling you to interpret scientific findings, identify significant patterns, and contribute to a deeper understanding of our planet's evolving climate system.

12.2 Climate Data Records and Long-Term Homogeneous Data Sets

In the study of climate, unlike short-term weather forecasting, the focus is on long-term patterns, trends, and variability. To achieve this, scientists rely on exceptionally high-quality, stable, and consistent datasets that span decades, centuries, or even millennia. These are known as **Climate Data Records (CDRs)**, and a critical characteristic of their utility is their **homogeneity**.

12.2.1 Climate Data Records (CDRs)

What are Climate Data Records (CDRs)? Climate Data Records (CDRs) are defined as long-term, consistently processed, and quality-controlled datasets of Earth system variables that are suitable for climate monitoring, trend detection, and climate model validation (Fig. 12.1). They are designed to provide a stable, accurate, and continuous record of essential climate variables (ECVs) over extended periods, minimizing the influence of non-climatic factors.

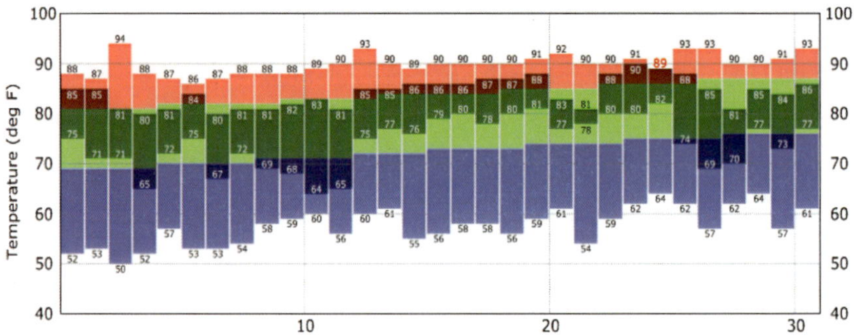

Fig. 12.1: Climate data records, Galveston TX (May, 2025)

Why are CDRs Essential for Climate Analysis? CDRs are the bedrock of climate science for several crucial reasons:

- **Detecting Climate Change:** To confidently identify long-term climate trends (like global warming) and distinguish them from natural climate variability, we need records that are long enough and free from artificial shifts caused by changes in observation methods.

- **Understanding Climate Variability:** CDRs allow scientists to study natural climate cycles and patterns (e.g., El Niño-Southern Oscillation, Pacific Decadal Oscillation) over multiple cycles, understanding their frequency, amplitude, and impacts.
- **Climate Model Validation:** CDRs provide the "ground truth" against which climate models are tested and evaluated. A model's ability to reproduce observed CDRs gives confidence in its projections of future climate.
- **Impact Assessment:** Accurate historical climate data from CDRs is vital for assessing the impacts of past climate events and for projecting future impacts on sectors like agriculture, water resources, and ecosystems.
- **Attribution Studies:** CDRs are used to attribute observed climate changes to specific natural or anthropogenic (human-caused) factors.
- **Key Characteristics of CDRs**
  - **Long Duration:** Typically span multiple decades, often 50 years or more, to capture long-term trends and multi-decadal variability.
  - **Consistency:** Processed using a consistent methodology throughout their entire period to avoid artificial shifts.
  - **Quality Control:** Undergo rigorous checks to identify and remove erroneous data points.
  - **Accuracy:** Strive for the highest possible accuracy, with uncertainties well-quantified.
  - **Homogeneity:** Free from non-climatic biases or shifts (discussed in detail below).
  - **Metadata:** Accompanied by comprehensive documentation (metadata) detailing their origin, processing steps, quality flags, and known limitations.

12.2.2 Long-Term Homogeneous Data Sets:

What does "Homogeneous" Mean in Climate Data? A **homogeneous climate data set** is one where variations in the data are primarily due to changes in the actual climate, rather than changes in the way the data was collected or processed. In other words, it's a dataset where non-climatic factors have been identified and, ideally, removed or adjusted for.

Why is Homogeneity Crucial for Climate Data? Imagine you're trying to determine if a city is getting warmer over a century. If the weather station was moved from a rural area to a city center in the middle of that century, or if the thermometer was changed from an old liquid-in-glass type to a modern electronic sensor, the observed temperature record might show an artificial warming trend that has nothing to do with the actual climate. This is an example of **inhomogeneity**.

Inhomogeneities can easily mimic or mask real climate signals, leading to incorrect conclusions about climate trends or variability. Therefore, ensuring homogeneity is paramount for reliable climate analysis.

- **Common Sources of Inhomogeneity in Climate Data**
    - **Instrument Changes:** Replacing older instruments with newer, more accurate, or different types of sensors (e.g., mercury thermometer to electronic thermistor, different types of rain gauges).
    - **Station Relocations:** Moving a weather station even a short distance can change its microclimate (e.g., moving from an open field to near a building or trees).
    - **Changes in Observing Practices:** Alterations in observation time, shielding of instruments, or manual vs. automated readings.
    - **Urbanization Effects:** For temperature records, the growth of urban areas around a station can lead to an artificial warming trend (the Urban Heat Island effect) that is not representative of regional climate change.
    - **Changes in Data Processing/Algorithms:** Updates to the algorithms used to process raw satellite data or to calculate derived variables can introduce artificial shifts.
    - **Environmental Changes Around the Station:** Growth of vegetation, construction of buildings, or changes in land use immediately surrounding a station.
    - **How are Homogeneous Data Sets (CDRs) Created and Ensured?**

The creation of homogeneous CDRs is a complex, multi-step process that often involves significant effort and sophisticated statistical techniques:

- **Data Rescue**
    - **Process:** Many valuable historical climate observations exist only in paper records (logbooks, charts, microfilms). Data rescue involves digitizing these records, often through manual transcription or optical character recognition.
    - **Importance:** Extends the length of climate records, providing crucial context for modern trends.

- **Quality Control (QC)**
  - **Process:** Initial checks to identify and flag obvious errors in the raw data, such as outliers (values far outside expected ranges), impossible values (e.g., negative precipitation), or sudden jumps. This often involves automated algorithms combined with manual review.
  - **Importance:** Removes gross errors that could severely distort analysis.
- **Homogenization Techniques**
  - **Concept:** The core process of identifying and adjusting for non-climatic shifts (breakpoints) in a time series. This is typically done by comparing a "candidate" station's record with records from nearby, highly correlated "reference" stations that are assumed to be more homogeneous.
- **Statistical Methods**
  - **Breakpoint Detection:** Algorithms (e.g., standard normal homogeneity test (SNHT), penalized maximal t-test) are used to statistically identify points in time where a sudden, non-climatic shift likely occurred.
  - **Relative Homogenization:** Once a breakpoint is detected, the candidate series is adjusted relative to the reference series to remove the detected shift. This often involves calculating differences or ratios between the candidate and reference series.
  - **Metadata Integration:** Known metadata (e.g., "station moved on YYYY-MM-DD,""instrument changed on YYYY-MM-DD") is crucial for confirming statistically detected breakpoints and guiding adjustments.
  - **Urban Heat Island (UHI) Adjustment:** For temperature records, specific methods are used to estimate and remove the UHI effect, often by comparing urban stations to nearby rural stations or using satellite-derived land cover information.
  - **Example:** If a station's temperature record shows a sudden 0.5^0C increase that correlates with a known instrument change but not with changes at surrounding stations, the pre-change data might be adjusted upwards by 0.5^0C to make it consistent with the post-change data.

- **Reanalysis Products (as a form of homogeneous data)**
  - **Concept:** As discussed in the NWP chapter, reanalysis products are created by running a single, fixed numerical weather prediction model and data assimilation system over a long historical period, assimilating all available observations.
  - **Importance:** Because the model and assimilation system are "frozen," any changes or trends in the reanalysis product are primarily due to actual changes in the atmosphere, not changes in the observing or modeling system. This makes reanalysis products inherently more homogeneous than raw observational archives. They provide a spatially and temporally complete, physically consistent, and largely homogeneous view of the past atmosphere.
- **Uncertainty Quantification**
  - **Process:** CDRs are not perfect. Scientists also work to quantify the remaining uncertainties in the homogenized data, often by providing confidence intervals or error bounds.
  - **Importance:** Allows users to understand the reliability of the data and the robustness of any conclusions drawn from it.
- **Examples of Key Climate Data Records and Homogeneous Datasets**
- **Global Surface Temperature Datasets:** These are perhaps the most well-known CDRs, crucial for tracking global warming. They are composites of millions of homogenized land and ocean temperature observations.
  - **HadCRUT (Hadley Centre/CRU Temperature):** From the UK Met Office Hadley Centre and the Climatic Research Unit at the University of East Anglia.
  - **GISS Surface Temperature Analysis (GISTEMP):** From NASA's Goddard Institute for Space Studies.
  - **NOAA GlobalTemp:** From the U.S. National Oceanic and Atmospheric Administration.
- **Global Precipitation Datasets:** Homogenized records of rainfall and snowfall.
- **Satellite-Derived CDRs:** Long-term, consistent records derived from satellite observations, such as:
  - **Sea Level Height:** From satellite altimetry.
  - **Sea Ice Extent and Volume:** From passive microwave sensors.

- **Ocean Heat Content:** Derived from various ocean observations.
- **Outgoing Longwave Radiation (OLR):** A measure of energy emitted by the Earth.
- **Reanalysis Products:** (e.g., ERA5, NCEP/NCAR Reanalysis, JRA-55) provide a consistent, gridded view of many atmospheric variables over decades.

Climate Data Records, underpinned by the rigorous process of ensuring data homogeneity, are the indispensable foundation for understanding Earth's climate system. By meticulously identifying and adjusting for non-climatic influences, scientists transform raw, heterogeneous observations into reliable, long-term records that allow us to confidently detect climate trends, characterize natural variability, validate climate models, and assess the profound impacts of a changing climate. The ongoing effort to extend, improve, and maintain these high-quality datasets is central to advancing climate science and informing global climate action.

12.3 Climate Indices and Teleconnections: Unraveling Global Climate Patterns

The Earth's climate system is a vast and intricately connected web, where atmospheric and oceanic phenomena in one part of the world can exert a profound influence on weather patterns thousands of kilometers away. To decipher these complex, far-reaching relationships, meteorologists and climate scientists employ two fundamental concepts: **climate indices** and **teleconnections**. These two ideas are deeply intertwined, with climate indices serving as the quantitative shorthand for the large-scale patterns that orchestrate these global **teleconnections** (Fig. 12.2).

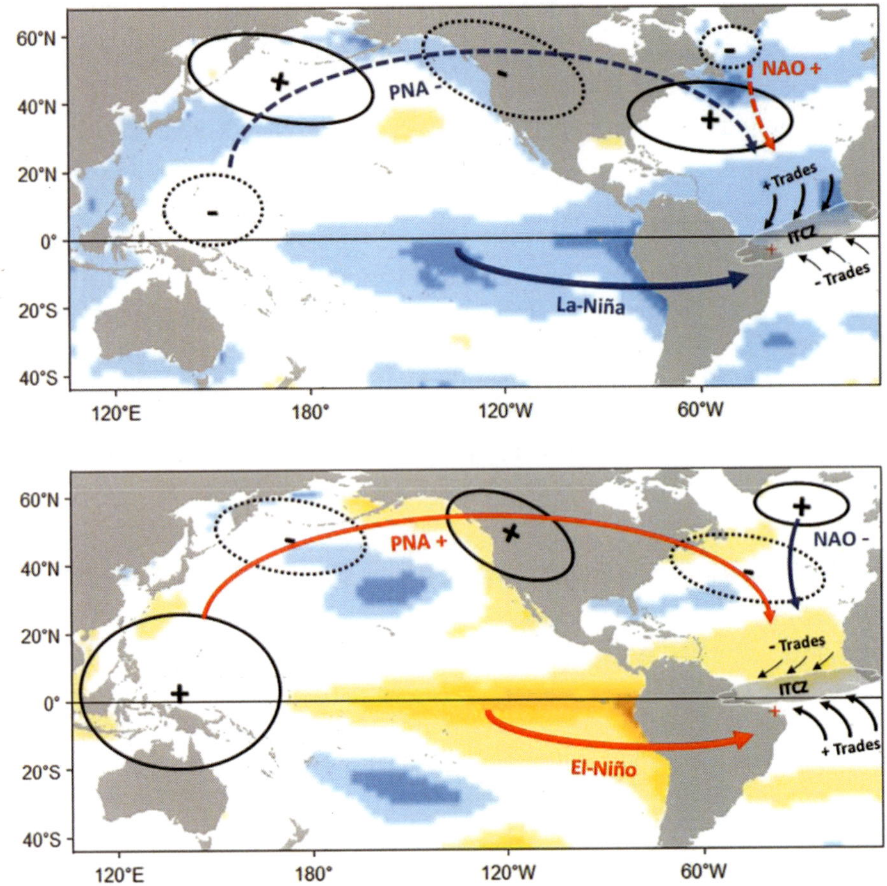

Fig. 12.2: Teleconnections pattern and weather conditions

12.3.1 Climate Indices: The Pulse of Global Climate Patterns

What are Climate Indices? Imagine trying to describe the intricate dance of a massive orchestra by listing every single note played by every instrument. It would be overwhelming. Climate indices are like the conductor's score or the main themes of the symphony. They are **simplified, standardized numerical values** that encapsulate the strength and phase of dominant, recurring patterns of climate variability within the atmosphere or oceans. Instead of analyzing complex maps of temperature or pressure anomalies across an entire ocean basin or hemisphere, an index condenses this vast information into a single, easily trackable number that summarizes the state of a major climate pattern.

- **Why are Climate Indices So Important?** Climate indices are indispensable tools for several reasons:

- **Simplification and Clarity:** They reduce immense amounts of complex spatial data into a single, digestible number, making it easier to monitor, track, and communicate the current state of a climate pattern.
- **Tracking Variability:** They provide a continuous time series that allows scientists to observe the evolution, strength, and duration of major climate phenomena over decades, centuries, or even longer. This helps distinguish natural fluctuations from long-term climate change.
- **Correlation with Regional Impacts:** Crucially, many indices correlate strongly with specific regional weather anomalies (e.g., droughts, floods, heatwaves, cold snaps). This makes them invaluable for understanding and assessing the impacts of climate variability.
- **Diagnostic Tools:** They help scientists diagnose the current state of the global climate system and understand the primary drivers behind observed weather patterns.
- **Predictability:** The evolution of some indices can be predicted weeks or even months in advance, offering valuable lead time for **sub-seasonal to seasonal (S2S) forecasts** – predictions for periods ranging from two weeks to several months ahead.
- **How are Climate Indices Derived?** Indices are typically constructed using various statistical methods applied to large datasets of meteorological and oceanic variables:
    - **Standardized Anomalies:** Calculating how much a variable deviates from its long-term average (the anomaly) and then scaling it by its typical variability (standard deviation).
    - **Principal Component Analysis (PCA) / Empirical Orthogonal Functions (EOFs):** As discussed in previous chapters, EOFs identify the dominant spatial patterns of variability. The time series associated with the leading EOFs can then serve as a robust climate index.
    - **Simple Differences/Ratios:** For some indices, it's as straightforward as calculating the difference in pressure or temperature between two key geographical locations.
- ***Examples of Prominent Climate Indices***
- **El Niño-Southern Oscillation (ENSO)**
    - **Description:** The most powerful and well-known driver of interannual (year-to-year) climate variability globally. It originates in the tropical Pacific Ocean and involves a fundamental interaction between the ocean and the atmosphere.

- **Phases**
- ***El Niño:*** *Characterized by warmer-than-average sea surface temperatures (SSTs) in the central and eastern equatorial Pacific. This shifts rainfall patterns and atmospheric circulation.*
- ***La Niña:*** *Characterized by cooler-than-average SSTs in the central and eastern equatorial Pacific, with opposite impacts to El Niño.*
- **Neutral:** Conditions are near average.

- **Key Indices**
    - **Southern Oscillation Index (SOI):** A standardized index based on the difference in sea level pressure between Tahiti (in the central Pacific) and Darwin, Australia (in the Indian Ocean). Negative SOI values typically indicate El Niño, while positive values indicate La Niña.
    - **Niño 3.4 Index:** This is a direct measure of SST anomalies in a specific, crucial region of the central equatorial Pacific (5°N−5°S,170°W−120°W). Positive values indicate El Niño, and negative values indicate La Niña.
    - **Importance:** ENSO has profound global teleconnections, influencing rainfall, temperature, and tropical cyclone activity across continents.
- **North Atlantic Oscillation (NAO)**
    - **Description:** A major mode of atmospheric variability in the North Atlantic region, characterized by a "seesaw" in atmospheric pressure between the Icelandic Low (near Iceland) and the Azores High (near the Azores islands).
- **Phases**
    - **Positive NAO:** A stronger-than-average pressure difference (deeper Icelandic Low, stronger Azores High). This typically steers storms more northward, bringing milder, wetter, and stormier winters to Northern Europe and drier, often colder conditions to Southern Europe and the Mediterranean. In eastern North America, it often leads to milder winters.
    - **Negative NAO:** A weaker-than-average pressure difference (shallower Icelandic Low, weaker Azores High). This often allows cold Arctic air to penetrate southward, resulting in colder, snowier winters in Northern Europe and warmer, wetter conditions in Southern Europe. Eastern North America can experience colder, snowier conditions.

   - **Index:** Typically calculated as the normalized difference in sea level pressure between a station in Iceland (e.g., Stykkisholmur) and one in Portugal (e.g., Lisbon) or the Azores.
- **Arctic Oscillation (AO)**
- **Description:** A large-scale atmospheric circulation pattern over the Arctic that influences weather across the Northern Hemisphere. It describes the strength and shape of the polar vortex (a large area of low pressure and cold air over the poles).
- **Phases**
- **Positive AO:** A stronger, more compact polar vortex, with lower-than-average pressure over the Arctic and higher-than-average pressure at mid-latitudes. This tends to keep cold air "bottled up" in the Arctic, leading to milder winters in much of mid-latitude North America, Europe, and Asia.
- **Negative AO:** A weaker, more distorted polar vortex, with higher-than-average pressure over the Arctic and lower-than-average pressure at mid-latitudes. This allows cold Arctic air to "spill" southward into mid-latitudes, leading to colder and snowier conditions in various regions.
- **Index:** Often derived from the leading EOF of sea level pressure anomalies north of 20°N.
- **Pacific Decadal Oscillation (PDO)**
   - **Description:** A long-lived (decadal-scale, typically 20-30 years per phase) pattern of sea surface temperature (SST) variability in the North Pacific Ocean. It's characterized by a horseshoe-shaped pattern of SST anomalies.
- **Phases**
   - **Positive PDO (Warm Phase):** Warmer-than-average SSTs along the North American west coast and cooler SSTs in the central North Pacific.
   - **Negative PDO (Cool Phase):** Cooler-than-average SSTs along the North American west coast and warmer SSTs in the central North Pacific.
   - **Importance:** Influences long-term patterns of precipitation and temperature across North America and the Pacific Rim, affecting phenomena like salmon populations, drought cycles, and wildfire risk over multi-year periods.

- **Madden-Julian Oscillation (MJO)**
  - **Description:** The largest element of intra-seasonal (30-90 day) variability in the tropical atmosphere. It manifests as an eastward-propagating pulse of enhanced clouds and rainfall that circles the globe near the equator.
  - **Phases:** Typically divided into 8 phases based on the longitude of the enhanced convective activity.
  - **Importance:** While originating in the tropics, the MJO has significant global teleconnections, influencing tropical cyclone formation and intensity, mid-latitude weather patterns (e.g., cold air outbreaks, precipitation events), and even the onset and break periods of monsoons.

12.3.2 Teleconnections: The Global Climate's Interconnected Web

What are Teleconnections?Teleconnections are the actual **remote atmospheric and oceanic linkages** where climate anomalies (deviations from the average) in one geographical region are statistically correlated with climate anomalies in geographically distant regions. These are not random coincidences but rather physically mediated connections, often involving the propagation of atmospheric waves or the influence of large-scale ocean temperature patterns. They represent the "cause-and-effect" (or at least strong correlation) across vast distances in the climate system.

- **How Do Teleconnections Work? (Underlying Mechanisms)** Teleconnections typically operate through a few key physical mechanisms:
  - **Atmospheric Wave Propagation (Rossby Waves):** Large-scale heating anomalies (e.g., from the unusually warm Pacific waters during El Niño) or the presence of major mountain ranges can generate atmospheric waves known as **Rossby waves**. These waves can propagate through the atmosphere, influencing pressure patterns and the position of the jet stream far downstream, leading to altered weather conditions in distant regions.
  - **Ocean-Atmosphere Coupling:** Changes in large-scale ocean temperatures (like the SST anomalies during ENSO) directly influence the overlying atmosphere. Warmer ocean waters can lead to increased evaporation and convection, altering atmospheric circulation patterns. The atmosphere then feeds back to the ocean, reinforcing the initial anomaly.

- **Moisture Transport:** Large-scale atmospheric circulation patterns, often influenced by climate indices, can steer "atmospheric rivers" or other concentrated moisture transport pathways, leading to remote impacts on precipitation (e.g., heavy rainfall in one region, drought in another).
- **Changes in Storm Tracks:** Teleconnection patterns can shift the preferred paths of mid-latitude storm systems. A shifted storm track can bring more frequent storms and precipitation to one region while leaving another drier than usual.

- **Why are Teleconnections So Crucial?**
  - **Enhanced Predictability:** Understanding teleconnections is a cornerstone of **sub-seasonal to seasonal (S2S) forecasting**. If meteorologists can predict the phase and strength of a major climate index (like ENSO or NAO) weeks or months in advance, they can then issue probabilistic outlooks for the *likely* temperature and precipitation anomalies in distant, teleconnected regions.
  - **Global Interconnectedness:** Teleconnections highlight the profound interconnectedness of the Earth's climate system, demonstrating that no region's weather is truly isolated from global influences.
  - **Climate Change Context:** They help us understand how climate change might alter these fundamental linkages, potentially leading to new or intensified regional impacts (e.g., if a teleconnection pattern becomes more frequent or intense).
  - **Attribution:** They help attribute observed regional weather extremes to large-scale climate drivers, distinguishing them from purely local phenomena.
  - **Examples of Teleconnections in Action (Linking Indices to Impacts):**
- **ENSO Teleconnections**
  - **North America:** During El Niño, the southern U.S. often experiences wetter-than-average and cooler conditions, while the Pacific Northwest and northern U.S. tend to be warmer and drier. During La Niña, these patterns often reverse.
  - **Asia:** El Niño can weaken the Indian and East Asian monsoons, leading to drought.
  - **Australia:** El Niño typically brings drier conditions and increased wildfire risk to eastern Australia.

- **Tropical Cyclones:** El Niño tends to suppress Atlantic hurricane activity while enhancing activity in the central and eastern Pacific.
- **NAO Teleconnections**
  - **Europe:** As mentioned, a strong positive NAO typically brings mild, wet, and stormy winters to Northern Europe (e.g., UK, Scandinavia) and drier, often colder conditions to Southern Europe (e.g., Spain, Mediterranean). A negative NAO reverses these patterns.
  - **Eastern North America:** A positive NAO often correlates with milder winters, while a negative NAO can bring colder, snowier conditions.
- **AO Teleconnections**
  - **Eurasia/North America:** A strong negative AO phase is often linked to severe cold air outbreaks and heavy snowfall in parts of Siberia, eastern North America, and sometimes Europe, as the polar vortex weakens and allows cold air to spill south.
- **MJO Teleconnections**
  - **Tropical Cyclones:** The MJO's phase can significantly modulate tropical cyclone activity in various basins, enhancing activity when the MJO's convective phase is in certain regions.
  - **Mid-Latitudes:** Can influence the timing and intensity of cold air outbreaks or precipitation events in mid-latitudes through its modulation of global circulation, even though it originates in the tropics.
- **The Intrinsic Relationship: Indices Quantify Teleconnections**

Climate indices are the **quantification** of the large-scale climate patterns that *drive* teleconnections. An index provides a single numerical value that tells us the current state and strength of a specific climate pattern (e.g., how strong El Niño is this month). Teleconnections, on the other hand, describe the **physical pathways and statistical relationships** through which that climate pattern (represented by the index) influences weather and climate in distant regions.

12.4 Climate Change Detection and Attribution: Analyzing Trends and Extreme Events

The Earth's climate is constantly changing, influenced by both natural processes and human activities. In the context of modern climate science, **climate change detection and attribution** is a critical field that aims to objectively determine if the climate is changing beyond natural variability, and if so, to identify the causes of those changes. This involves meticulous analysis of long-term climate data, with particular emphasis on identifying **trends** in average conditions and changes in the frequency or intensity of **extreme events**.

12.4.1 Climate Change Detection

What is Climate Change Detection? Climate change detection is the process of demonstrating that the climate has changed in a statistically significant way, beyond the range of natural internal variability. It's about identifying a "signal" of change that stands out from the "noise" of natural fluctuations.

- **Why is Detection Challenging?**
    - **Natural Variability:** The Earth's climate naturally fluctuates on various timescales (e.g., El Niño cycles, volcanic eruptions, solar cycles). Distinguishing a long-term, forced change from these natural ups and downs requires long, homogeneous datasets.
    - **Data Limitations:** Historical data can be sparse, incomplete, or contain inhomogeneities (e.g., changes in instruments or station locations).
    - **Complex System:** The climate system is highly complex, with many interacting components, making it difficult to isolate individual causes without sophisticated models.
    - **How is Climate Change Detected?** Detection primarily relies on analyzing long-term, high-quality **Climate Data Records (CDRs)** for consistent trends that cannot be explained by known natural variability alone.

12.4.2 Analyzing Trends in Climate Data

What are Climate Trends? A climate trend refers to a sustained, directional change in a climate variable (e.g., temperature, precipitation, sea level) over a significant period, typically decades or longer. Detecting a trend means identifying a statistically significant upward or downward movement in the data.

- **Methodologies for Analyzing Trends**
    - **Visual Inspection:** The first step often involves simply plotting the time series of a climate variable (e.g., global average temperature) and visually inspecting for a general upward or downward slope.
- **Linear Regression**
    - **Concept:** The most common statistical method. It fits a straight line through the data points, representing the average rate of change per unit of time (e.g., degrees Celsius per decade).

- **Statistical Significance:** Statistical tests (e.g., t-tests) are used to determine if the slope of the trend line is significantly different from zero, meaning the observed trend is unlikely to have occurred by random chance.

- **Non-parametric Tests (e.g., Mann-Kendall Test)**
  - **Concept:** These tests are used when the data may not follow a normal distribution or when there are outliers. They assess the monotonic (consistently increasing or decreasing) nature of a trend without assuming a specific linear relationship.
  - **Robustness:** They are less sensitive to extreme values and don't require assumptions about the data's distribution.
- **Moving Averages and Smoothing**
  - **Concept:** Applying a moving average (e.g., a 10-year or 30-year average) to a time series helps to smooth out short-term fluctuations and highlight underlying long-term trends.
  - **Visualization:** Often used in conjunction with linear regression to visually emphasize the trend.
- **Spatial Trend Analysis**
  - **Concept:** Analyzing trends not just for global averages, but for specific regions or grid cells on a map. This reveals geographical patterns of warming, cooling, or precipitation changes.
  - **Visualization:** Often presented as maps showing the magnitude and statistical significance of trends across different regions.
- **Challenges in Trend Analysis**
  - **Natural Variability:** Short-term trends can be dominated by natural cycles (e.g., a decade of warming might be part of an El Niño cycle). Long records are needed to separate forced trends.
  - **Inhomogeneities:** Non-climatic shifts in data (e.g., station moves, instrument changes) can create artificial trends or mask real ones. Homogenization is critical.
  - **Choice of Start/End Points:** Trends can appear different depending on the specific start and end years chosen for analysis, especially for shorter periods.

12.4.3 Climate Change Attribution

What is Climate Change Attribution? Climate change attribution is the process of **determining the causes** of detected climate changes. It aims to quantify the relative contributions of different factors (both natural and human-induced) to observed changes in climate. It moves beyond "is the climate changing?" to "why is the climate changing?"

- **Why is Attribution Challenging?**
  - **Multiple Forcings:** The climate system is influenced by many factors simultaneously (greenhouse gases, aerosols, volcanic eruptions, solar variations, land-use changes).
  - **Complex Interactions:** These factors interact in complex, often non-linear ways.
  - **Uncertainty:** There are uncertainties in quantifying past emissions, natural forcings, and the climate's response to them.
  - **How is Climate Change Attributed? (Fingerprinting)** Attribution primarily relies on a "fingerprinting" approach, combining observations with sophisticated climate models:
- **Climate Models as Laboratories**
  - **Concept:** Climate models are used to simulate the Earth's climate under different scenarios.
  - **All Forcings Run:** Models are run with all known historical natural and human-induced forcings (greenhouse gases, aerosols, solar variations, volcanic eruptions).
  - **Natural Forcings Only Run:** Models are run again, but *only* with natural forcings (e.g., solar and volcanic activity), keeping human-induced forcings constant.
  - **Comparison:** The results from these two sets of model runs are then compared to each other and to observations.
- **Fingerprinting**
  - **Concept:** Different climate forcings are expected to produce distinct "fingerprints" (spatial and temporal patterns) of climate change. For example, greenhouse gas warming is expected to be stronger at poles and in the troposphere, while solar forcing might have a different vertical profile.
  - **Matching Patterns:** Scientists look for statistical matches between the observed climate change patterns and the patterns simulated by models under different forcing scenarios.
  - **Example:** If observed global warming patterns (e.g., more warming over land than ocean, stronger warming at poles) closely match model simulations that include human-caused greenhouse gas emissions,

but *not*simulations that only include natural factors, it strengthens the attribution to human influence.

- **Statistical Methods:** Advanced statistical techniques are used to separate the signals of different forcings from the background noise of natural variability.

12.4.4 Analyzing Extreme Events

What are Extreme Events in Climate Change? Extreme events are rare occurrences at the tails of the distribution of weather and climate variables (e.g., unusually high temperatures, very heavy rainfall, prolonged droughts). Climate change is not just about changes in averages; it's also about changes in the **frequency, intensity, duration, and spatial extent** of these extreme events.

- **Why is Analyzing Extreme Events Important?**
  - **High Impact:** Extreme events cause the most significant societal and economic damage (e.g., heat-related deaths, flood damage, crop failures).
  - **Non-linear Response:** Even small changes in average conditions can lead to disproportionately large changes in the frequency of extremes.
  - **Vulnerability:** Understanding changes in extremes is crucial for adaptation planning and building resilience.
- **Methodologies for Analyzing Extreme Events**
- **Trend Analysis of Extremes**
  - **Concept:** Instead of looking at average temperature, analyze trends in metrics like the number of "hot days" (days above a certain threshold), or the maximum 1-day precipitation total per year.
  - **Indices of Extremes:** Climate scientists use a suite of standardized indices for extremes (e.g., those defined by the Expert Team on Climate Change Detection and Indices - ETCCDI), such as:
  - **TXx:** Annual maximum value of daily maximum temperature.
  - **R95p:** Annual total precipitation from days exceeding the 95th percentile of daily precipitation.
  - **Drought Indices:** (e.g., Standardized Precipitation Index - SPI, Palmer Drought Severity Index - PDSI) to track drought severity and duration.

- **Attribution of Extreme Events (Event Attribution)**
  - **Concept:** A specialized field that attempts to quantify how human-induced climate change influenced the probability or intensity of *specific* extreme weather events (e.g., "Was this heatwave made more likely by climate change?").
  - **Methodology:** Involves comparing the probability/intensity of an event in the current climate (with human influence) to its probability/intensity in a hypothetical "pre-industrial" or "natural" climate (without human influence), often using climate models.
- **Changes in Return Periods**
  - **Concept:** Analyzing how the frequency of extreme events changes. A "1-in-100-year flood" in the past might become a "1-in-50-year flood" in a changed climate.
  - **Statistical Models:** Using statistical models (e.g., Extreme Value Theory) to estimate changes in the likelihood of very rare events.
- **Challenges in Analyzing Extreme Events**
  - **Rarity:** By definition, extreme events are rare, making robust statistical analysis challenging due to limited data points.
  - **Data Quality:** Extremes are highly sensitive to data quality and inhomogeneities.
  - **Model Resolution:** Climate models often need very high resolution to accurately simulate the physical processes that generate extremes.
  - **Complex Interactions:** Extremes are often the result of complex interactions between multiple meteorological factors.

Climate change detection and attribution, underpinned by rigorous analysis of trends and extreme events, are fundamental to modern climate science. By meticulously identifying statistically significant changes in long-term climate records and employing sophisticated climate models, scientists can confidently demonstrate that the Earth's climate is indeed changing beyond natural variability and, crucially, attribute a substantial portion of these changes to human activities. The focus on extreme events highlights the tangible and often devastating impacts of a warming world, providing critical information for adaptation strategies and informing the urgent need for climate action. This ongoing scientific endeavor provides the essential evidence base for global policy and societal resilience in the face of our evolving climate.

12.5 Paleoclimatic Data: Proxy Records and Their Analysis

While instrumental records (thermometers, rain gauges, etc.) provide invaluable climate data for the past few centuries, understanding Earth's climate history over longer timescales – millennia, hundreds of thousands of years, or even millions of years – requires a different approach. This is where **paleoclimatic data** comes into play. Paleoclimatology is the study of past climates, and since direct instrumental measurements don't exist for these ancient periods, scientists rely on **proxy records** to reconstruct past climate conditions.

12.5.1 Paleoclimatic Data and Proxy Records

- **What are Paleoclimatic Data?** Paleoclimatic data refers to any natural or historical archive that preserves information about past climate conditions. These are not direct measurements of temperature or precipitation, but rather indirect indicators that can be interpreted to infer past climate states.
- **What are Proxy Records?Proxy records** are natural archives that contain physical, chemical, or biological information that can be correlated with climate variables. They are essentially "natural recorders" of past environmental conditions. Scientists study these proxies to reconstruct past temperatures, precipitation, atmospheric composition, ocean currents, ice volume, and other climate parameters.
- **Why are Proxy Records Essential?**
  - **Extend Climate Records:** They allow scientists to extend climate records far beyond the instrumental period (which is only a few centuries long), providing context for current climate change.
  - **Understand Natural Variability:** They reveal the full range of Earth's natural climate variability, including past warm periods, ice ages, and abrupt climate shifts. This helps distinguish human-caused change from natural cycles.
  - **Validate Climate Models:** Past climate states reconstructed from proxies provide independent targets for climate models, allowing scientists to test if models can accurately simulate past climates, thus building confidence in their projections of future climate.
  - **Understand Climate Sensitivity:** They provide insights into how the Earth system responds to different climate forcings (e.g., changes in greenhouse gases, solar radiation) under different past conditions.

12.5.2 Types of Proxy Records and Their Analysis

Paleoclimatologists use a diverse array of proxy records, each offering unique insights into different aspects of past climate and covering different timescales.

- **Ice Cores**
  - **Description:** Cylinders of ice drilled from ice sheets (Antarctica, Greenland) or high-altitude glaciers. As snow falls and compacts into ice, it traps atmospheric gases, dust, volcanic ash, and other impurities.
- **Climate Information**
  - **Trapped Air Bubbles:** Analyze the composition of ancient air, providing direct measurements of past atmospheric greenhouse gas concentrations (e.g., CO_2 , methane) over hundreds of thousands of years.
  - **Stable Isotopes of Water (δ18O, δD):** The ratio of heavy to light isotopes of oxygen and hydrogen in the ice is highly correlated with the temperature at which the snow originally fell. This provides a direct temperature proxy.
  - **Dust/Aerosols:** Indicate past atmospheric circulation patterns, aridity, and volcanic activity.
  - **Ionic Concentrations:** Reveal past atmospheric chemistry, sea salt, and pollutant levels.
  - **Timescale:** Up to 800,000 years for Antarctic ice cores (e.g., EPICA Dome C), tens of thousands of years for Greenland cores.
  - **Analysis:** Requires highly specialized laboratories for precise isotopic analysis, gas chromatography, and impurity counting.
- **Tree Rings (Dendroclimatology)**
  - **Description:** Trees grow annual rings, with the width and density of the ring influenced by environmental conditions during the growing season (e.g., temperature, precipitation, sunlight).
- **Climate Information**
  - **Ring Width:** Wider rings generally indicate favorable growing conditions (e.g., warmer temperatures, more precipitation), while narrower rings indicate stress (e.g., drought, cold).
  - **Wood Density:** Denser wood can indicate longer growing seasons or specific temperature conditions.

- **Stable Isotopes:** Isotopic ratios within the wood can provide additional information about water availability and temperature.
- **Timescale:** Up to a few thousand years, limited by the lifespan of trees and the preservation of ancient wood.
- **Analysis:** Involves careful cross-dating (matching patterns of wide/narrow rings across multiple trees to ensure accurate dating) and statistical calibration against instrumental records.

- **Sediment Cores (Ocean, Lake, and Terrestrial)**
  - **Description:** Layers of accumulated sediment at the bottom of oceans, lakes, or in peat bogs and caves. These layers preserve a record of past environmental conditions.
- **Climate Information**
  - **Microfossils (Foraminifera, Pollen):** Different species of tiny marine organisms (foraminifera) thrive in specific ocean temperatures, and their shells are preserved in sediments. Pollen grains from terrestrial plants indicate past vegetation types and, by extension, climate (e.g., presence of cold-tolerant species).
  - **Isotopic Ratios in Shells/Skeletons (δ18O, δ13C):** Oxygen isotopes in the calcium carbonate shells of marine organisms reflect past ocean temperatures and global ice volume. Carbon isotopes can indicate ocean circulation and productivity.
  - **Sediment Composition:** Changes in sediment type (e.g., clay vs. sand), grain size, and mineral content can indicate past ocean currents, sea level, aridity, or glacial activity.
  - **Biomarkers:** Organic molecules (e.g., lipids) from ancient organisms can provide direct temperature reconstructions.
  - **Timescale:** From thousands to millions of years, depending on sedimentation rates and preservation. Ocean sediment cores can go back tens of millions of years.
  - **Analysis:** Requires microscopy, chemical analysis (mass spectrometry for isotopes), and careful dating using radiometric methods (e.g., Carbon-14, Uranium-Thorium).
- **Speleothems (Cave Formations - Stalagmites, Stalactites)**
  - **Description:** Deposits of calcium carbonate (calcite) that form in caves as water drips from the ceiling. Growth rings (laminae) can form annually or sub-annually.

- **Climate Information**
  - **Growth Rate/Density:** Can reflect past rainfall amounts.
  - **Stable Isotopes (δ18O, δ13C):** Oxygen isotopes in the calcite reflect the temperature and isotopic composition of the rainfall above the cave. Carbon isotopes can indicate vegetation changes above the cave.
  - **Trace Elements:** Concentrations of elements like magnesium can reflect past aridity or temperature.
  - **Timescale:** From thousands to hundreds of thousands of years, depending on the speleothem's age and continuous growth.
  - **Analysis:** High-resolution sampling, isotopic analysis, and precise Uranium-Thorium dating.
- **Corals**
  - **Description:** Tropical corals grow annual bands, similar to tree rings, with their chemistry influenced by ocean conditions.
  - Climate Information:
  - **Growth Rate:** Can reflect ocean temperature and light availability.
  - **Stable Isotopes (δ18O):** Oxygen isotopes in coral skeletons are sensitive to past sea surface temperature (SST) and salinity.
  - **Trace Elements (e.g., Sr/Ca ratios):** Strontium to calcium ratios in coral skeletons are also a very robust proxy for past SST.
  - **Timescale:** Up to a few centuries to a few thousand years for long-lived corals.
  - **Analysis:** High-resolution sampling, chemical analysis, and precise dating.
- **Historical Documents and Archaeological Evidence**
  - **Description:** Written records (ship logs, harvest records, diaries, government documents), paintings, and archaeological findings.
  - **Climate Information:** Provide qualitative or semi-quantitative information about past weather events (e.g., severe winters, droughts, floods), harvest successes/failures, and sea ice conditions.
  - **Timescale:** Limited to the period of written history, typically the last few centuries to millennia.
  - **Analysis:** Requires careful historical research, cross-referencing, and critical evaluation of subjective accounts.

12.5.3 Analysis of Paleoclimatic Data

Analyzing proxy records is a multi-step process that requires a strong interdisciplinary approach:

- **Dating the Proxy:** This is paramount. Accurate chronological control is essential to ensure that climate changes are placed in their correct temporal sequence. Methods include:
- **Radiometric Dating:** Carbon-14, Uranium-Thorium, Potassium-Argon.
- **Dendrochronology:** Tree-ring cross-dating.
- **Ice Core Layer Counting:** Counting annual layers in ice.
- **Tephrochronology:** Dating layers using volcanic ash.
- **Orbital Tuning:** Aligning long climate records with known changes in Earth's orbit (Milankovitch cycles).

- **Calibration**
  - **Concept:** Establishing a quantitative relationship between the proxy signal and the climate variable it represents. This is done by comparing the proxy's modern-day signal with instrumental climate data from the same period.
  - **Statistical Methods:** Regression analysis is commonly used to build transfer functions (e.g., a regression equation that converts δ18O values in ice into temperature).
  - **Assumption:** Assumes that the relationship between the proxy and the climate variable remains constant over time.
- **Reconstruction**
  - **Concept:** Applying the calibrated relationship to the entire proxy record to reconstruct the past climate variable (e.g., reconstructing past temperatures from ice core isotopes).
  - **Uncertainty Quantification:** Reconstructions always have uncertainties. Paleoclimatologists strive to quantify these uncertainties (e.g., by providing error bars or confidence intervals) to reflect the limitations of the proxy and the calibration.
- **Integration and Synthesis**
  - build a more comprehensive and robust picture of past climate. Different proxies provide different types of information and cover different timescales.

- **Multi-proxy Networks:** Large databases exist that compile and synthesize data from thousands of proxy records (e.g., PAGES 2k Network).
- **Model-Data Comparison:** Comparing paleoclimate reconstructions with simulations from climate models to test model performance and understand past climate dynamics.

Paleoclimatic data, derived from a rich array of natural proxy records, provides the essential deep-time perspective needed to understand Earth's climate system. By meticulously analyzing these archives – from the bubbles in ancient ice to the rings of long-lived trees and the shells of microscopic marine organisms – scientists can reconstruct past temperatures, atmospheric composition, and environmental conditions over vast timescales. This invaluable information not only reveals the full spectrum of Earth's natural climate variability but also provides critical context for current climate change, validates the projections of climate models, and deepens our fundamental understanding of how our planet's climate system responds to various forcings. The study of paleoclimate is thus central to placing present and future climate change in its proper historical context.

12.6 Climate Prediction Models: Uses and Limitations

In the study of climate, understanding the past and present is crucial, but equally vital is the ability to project future climate conditions. This is the domain of **Climate Prediction Models**, often referred to as **Global Climate Models (GCMs)** or **Earth System Models (ESMs)**. These are highly sophisticated computer programs designed to simulate the physical, chemical, and biological processes that govern the Earth's climate system.

12.6.1 What are Climate Prediction Models?

Climate prediction models are complex mathematical representations of the Earth's climate system. Unlike Numerical Weather Prediction (NWP) models, which focus on short-term, deterministic forecasts of atmospheric states (days to weeks), climate models are designed to simulate the **long-term evolution of the climate system** (decades to centuries or even millennia). They are built upon the same fundamental physical laws as weather models but are configured differently to address longer timescales and broader interactions.

- **Key Components of Climate Models:** Modern climate models are truly "Earth System Models" because they integrate multiple interacting components:
- **Atmosphere:** Simulates temperature, pressure, winds, humidity, clouds, precipitation, and radiation.

- **Ocean:** Simulates ocean currents, temperature, salinity, sea ice, and ocean heat content.
- **Land Surface:** Represents soil moisture, vegetation, snow cover, and land-atmosphere interactions.
- **Sea Ice:** Simulates the formation, movement, and melting of sea ice.
- **Carbon Cycle (in ESMs):** Simulates the exchange of carbon dioxide and other greenhouse gases between the atmosphere, ocean, and land, including biological processes.
- **Aerosols and Atmospheric Chemistry (in ESMs):** Represents the formation and effects of atmospheric particles and chemical reactions.

These components are coupled together, meaning they exchange energy, momentum, and matter, allowing the model to simulate the complex feedbacks within the Earth system.

12.6.2 How Climate Models Work (Briefly)

At their core, climate models divide the Earth (atmosphere, ocean, land) into a 3D grid of cells. Within each cell, and at each time step, the models numerically solve a set of differential equations that describe the conservation of energy, momentum, mass, and water.

- **Physical Laws:** They are based on fundamental laws of physics (e.g., Newton's laws of motion, laws of thermodynamics, conservation of mass).
- **Numerical Approximation:** Since the equations are too complex for analytical solutions, they are approximated and solved numerically on a discrete grid.
- **Parameterizations:** Processes that occur at scales smaller than the model's grid cells (e.g., individual clouds, turbulence, vegetation processes) cannot be explicitly resolved. Their average effects are represented by **parameterization schemes** – simplified mathematical formulas derived from observations and theoretical understanding.
- **Forcings:** Models are "forced" with historical and projected changes in factors that influence climate, such as:
- **Greenhouse Gas Concentrations:** Historical observations and future emission scenarios (e.g., Shared Socioeconomic Pathways - SSPs).
    - **Solar Radiation:** Variations in solar output.
    - **Volcanic Eruptions:** Injection of aerosols into the stratosphere.
    - **Land Use Changes:** Deforestation, urbanization, etc.

By integrating these equations forward in time for decades or centuries under different forcing scenarios, climate models project how the climate system might evolve.

12.6.3 Uses of Climate Prediction Models

Climate prediction models are indispensable tools for a wide range of scientific, policy, and societal applications:

- **Projecting Future Climate Change:**
  - **Core Use:** This is the primary purpose. Models are run under various future greenhouse gas emission scenarios (e.g., SSPs) to project how global and regional temperatures, precipitation patterns, sea levels, and extreme events might change over the 21st century and beyond.
  - **Intergovernmental Panel on Climate Change (IPCC):** The IPCC reports heavily rely on climate model projections from multiple international modeling centers (e.g., CMIP - Coupled Model Intercomparison Project).
- **Understanding Past Climate Variability and Change**
  - **Attribution Studies:** By running models with and without specific historical forcings (e.g., human-caused greenhouse gases vs. natural factors), models help attribute observed climate changes to their likely causes.
  - **Paleoclimate Simulations:** Models can simulate past climates (e.g., ice ages, warm periods) to test their ability to reproduce known conditions and understand the climate system's response to different forcings.
- **Investigating Climate System Processes and Feedbacks**
  - **"What If" Scenarios:** Models allow scientists to conduct experiments that are impossible in the real world (e.g., "What if CO_2 concentrations doubled?", "What if there were no aerosols?"). This helps understand complex interactions and feedbacks (e.g., cloud feedbacks, ice-albedo feedback).
  - **Sensitivity Studies:** Assessing how sensitive the climate system is to changes in different parameters or forcings.
- **Assessing Climate Change Impacts**
  - **Input for Impact Models:** Climate model outputs (e.g., projected temperature, precipitation) serve as crucial inputs for a wide array of impact models in other sectors:

- **Hydrological Models:** To assess future water availability and flood risk.
- **Agricultural Models:** To project crop yields and food security.
- **Ecological Models:** To understand shifts in ecosystems and biodiversity.
- **Public Health Models:** To assess heat-related illnesses or vector-borne disease spread.
- **Vulnerability Assessment:** Identifying regions and communities most vulnerable to future climate impacts.

- **Informing Policy and Adaptation Strategies**
    - **Mitigation Targets:** Model projections inform international climate agreements and national emission reduction targets.
    - **Adaptation Planning:** Providing region-specific climate change information to help communities and industries plan for adaptation (e.g., building sea walls, developing drought-resistant crops).
- **Developing and Testing New Scientific Hypotheses**
    - Models serve as a dynamic laboratory for testing new theories about how the Earth system works.

12.6.4 Limitations of Climate Prediction Models

Despite their sophistication and invaluable contributions, climate prediction models have inherent limitations that are important to understand:

- **Resolution Limitations**
    - **Coarse Grids:** Even the highest-resolution global climate models still operate at resolutions (tens to hundreds of kilometers) that are too coarse to explicitly resolve many important local-scale processes (e.g., individual thunderstorms, complex terrain effects, urban heat islands).
    - **Parameterization Reliance:** These sub-grid scale processes must be represented by parameterizations, which are simplifications and a major source of uncertainty. This means models cannot perfectly simulate local weather, only its statistical properties over time.
- **Uncertainties in Future Emissions Scenarios**
    - **Human Behavior:** Model projections depend on future greenhouse gas emission pathways (SSPs), which are based on assumptions about human population growth, economic development, technological advancements, and policy choices. These are inherently uncertain.

    - **Not Predictions of Human Decisions:** Models predict *what might happen* under a given scenario, not *what will happen*, because human actions are not predetermined.
- **Uncertainties in Physical Processes and Feedbacks**
    - **Clouds:** Clouds are one of the largest sources of uncertainty. They both reflect sunlight (cooling effect) and trap heat (warming effect), and how their properties and distribution will change in a warmer world is complex and not fully understood.
    - **Aerosols:** The direct and indirect effects of aerosols (tiny particles in the atmosphere) on climate are still highly uncertain.
    - **Carbon Cycle Feedbacks:** How much carbon the land and ocean will continue to absorb as the climate warms is uncertain. For example, thawing permafrost could release large amounts of methane, a strong greenhouse gas, creating a positive feedback loop.
- **Computational Constraints**
    - Running high-resolution, complex Earth System Models for centuries requires immense supercomputing power, limiting the number of simulations that can be performed and the resolution achievable.
    - This also means that some very slow processes (e.g., deep ocean circulation, ice sheet dynamics over millennia) might not be fully captured in shorter simulations.
- **Initial Conditions for Climate Projections**
    - Unlike weather forecasts that need precise initial conditions, climate projections are less sensitive to the exact initial state (the "butterfly effect" averages out over long timescales). However, the initial state can still influence the first few decades of a projection.
- **Model Biases**
    - Despite continuous improvement, all climate models have systematic biases (e.g., a tendency to be too warm in certain regions, or to over-predict certain types of precipitation). These biases are often quantified through comparison with historical observations.
    - **Ensemble Approach:** To account for these uncertainties and model differences, climate scientists typically use **multi-model ensembles** (running many different models from various institutions) and **single-model initial-condition ensembles** (running the same model multiple times with slightly different starting points). The spread among these ensemble members provides an estimate of the overall uncertainty.

- **Emergent Constraints**
  - Scientists are actively working on "emergent constraints," where relationships observed in current climate data are used to narrow down the range of future projections from models.

Climate prediction models are extraordinary scientific achievements, serving as our primary tools for understanding the complex dynamics of the Earth's climate system and projecting its future under various scenarios. They are indispensable for climate change research, impact assessment, and informing global policy decisions. While inherent uncertainties related to future human behavior, the complexity of physical processes, and computational limitations mean that model projections are not perfect predictions, the continuous advancements in model development, coupled with the power of multi-model ensembles, provide increasingly robust and reliable insights into the challenges and opportunities of our changing climate. These models are our best scientific guides for navigating the future of our planet.

12.7 Representative Concentration Pathways (RCPs): Applications and Limitations

In the realm of climate change research, particularly when discussing future climate projections, the term **RPCs** almost certainly refers to **Representative Concentration Pathways**. These are a set of greenhouse gas emission scenarios developed by the climate modeling community to provide a consistent framework for climate model simulations and impact assessments. They are not predictions of the future, but rather plausible pathways of future greenhouse gas concentrations, aerosols, and land-use changes.

12.7.1 What are Representative Concentration Pathways (RCPs)?

Representative Concentration Pathways (RCPs) are a set of four greenhouse gas concentration trajectories adopted by the Intergovernmental Panel on Climate Change (IPCC) for its Fifth Assessment Report (AR5) and subsequent research. They describe different plausible future climate outcomes based on various assumptions about economic growth, technological development, energy use, population growth, and climate policy.

The "concentration" in their name is key: unlike earlier scenarios that focused on emissions, RCPs specify the **concentration of greenhouse gases** in the atmosphere over time. Climate models then use these concentration pathways as input to simulate the Earth's climate response.

Each RCP is named after a possible range of **radiative forcing** in the year 2100 relative to pre-industrial values. Radiative forcing is a measure of the change in energy balance of the Earth's system due to a particular factor (e.g., increased greenhouse gas concentrations). It's measured in Watts per square meter (W/m^2).

The four main RCPs are

- **RCP2.6 (Very Low Emissions)**
    - **Radiative Forcing by 2100:** 2.6 W/m^2.
    - **Description:** Represents a "peak and decline" scenario where greenhouse gas emissions peak around 2020 and then decline substantially. This pathway aims to keep global warming below 2°C relative to pre-industrial levels, requiring aggressive mitigation efforts.
    - **Interpretation:** A very ambitious mitigation scenario.
- **RCP4.5 (Intermediate Emissions)**
    - **Radiative Forcing by 2100:** 4.5 W/m^2.
    - **Description:** A stabilization scenario where emissions peak around 2040 and then decline. This pathway suggests a moderate level of climate change.
- **RCP6.0 (Intermediate to High Emissions)**
    - **Radiative Forcing by 2100:** 6.0 W/m^2.
    - **Description:** Another stabilization scenario, but with emissions peaking around 2080 and then declining. It represents a higher level of climate change than RCP4.5.
- **RCP8.5 (Very High Emissions / Business-as-Usual):**
    - **Radiative Forcing by 2100:** 8.5 W/m^2.
    - **Description:** Represents a "high-end" or "business-as-usual" scenario with continually increasing greenhouse gas emissions throughout the 21st century. It leads to the most significant warming among the RCPs.
    - **Interpretation:** Often considered a worst-case scenario or a high-emissions baseline without aggressive climate policy.

12.7.2 Applications of RCPs in Climate Research and Policy

RCPs are fundamental tools in climate science and policy for several reasons:

- **Standardized Input for Climate Models**
    - **Consistency:** RCPs provide a common, consistent set of future greenhouse gas concentrations, land use changes, and aerosol emissions. This allows different climate models from various research

institutions around the world to run simulations using the same inputs, making their outputs comparable (e.g., through the Coupled Model Intercomparison Project - CMIP).

- **Intercomparison:** This standardization is crucial for understanding model uncertainties and identifying robust projections across multiple models.

- **Projecting Future Climate Scenarios**
  - **Range of Futures:** By running models under the four RCPs, scientists can explore a wide range of plausible future climate outcomes, from ambitious mitigation (RCP2.6) to high emissions (RCP8.5). This provides a comprehensive picture of potential warming, precipitation changes, sea level rise, and extreme event shifts.
  - **Impact Assessment:** The climate projections derived from RCPs are then used as input for various impact models to assess the consequences of climate change on sectors like agriculture, water resources, human health, ecosystems, and infrastructure.
- **Informing Policy and Mitigation Efforts**
  - **Policy Targets:** RCPs help policymakers understand what level of emissions reduction (mitigation) is required to achieve specific climate goals (e.g., limiting warming to 1.5°C or 2°C). RCP2.6, for instance, illustrates the scale of effort needed for ambitious climate targets.
  - **Risk Assessment:** They allow governments and businesses to assess the risks associated with different emission pathways, informing adaptation strategies and investments.
  - **Decision Support:** By presenting a range of possible futures, RCPs support robust decision-making under uncertainty.
- **Communicating Climate Science**
  - **Simplified Communication:** RCPs provide a clear, numerical shorthand for complex climate scenarios, aiding in the communication of climate change projections to the public & non-expert stakeholders.
  - **IPCC Reports:** They are central to the IPCC Assessment Reports, which synthesize the latest climate science for policymakers.
- **Attribution Studies**
  - While primarily for future projections, the underlying methodologies that build RCPs also contribute to understanding the drivers of historical climate change.

12.7.3 Limitations of RCPs

Despite their widespread use and utility, RCPs, like any scenario framework, have important limitations:

- **Not Predictions**
  - **Misinterpretation:** The most common misunderstanding is that RCPs are forecasts or predictions of what *will* happen. They are not. They are "if-then" scenarios: *if* emissions follow this pathway, *then* this climate outcome is plausible.
  - **Human Behavior:** The future development pathways that underpin RCPs (population, technology, policy) are uncertain and depend on human choices, which cannot be predicted.
- **Fixed Number of Pathways**
  - **Limited Granularity:** Having only four main pathways means they don't capture the full spectrum of possible future socio-economic developments or mitigation efforts.
  - **Interpolation Issues:** Users sometimes linearly interpolate between RCPs, which may not be physically realistic as the underlying socio-economic drivers are non-linear.
- **"Business-as-Usual" Misconception (RCP8.5)**
  - **High-End Scenario:** RCP8.5 was initially labeled "business-as-usual," but as climate policies have developed and renewable energy costs have fallen, many researchers now view it as a less likely "high-end" or "worst-case" scenario, rather than the most probable business-as-usual path. It typically requires very high coal consumption and slower technological progress.
  - **Still Relevant:** Despite this, it remains important for understanding the potential *magnitude* of climate change risks if mitigation efforts are insufficient.
- **Assumptions within Pathways**
  - Each RCP is built on specific assumptions about population, economic growth, energy use, and technological progress. These assumptions might evolve or be challenged over time.
  - **Negative Emissions (for RCP2.6):** The RCP2.6 pathway relies significantly on achieving large-scale **net negative emissions** in the latter half of the 21st century (i.e., removing more CO_2 from the atmosphere than is emitted). The feasibility and scale of such technologies are still uncertain.

- **Successor Scenarios (SSPs)**
  - While RCPs define the *concentration* pathways, the scientific community has moved towards **Shared Socioeconomic Pathways (SSPs)** as the successor framework for integrated assessment modeling. SSPs describe alternative future socioeconomic developments (e.g., sustainable development, regional rivalry, fossil-fueled development) that then lead to different emission levels, which are *then* linked to RCPs. This provides a more comprehensive narrative backdrop for the emission scenarios.
  - While CMIP5 used RCPs directly, CMIP6 and subsequent research increasingly use the combined SSP-RCP framework (e.g., SSP2-4.5 represents a specific socioeconomic pathway leading to a 4.5 W/m^2 radiative forcing).

Representative Concentration Pathways (RCPs) have been an indispensable framework for standardizing climate model projections, enabling intercomparisons, and informing global climate policy for over a decade. They provide a crucial set of plausible future climate scenarios, ranging from ambitious mitigation efforts to high-emission pathways, allowing scientists to assess the potential scale of future warming and its impacts. While their interpretation requires careful understanding (they are not predictions) and they have some inherent limitations (especially concerning the "business-as-usual" interpretation of RCP8.5 and the need for negative emissions in RCP2.6), RCPs have significantly advanced our ability to explore the future of our climate. Their evolution into the more comprehensive SSP-RCP framework continues to provide the essential scientific basis for global efforts to understand, mitigate, and adapt to climate change.

12.8 Conclusion

Concluding our extensive exploration of Climate Data and Analysis, it is evident that understanding Earth's long-term climate patterns is a monumental scientific endeavor, distinct yet complementary to short-term weather forecasting. We've traversed the intricate landscape of **Climate Data Records (CDRs)**, recognizing the paramount importance of their homogeneity to confidently discern true climate signals from observational noise. The journey revealed how **climate indices** distill vast atmospheric and oceanic complexities into actionable metrics, unveiling the powerful influence of **teleconnections** that link distant regions of our interconnected planet. Furthermore, we delved into the rigorous methodologies of **climate change detection and attribution**, meticulously analyzing trends in average conditions and shifts in extreme events to unequivocally establish human influence on our warming world.

Finally, we examined the indispensable role of **Climate Prediction Models (GCMs/ESMs)**, which, guided by frameworks like the **Representative Concentration Pathways (RCPs)** and their successors, allow us to project plausible future climate scenarios. While these models possess inherent limitations and uncertainties, their continuous refinement, coupled with the insights gained from historical data analysis, provides our best scientific guidance for navigating the challenges of a changing climate. The meticulous collection, curation, and analysis of climate data, from ancient proxy records to cutting-edge model simulations, form the bedrock of our understanding, empowering informed decisions for adaptation and mitigation in the face of one of humanity's most pressing challenges.

13

Specialized Applications and Future Trends

13.1 Introduction

Having navigated the foundational statistical methods, advanced analytical techniques, visualization strategies, and the intricate world of Numerical Weather Prediction and Climate Data Analysis, we now arrive at a pivotal point in the chapter **"Specialized Applications and Future Trends."** This chapter is dedicated to exploring how the comprehensive understanding and methodologies developed in previous chapters are being applied to address some of the most pressing environmental and societal challenges of our time. Furthermore, we will look ahead to the emerging technologies and interdisciplinary approaches that are shaping the future of atmospheric and climate science.

The atmosphere is not an isolated system; it profoundly influences, and is influenced by, human activities, energy systems, water resources, and urban environments. Therefore, thc insights derived from meteorological and climate data are increasingly being tailored and integrated into highly specialized applications that directly impact public health, economic sustainability, and urban planning. This goes beyond traditional weather forecasting to provide actionable intelligence for complex, real-world problems.

In this chapter, we will delve into several key areas where atmospheric science is making critical contributions:

- **Air Quality Analysis and Pollutant Measurements:** We will explore how meteorological principles are applied to understand the transport, dispersion, and transformation of atmospheric pollutants. This includes the methods for measuring various pollutants and the sophisticated **dispersion modeling** techniques used to predict air pollution levels and their spread, crucial for public health warnings and environmental regulations.
- **Hydrometeorological Data and Water Resource Management:** Recognizing the intimate link between the atmosphere and the

Earth's water cycle, we will examine the collection and analysis of hydrometeorological data (precipitation, evaporation, soil moisture, river flow). This section will highlight its vital role in flood forecasting, drought monitoring, and sustainable water resource management, especially in a changing climate.

- **Renewable Energy Integration:** As the world transitions to cleaner energy sources, accurate meteorological and climate data become paramount. We will discuss how specialized atmospheric science supports the optimal siting, forecasting, and operation of wind and solar energy farms, ensuring grid stability and efficiency.
- **Urban Climatology and Smart Cities:** With a growing global urban population, understanding the unique atmospheric conditions within cities is increasingly important. This section will explore how meteorological principles are applied to study urban heat islands, ventilation, and local weather phenomena, informing sustainable urban planning and the development of "smart city" initiatives.

Finally, we will look to the **future trends** that are poised to revolutionize these specialized applications and the broader field of atmospheric science. This includes the accelerating integration of **Artificial Intelligence and Machine Learning**, the development of **hyper-resolution models** that bridge weather and climate scales, and the increasing reliance on **integrated Earth system observations**. This chapter will demonstrate how atmospheric science is not only adapting to new challenges but is actively innovating to provide the essential knowledge and tools for a more sustainable and resilient future.

13.2 Air Quality Analysis

Air quality analysis is a critical interdisciplinary field that combines principles of atmospheric science, chemistry, public health, and environmental engineering to understand the composition of the atmosphere, identify sources of pollution, monitor pollutant concentrations, predict their dispersion, and assess their impacts on human health and ecosystems. In an increasingly urbanized and industrialized world, where air pollution remains a significant global health concern, robust air quality analysis is essential for informing policy, issuing public health warnings, and developing effective mitigation strategies (Fig. 13.1)

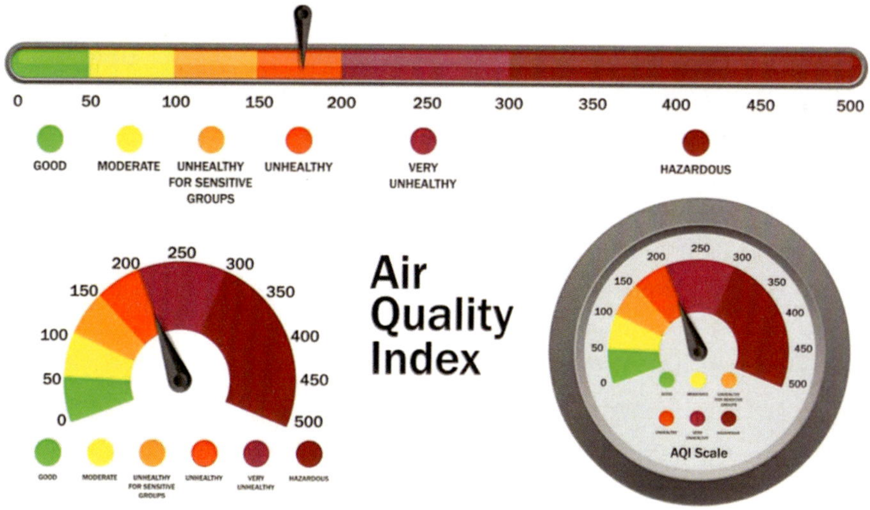

Fig. 13.1: Air quality analysis

13.2.1 What is Air Quality?

Air quality refers to the condition of the air around us, specifically the presence and concentration of various substances that can harm human health, damage ecosystems, or affect the climate. These substances are broadly categorized as **air pollutants**.

13.2.2 Key Air Pollutants Measured and Analyzed

While many substances can pollute the air, several are commonly monitored due to their widespread presence and significant impacts:

- **Particulate Matter (PM)**
    - **Description:** Tiny solid particles or liquid droplets suspended in the air. Categorized by size:
    - **PM2.5:** Fine particulate matter, less than 2.5 micrometers in diameter. These are particularly dangerous because they can penetrate deep into the lungs and even enter the bloodstream.
    - **PM10:** Coarse particulate matter, less than 10 micrometers in diameter.
    - **Sources:** Combustion (vehicles, power plants, industrial processes, wildfires, residential heating), dust (construction, roads), agriculture.
    - **Health Impacts:** Respiratory and cardiovascular diseases, premature death.

- **Ozone (O_3)**
  - **Description:** A highly reactive gas. In the stratosphere, it forms the protective ozone layer. At ground level (tropospheric ozone), it is a harmful air pollutant and a greenhouse gas.
  - **Sources:** Not directly emitted, but formed in the atmosphere through chemical reactions between nitrogen oxides (NOx) and volatile organic compounds (VOCs) in the presence of sunlight and heat.
  - **Health Impacts:** Respiratory problems (asthma, bronchitis), lung damage.
- **Nitrogen Dioxide (NO_2)**
  - **Description:** A reddish-brown gas with a pungent odor.
  - **Sources:** Primarily from the combustion of fossil fuels (vehicles, power plants, industrial boilers).
  - **Health Impacts:** Respiratory problems, contributes to acid rain and smog formation.
- **Sulfur Dioxide (SO_2)**
  - **Description:** A colorless gas with a sharp, irritating odor.
  - **Sources:** Primarily from the burning of fossil fuels (especially coal) in power plants and industrial facilities.
  - **Health Impacts:** Respiratory problems, contributes to acid rain and particulate matter formation.
- **Carbon Monoxide (CO)**
  - **Description:** A colorless, odorless, poisonous gas.
  - **Sources:** Incomplete combustion of carbon-containing fuels (vehicles, industrial processes, residential heating, wildfires).
  - **Health Impacts:** Reduces oxygen delivery to organs and tissues, can be fatal at high concentrations.

13.2.3 Methods of Air Quality Analysis

Air quality analysis relies on a combination of observation, modeling, and statistical techniques:

- **Pollutant Measurements (Monitoring Networks)**
  - **Ground-Based Stations:** Networks of fixed monitoring stations equipped with specialized instruments to continuously measure pollutant concentrations in ambient air. These provide highly accurate, real-time data at specific locations.

- **Mobile Monitoring:** Using vehicles or portable sensors to measure pollution across a larger area or in specific "hotspots."
- **Satellite Remote Sensing:** Satellites equipped with sensors can measure column concentrations of certain pollutants (NO_2, SO_2, O_3, aerosols) over large regions, providing broad spatial coverage, especially in areas with sparse ground monitoring.
- **Low-Cost Sensors:** A growing trend involves deploying networks of smaller, less expensive sensors, often in urban areas, to provide higher spatial resolution data, though with varying levels of accuracy.
- **Data Types:** Measurements typically include hourly or daily average concentrations, often reported in micrograms per cubic meter ($\mu g/m^3$) or parts per billion (ppb).

- **Air Quality Index (AQI)**
  - **Concept:** A standardized, color-coded index developed by environmental agencies (e.g., U.S. EPA, European Environment Agency) to communicate daily air quality levels to the public in a simple, understandable way.
  - **How it works:** The AQI converts the concentrations of various pollutants into a single number on a scale, with corresponding health implications and cautionary statements. The highest pollutant concentration determines the overall AQI for that day or hour.
  - **Interpretation:** Different color categories (e.g., Green: Good, Yellow: Moderate, Orange: Unhealthy for Sensitive Groups, Red: Unhealthy, Purple: Very Unhealthy, Maroon: Hazardous) provide immediate visual cues about potential health risks.
  - **Utility:** Essential for public health warnings, informing daily activities, and tracking air quality trends.
- **Dispersion Modeling**
  - **Concept:** Uses mathematical models to simulate how air pollutants are transported, dispersed, and chemically transformed in the atmosphere after they are emitted from a source.
  - **Inputs:** Requires detailed information on:
  - **Emission Sources:** Location, height, emission rate, and chemical composition of pollutants from factories, vehicles, etc.
  - **Meteorological Data:** Crucial inputs include wind speed and direction, atmospheric stability (how easily air mixes vertically),

temperature, humidity, and mixing height (the height of the well-mixed layer of the atmosphere).

- **Topography and Land Use:** Terrain features and surface characteristics influence airflow and dispersion.
- **How it works:** Models apply principles of fluid dynamics, atmospheric chemistry, and turbulence to calculate pollutant concentrations at various downwind locations.

- **Types of Models**
  - **Gaussian Plume Models:** Simple models for continuous emissions from single sources in relatively flat terrain.
  - **Lagrangian Particle Models:** Track individual "parcels" of pollutants as they move and disperse.
  - **Eulerian Grid Models (Chemical Transport Models - CTMs):** Divide the atmosphere into a 3D grid and solve equations for transport, diffusion, and chemical reactions of pollutants within each grid cell. These are complex, computationally intensive models that can simulate regional or global air quality.
- **Applications**
  - **Regulatory Compliance:** Assessing whether new industrial facilities will meet air quality standards.
  - **Forecasting:** Predicting future air pollution levels for public advisories.
  - **Source Apportionment:** Identifying the contribution of different emission sources to observed pollution levels.
  - **Emergency Response:** Predicting the spread of accidental chemical releases.
- **Statistical Analysis and Machine Learning**
  - **Trend Analysis:** Identifying long-term increases or decreases in pollutant concentrations.
  - **Source Apportionment (Statistical):** Using statistical techniques (e.g., Positive Matrix Factorization) to identify and quantify the contributions of different source types (e.g., traffic, industry, residential burning) to ambient particulate matter.
  - **Forecasting:** Applying machine learning algorithms (e.g., neural networks, random forests) to predict future air quality based on historical pollutant data, meteorological variables, and other factors.

- **Relationship Discovery:** Exploring correlations between meteorological conditions and pollutant levels (e.g., stagnation events leading to high pollution).

13.2.4 Applications of Air Quality Analysis in Meteorology and Beyond

- **Public Health Protection:** Informing health advisories, developing policies to reduce exposure, and understanding the health burden of pollution.
- **Environmental Regulation:** Setting and enforcing air quality standards, evaluating the effectiveness of emission control strategies.
- **Urban Planning:** Designing cities to promote better air circulation, locating industrial zones appropriately, and planning green spaces.
- **Climate Change Research:** Understanding the role of aerosols (which can cool or warm the planet) and short-lived climate pollutants (like black carbon, methane) in the climate system.
- **Forecasting and Warning Systems:** Providing real-time and forecast air quality information to the public and emergency services.
- **Interdisciplinary Research:** Linking air quality with meteorology, hydrology (e.g., acid rain effects on water bodies), and ecology (e.g., ozone damage to crops).

Air quality analysis is a dynamic and increasingly vital field at the intersection of atmospheric science and societal well-being. By meticulously measuring pollutants, leveraging sophisticated dispersion models, and applying advanced statistical techniques, scientists and policymakers gain crucial insights into the complex interplay between emissions, atmospheric processes, and environmental impacts. This comprehensive understanding is fundamental for developing effective strategies to mitigate air pollution, protect public health, and contribute to a healthier and more sustainable planet.

13.3 Hydrometeorological Data and Water Management

The Earth's climate and water cycle are intrinsically linked. The atmosphere dictates where, when, and how much water falls as precipitation, while the land surface and oceans govern its storage, movement, and eventual return to the atmosphere. **Hydrometeorology** is the interdisciplinary science that studies the transfer of water and energy between the land surface and the lower atmosphere. Consequently, **hydrometeorological data** forms the bedrock for effective **water management**, a critical challenge in a world facing increasing water scarcity, extreme floods, and prolonged droughts.

13.3.1 What is Hydrometeorological Data?

Hydrometeorological data refers to the collection of measurements and observations related to both atmospheric (meteorological) and hydrological (water-related) components of the water cycle. It provides a comprehensive picture of water's journey through the Earth system.

13.3.2 Key Types of Hydrometeorological Data

- **Precipitation**
  - **Measurements:** Rain gauges (manual, automated tipping bucket, weighing gauges), weather radar (estimating rainfall intensity and accumulation over areas), satellite precipitation estimates (e.g., from GPM, TRMM missions).
  - **Importance:** The primary input of water into terrestrial systems. Crucial for understanding water availability, flood potential, and drought severity.
- **Temperature**
  - **Measurements:** Thermometers (air, soil, water), satellite-derived surface temperatures.
  - **Importance:** Influences evaporation rates, snowmelt, and the phase of precipitation (rain vs. snow).
- **Evaporation and Evapotranspiration (ET)**
  - **Measurements:** Evaporation pans, lysimeters, eddy covariance systems, satellite-derived ET estimates.
  - **Importance:** Represents water loss from land and water surfaces to the atmosphere. Critical for water balance calculations, irrigation scheduling, and drought monitoring.
- **Humidity (e.g., Dew Point, Relative Humidity)**
  - **Measurements:** Hygrometers, radiosondes, satellite sensors.
  - **Importance:** Influences evaporation, cloud formation, and atmospheric stability.
- **Wind Speed and Direction**
  - **Measurements:** Anemometers, Doppler radar, satellite scatterometers.
  - **Importance:** Drives advection of moisture, influences evaporation rates, and affects snow redistribution.
- **Solar Radiation**
  - **Measurements:** Pyranometers, satellite sensors.
  - **Importance:** Provides the energy for evaporation and snowmelt.

- **Snowpack and Snow Water Equivalent (SWE)**
  - **Measurements:** Snow depth sensors, snow pillows (measuring weight), ground-penetrating radar, satellite remote sensing (e.g., passive microwave, lidar).
  - **Importance:** Snowpack acts as a natural reservoir, storing water for later release during spring melt. SWE quantifies the amount of water contained within the snow. Crucial for seasonal water supply forecasts and flood risk.
- **Soil Moisture**
  - **Measurements:** Soil moisture sensors (in-situ), satellite remote sensing (e.g., SMAP, SMOS missions).
  - **Importance:** Influences runoff generation, plant growth, and land-atmosphere feedbacks. A key indicator of drought conditions.
- **River Flow (Streamflow/Discharge)**
  - **Measurements:** Stream gauges (measuring water level, then converted to flow using rating curves).
  - **Importance:** Direct measure of water moving through river systems. Essential for flood forecasting, water supply allocation, and hydropower generation.
- **Lake and Reservoir Levels/Volumes**
  - **Measurements:** Staff gauges, pressure transducers, satellite altimetry.
  - **Importance**: Indicates water storage for human consumption, irrigation, and hydropower.

13.3.3 The Role of Hydrometeorological Data in Water Management:

Effective water management involves planning, developing, distributing, and managing the optimum use of water resources. Hydrometeorological data is fundamental to every aspect of this process:

- **Flood Forecasting and Warning**
  - **Data Use:** Real-time precipitation (radar, rain gauges), river gauge levels, soil moisture, snowmelt forecasts, and NWP precipitation forecasts.
  - **Application:** Hydrological models ingest this data to predict river discharge and inundation areas. This allows for timely flood warnings, evacuation orders, and deployment of emergency resources, minimizing loss of life and property.

- **Drought Monitoring and Management**
  - **Data Use:** Long-term precipitation records, temperature, soil moisture, evaporation, and streamflow.
  - **Application:** Used to calculate various drought indices (e.g., Standardized Precipitation Index - SPI, Palmer Drought Severity Index - PDSI) to assess drought severity, duration, and spatial extent. This informs water use restrictions, agricultural aid, and long-term water conservation planning.
- **Water Supply Forecasting**
  - **Data Use:** Snowpack (SWE), historical precipitation, temperature forecasts, and streamflow data.
  - **Application:** Particularly vital in regions reliant on snowmelt. Forecasts of seasonal water supply help water managers allocate resources for irrigation, municipal use, hydropower generation, and environmental flows.
- **Reservoir Operations**
  - **Data Use:** Inflow forecasts (from precipitation and snowmelt), current reservoir levels, and downstream demands.
  - **Application:** Hydrometeorological data guides decisions on how much water to release from dams for hydropower, flood control, irrigation, and maintaining ecological flows, balancing competing needs.
- **Irrigation Scheduling**
  - **Data Use:** Evapotranspiration rates, soil moisture, precipitation forecasts, and crop water requirements.
  - **Application:** Helps farmers determine precisely when and how much to irrigate, optimizing water use, conserving resources, and improving crop yields.
- **Hydropower Generation**
  - **Data Use:** River flow forecasts, snowmelt forecasts, and reservoir levels.
  - **Application:** Enables power companies to optimize hydropower generation, balancing energy demand with water availability and other reservoir uses.
- **Water Quality Management**
  - **Data Use:** Precipitation (runoff), streamflow, temperature, and land use data.

    - **Application:** Understanding how rainfall events lead to pollutant runoff into rivers and lakes, informing strategies for water quality protection.

- **Climate Change Adaptation**
    - **Data Use:** Long-term hydrometeorological records, climate model projections of precipitation and temperature.
    - **Application:** Assessing how climate change will impact future water availability, flood frequency, and drought severity, informing long-term infrastructure planning and adaptation strategies.

13.3.4 Challenges in Hydrometeorological Data and Water Management

- **Data Scarcity:** Many regions, particularly in developing countries, lack sufficient hydrometeorological monitoring networks.
- **Data Quality and Homogeneity:** Ensuring the accuracy and consistency of long-term records is challenging due to instrument changes, station moves, and varying observation practices.
- **Spatial and Temporal Resolution:** Matching the resolution of atmospheric data (e.g., radar, satellite) with the needs of hydrological models (e.g., at catchment scale) is complex.
- **Model Coupling:** Accurately coupling atmospheric, land surface, and hydrological models remains a scientific challenge.
- **Uncertainty Quantification:** Effectively communicating the uncertainties in hydrometeorological forecasts and water resource projections.
- **Transboundary Water Issues:** Managing water resources across international or state borders often involves complex political and data-sharing challenges.

13.3.5 Future Trends

- **Remote Sensing:** Increased reliance on satellite observations for precipitation, soil moisture, snowpack, and surface water levels, providing broader and more consistent coverage.
- **Machine Learning and AI:** Application of ML for improving hydrometeorological forecasts, downscaling, bias correction, and identifying complex relationships in large datasets.
- **Integrated Modeling Systems:** Development of fully coupled Earth System Models that seamlessly integrate atmospheric, hydrological, and land surface processes.

- **Citizen Science:** Leveraging crowd-sourced observations (e.g., rainfall reports) to augment traditional monitoring networks.
- **Real-time Data Streams:** Enhanced infrastructure for real-time data collection, transmission, and assimilation into operational systems.
- **Decision Support Systems:** Development of user-friendly, interactive platforms that integrate hydrometeorological data and models to support water management decisions.

Hydrometeorological data forms the indispensable backbone of effective water management, providing the critical information needed to understand the water cycle, predict its behavior, and manage water resources sustainably. From issuing life-saving flood warnings to optimizing irrigation and planning for future water security in a changing climate, the meticulous collection, analysis, and application of these interconnected atmospheric and hydrological measurements are paramount. As global water challenges intensify, the continued advancement and integration of hydrometeorological data and modeling will be central to building resilient societies.

13.4 Renewable Energy Integration: The Role of Meteorological Data

The global transition towards a sustainable energy future heavily relies on the widespread adoption and effective integration of renewable energy sources, primarily **solar (photovoltaic) and wind power**. Unlike traditional fossil fuel-based power generation, which can be dispatched on demand, solar and wind energy are inherently **intermittent and variable**, directly dependent on weather conditions. This dependence makes accurate meteorological and climate data, along with specialized atmospheric science applications, absolutely crucial for the successful **integration of renewable energy** into national and global power grids.

13.4.1 Why is Meteorological Data Crucial for Renewable Energy?

The variability of solar and wind resources poses significant challenges for grid operators, who must maintain a constant balance between electricity supply and demand. Meteorological data provides the essential intelligence to address these challenges across the entire lifecycle of a renewable energy project:

- **Resource Assessment (Siting):** Identifying the best locations for new wind farms or solar power plants requires long-term, high-resolution data on wind speeds, solar radiation, and other relevant atmospheric variables.
- **Forecasting and Grid Management:** Predicting how much electricity will be generated by solar and wind farms in the coming hours, days, and

even weeks is vital for balancing the grid, managing energy markets, and ensuring reliability.

- **Operations and Maintenance:** Understanding weather impacts on existing infrastructure (e.g., strong winds affecting turbine operation, dust on solar panels) helps optimize maintenance schedules and minimize downtime.
- **Climate Change Impact Assessment:** Projecting how climate change might alter future solar and wind resources is crucial for long-term energy planning and investment.

13.4.2 Key Meteorological Data and Its Application in Renewable Energy

A wide range of meteorological data, from real-time observations to long-term climate projections, is utilized:

- **Wind Energy**
    - **Data Needed:** Wind speed, wind direction, turbulence, wind shear (change in wind with height), air density, temperature, icing potential.
- **Applications**
    - **Resource Assessment:** Long-term wind speed data (from meteorological towers, reanalysis, or climate models) at turbine hub height (e.g., 80–160 meters) is used to estimate annual energy production (AEP) and identify optimal wind farm locations.
    - **Power Forecasting:** Numerical Weather Prediction (NWP) models provide forecasts of wind speed and direction for the next hours to days. These are converted into power output forecasts using turbine power curves. This information helps grid operators predict how much wind power will be available, enabling them to ramp up or down other power sources.
    - **Turbine Design & Load Management:** Understanding extreme wind events and turbulence helps in designing robust turbines and managing their operational loads.
    - **Icing Forecasts:** Predicting ice accretion on turbine blades, which can reduce efficiency and cause damage.
- **Solar Energy (Photovoltaic - PV & Concentrated Solar Power - CSP)**
- **Data Needed**
    - **Solar Radiation:** Global Horizontal Irradiance (GHI), Direct Normal Irradiance (DNI), Diffuse Horizontal Irradiance (DHI).
    - **Cloud Cover:** Type, height, and movement of clouds.

- **Temperature:** Air temperature (influences PV panel efficiency).
- **Aerosols/Dust:** Can reduce solar radiation reaching the surface.

- **Applications**
  - **Resource Assessment:** Long-term solar radiation data (from ground stations, reanalysis, or satellite-derived products) is used to estimate the energy potential of a site.
  - **Power Forecasting:** NWP models and specialized solar forecasting models predict solar radiation and cloud cover. Short-term (minutes to hours) forecasts are critical for managing PV output fluctuations caused by passing clouds. Longer-term forecasts (days to weeks) help in grid scheduling.
  - **Panel Efficiency:** Temperature forecasts are used to estimate the efficiency of PV panels, which decreases with increasing temperature.
  - **Soiling:** Data on dust and precipitation helps predict soiling losses on solar panels and optimize cleaning schedules.
- **Hydroelectric Power (Indirectly):**
  - **Data Needed:** Precipitation (rain, snow), temperature (for snowmelt), evaporation, soil moisture.
  - **Applications:** Hydrometeorological forecasts of river flow and reservoir inflows are crucial for optimizing hydropower generation, balancing energy production with water management needs (e.g., flood control, irrigation).

13.4.3 Specialized Applications and Techniques

The integration of meteorological data into renewable energy operations has led to several specialized applications:

- **High-Resolution NWP Models:** Regional and even convection-permitting NWP models are increasingly used to provide very localized forecasts of wind and solar resources, especially in areas with complex terrain or coastal effects.
- **Ensemble Forecasting:** Ensemble NWP provides probabilistic forecasts of renewable energy output, quantifying the uncertainty in future generation. This allows grid operators to make more robust decisions and manage risks.

1. **Machine Learning (ML) and Artificial Intelligence (AI):**
   - Post-processing: ML models are used to statistically correct biases in raw NWP model output for wind and solar.

- Power Conversion: ML can learn complex relationships between meteorological variables and actual power output from specific farms, improving power forecasts.
- Short-Term Forecasting/Nowcasting: AI models can rapidly analyze satellite and sky camera images to predict cloud movement and solar output fluctuations in the very short term (0-6 hours).
- Anomaly Detection: Identifying unusual performance of wind turbines or solar panels based on meteorological conditions.

1. **Remote Sensing:** Satellite data is becoming increasingly vital for assessing solar resources, tracking cloud movement, and monitoring snowpack for hydropower, especially in data-sparse regions.
    - **Wind Lidar and Sodar:** Ground-based remote sensing instruments (Lidar, Sodar) provide high-resolution vertical profiles of wind speed and turbulence, crucial for wind resource assessment and turbine siting.
    - **Grid Integration Software:** Specialized software platforms integrate meteorological forecasts with grid management systems, enabling real-time optimization of energy dispatch, storage, and market operations.

13.4.5 Challenges in Renewable Energy Integration

- **Intermittency and Variability:** The inherent nature of solar and wind power makes balancing supply and demand challenging, requiring flexible conventional power sources or energy storage.
- **Forecast Uncertainty:** While improving, forecasts for renewable energy output still have uncertainties, especially for short-term fluctuations and extreme weather events.
- **Grid Infrastructure:** Existing grid infrastructure may not be designed to handle large amounts of variable renewable energy, requiring significant upgrades.
- **Data Availability and Quality:** High-resolution, long-term meteorological data at specific sites can be scarce or expensive.
- **Complex Interactions:** Understanding how large-scale weather patterns and local microclimates interact to affect energy generation is complex.
- **Climate Change Impacts:** Projecting how climate change will alter future wind and solar resources adds another layer of uncertainty to long-term energy planning.

13.4.6 Future Trends

- **"Digital Twins" of Renewable Energy Farms:** Creating highly detailed virtual models of wind and solar farms that integrate real-time meteorological data, operational data, and AI to optimize performance and predict maintenance needs.
- **Increased Use of AI for Grid Optimization:** AI will play an even larger role in real-time grid balancing, energy trading, and demand-side management, leveraging advanced meteorological forecasts.
- **Hybrid Renewable Energy Systems:** Combining solar and wind with energy storage (batteries, pumped hydro) to mitigate intermittency, requiring integrated meteorological forecasting for the entire system.
- **Atmospheric Boundary Layer (ABL) Modeling:** More accurate and higher-resolution modeling of the ABL (the lowest part of the atmosphere) is critical for wind energy, as most turbines operate within this layer.
- **Climate-Resilient Energy Systems:** Using climate change projections to design renewable energy infrastructure that can withstand future extreme weather events.
- **Global Energy System Models:** Integrating climate models with energy system models to project future energy pathways & their climate impacts.

The successful integration of renewable energy into our global power grids is inextricably linked to advancements in meteorological science and data analysis. From initial resource assessment and optimal siting to real-time power forecasting and grid management, accurate and timely meteorological information is paramount. As the world accelerates its transition to clean energy, the continued development of specialized atmospheric applications, leveraging high-resolution models, ensemble forecasting, and cutting-edge AI, will be central to overcoming the challenges of intermittency and variability, ultimately enabling a resilient, sustainable, and decarbonized energy future.

13.5 Urban Climatology and Smart Cities

As humanity increasingly congregates in urban centers, understanding the unique atmospheric environment of cities becomes paramount. **Urban climatology** is the specialized field that investigates the climate of urban areas, focusing on how built environments modify atmospheric conditions compared to surrounding rural areas. This discipline is not merely academic; its insights are crucial for informing **smart city initiatives**, which leverage technology and data to improve urban living, sustainability, and resilience in the face of environmental challenges (Fig. 13.2).

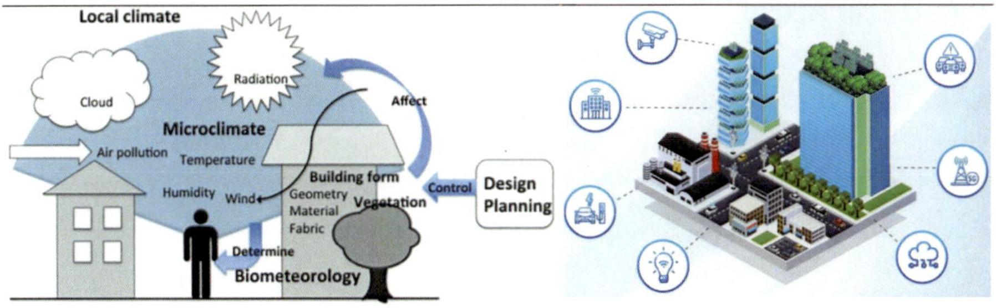

Fig. 13.2: Urban climate and smart city concept

13.5.1 Urban Climatology: The Climate of Cities

Cities are complex, three-dimensional structures that fundamentally alter the natural energy and water balances of the Earth's surface. They create their own distinct microclimates, often referred to as the **Urban Climate**.

13.5.2 Key Phenomena Studied in Urban Climatology:

1 **Urban Heat Island (UHI) Effect**

- **Description:** The most well-known urban climatic phenomenon. Urban areas are consistently warmer than their surrounding rural areas, particularly at night.

- **Causes**
  - **Reduced Evapotranspiration:** Less vegetation and more impervious surfaces (concrete, asphalt) mean less water evaporates, leading to less evaporative cooling.
  - **Increased Heat Storage:** Building materials and paved surfaces absorb and store more solar radiation during the day and release it slowly at night.
  - **Anthropogenic Heat:** Heat generated by human activities (vehicles, air conditioning, industrial processes).
  - **Reduced Airflow:** Buildings can block winds, reducing convective cooling and trapping heat.
  - **Impacts:** Increased energy consumption for cooling, higher heat-related morbidity and mortality (especially during heatwaves), altered local weather patterns (e.g., enhanced convection downwind of cities), and reduced air quality (chemical reactions are temperature-dependent).

- **Urban Boundary Layer (UBL)**
  - **Description:** The lowest part of the atmosphere directly influenced by the urban surface. It has a distinct structure compared to the rural boundary layer, characterized by increased turbulence, different mixing heights, and altered wind profiles.
  - **Impacts:** Affects the dispersion of pollutants, the transport of heat and moisture, and local cloud formation.
- **Altered Wind Patterns:**
  - **Description:** Buildings create drag, slowing down winds at street level. They also create complex flow patterns, including channeling effects (winds funneling down streets), vortexes around tall buildings, and increased turbulence.
  - **Impacts:** Influences pedestrian comfort, pollutant dispersion, and the efficiency of urban wind energy systems.
- **Changes in Precipitation:**
  - **Description:** Cities can sometimes enhance precipitation downwind due to the UHI (providing uplift) and increased aerosols (acting as cloud condensation nuclei).
  - **Impacts:** Can lead to localized heavy rainfall and increased flood risk, especially in combination with impervious surfaces.
- **Air Quality and Pollution Dispersion:**
  - **Description:** Cities are major sources of air pollutants. Urban climatology studies how these pollutants are transported and dispersed within the complex urban environment, influenced by UHI, wind patterns, and boundary layer dynamics.
  - **Impacts:** Direct impact on public health, visibility, and local ecosystems.

13.5.3 Data and Tools in Urban Climatology:

- **Dense Sensor Networks:** Micro-scale weather stations, low-cost air quality sensors.
- **Remote Sensing:** Satellite and drone imagery (thermal, visible) to map surface temperatures, land cover, and building heights.
- **Computational Fluid Dynamics (CFD) Models:** High-resolution numerical models that simulate airflow and pollutant dispersion around individual buildings or street canyons.

- **Urban Canopy Models (UCMs):** Represent the urban environment within larger meteorological models to simulate urban-atmosphere interactions.
- **Geographic Information Systems (GIS):** Essential for integrating spatial data on buildings, vegetation, impervious surfaces, and meteorological observations.

13.5.4 Smart Cities: Leveraging Data for Urban Resilience

- **What are Smart Cities?** A **smart city** is an urban area that uses various types of electronic Internet of Things (IoT) sensors to collect data. Insights gained from that data are used to manage assets, resources, and services efficiently. The goal is to improve the quality of life for residents, enhance urban operations, and promote sustainability.
- **The Nexus of Urban Climatology and Smart Cities:** Urban climatology provides the scientific understanding of urban atmospheric processes, while smart city initiatives provide the technological infrastructure (sensors, data platforms, communication networks) to monitor, analyze, and respond to these processes in real-time. This synergy is critical for building truly resilient and livable cities.

13.5.5 Applications of Urban Climatology in Smart City Initiatives:

- **Urban Heat Island Mitigation:**
  - **Smart City Role:** Deploying dense networks of temperature sensors to map UHI hotspots in real-time. Using predictive models to forecast heat stress.
  - **Climatology Integration:** Informing strategies like:
  - **Green Infrastructure:** Identifying optimal locations for parks, green roofs, and urban forests to increase evaporative cooling and shading.
  - **Cool Pavements/Roofs:** Promoting the use of reflective materials for roofs and roads to reduce heat absorption.
  - **Water Features:** Designing urban water bodies for evaporative cooling.
  - **Outcome:** Reduces energy consumption for cooling, improves public health during heatwaves, and enhances outdoor comfort.
- **Air Quality Management**
  - **Smart City Role:** Deploying networks of low-cost air quality sensors at street level, integrating with traffic data and meteorological forecasts.

- **Climatology Integration:** Using dispersion models (informed by urban wind patterns and boundary layer dynamics) to predict pollutant concentrations at high resolution. Identifying "pollution canyons" or areas of poor ventilation.
- **Outcome:** Real-time air quality advisories, dynamic traffic management (e.g., rerouting traffic during high pollution episodes), informing urban design to improve ventilation, and targeted interventions to reduce emissions.

- **Sustainable Urban Planning and Design**
  - **Smart City Role:** Using GIS and urban modeling tools to simulate the climatic impact of new developments or infrastructure projects.
  - Climatology Integration:
  - **Building Orientation:** Advising on optimal building orientation to maximize natural ventilation and minimize solar gain.
  - **Street Canyon Design:** Designing street widths and building heights to promote airflow and pollutant dispersion.
  - **Microclimate Mapping:** Creating detailed maps of thermal comfort, wind exposure, and air quality at neighborhood scales to guide development.
  - **Outcome:** Creates more comfortable, energy-efficient, and healthier urban spaces.
- **Water Management and Flood Resilience**
- **Smart City Role:** Deploying smart rain gauges, soil moisture sensors, and real-time river level monitors. Integrating with smart drainage systems.
- **Climatology Integration:** Using high-resolution hydrometeorological models to forecast localized heavy rainfall and urban flash floods. Understanding how impervious surfaces and urban drainage systems interact with extreme precipitation.
- **Outcome:** Improved flood warnings, optimized stormwater management (e.g., smart retention ponds), and enhanced urban flood resilience.
- **Renewable Energy Integration (Urban Scale)**
- **Smart City Role:** Monitoring solar panel output on buildings, managing microgrids.
- **Climatology Integration:** Providing highly localized solar radiation forecasts (accounting for urban shading and aerosols) and wind resource assessments for rooftop wind turbines.
- **Outcome:** Optimizing urban energy generation and consumption.

13.5.6 Future Trends in Urban Climatology and Smart Cities

- **Hyper-Resolution Urban Models:** Development of increasingly sophisticated urban climate models that can resolve atmospheric processes down to meter scales, explicitly simulating airflow around individual buildings.
- **Dense Sensor Networks and IoT:** Proliferation of low-cost, interconnected sensors providing unprecedented spatial and temporal resolution of urban meteorological and air quality data.
- **AI and Machine Learning for Urban Systems:**
  - **Predictive Modeling:** Using ML to forecast urban heat island intensity, air pollution hotspots, and localized extreme weather.
  - **Data Fusion:** AI for integrating heterogeneous sensor data (ground, satellite, mobile) to create comprehensive urban environmental maps.
  - **Optimization:** AI for optimizing urban design, traffic flow, and energy consumption based on real-time environmental conditions.
  - **Digital Twins of Cities:** Creating virtual replicas of cities that integrate real-time environmental data, infrastructure, and human activity to simulate scenarios and test interventions.
  - **Citizen Science:** Engaging urban residents in collecting environmental data using personal sensors, contributing to denser monitoring networks.
  - **Integration with Public Health Systems:** Direct links between urban environmental data and public health advisories, emergency services, and long-term health planning.
  - **Nature-Based Solutions:** Increased focus on using natural elements (trees, green spaces, water bodies) as climate adaptation and mitigation strategies, guided by urban climatology insights.

Urban climatology and smart city initiatives are converging to create more livable, sustainable, and resilient urban environments. By meticulously analyzing the unique atmospheric dynamics of cities and leveraging advanced technologies for real-time monitoring and data-driven decision-making, we can effectively address critical challenges like the urban heat island effect, air pollution, and urban flooding. This interdisciplinary fusion of atmospheric science, urban planning, and cutting-edge technology is essential for shaping the future of our rapidly urbanizing world, ensuring that cities are not just centers of population but also beacons of environmental intelligence and human well-being.

13.6 Emerging Technologies in Meteorology

The field of meteorology is undergoing a rapid transformation, driven by an explosion of new technologies that are revolutionizing how we observe, analyze, and predict the atmosphere. Beyond traditional instruments and supercomputing, a new wave of innovations is providing unprecedented data resolution, accessibility, and insights. This section explores some of the most impactful emerging technologies and concepts: the Internet of Things (IoT) sensors, drones, citizen science data, and the crucial role of Explainable AI (XAI).

13.6.1 Internet of Things (IoT) Sensors

What are IoT Sensors in Meteorology? The Internet of Things (IoT) refers to a network of physical objects ("things") embedded with sensors, software, and other technologies for the purpose of connecting and exchanging data with other devices and systems over the internet. In meteorology, IoT sensors are typically **low-cost, compact, and often wireless devices** that can measure various atmospheric parameters (temperature, humidity, pressure, rainfall, wind speed, air quality pollutants) and transmit data in real-time.

13.6.2 How they are Changing Meteorology:

- **Hyper-Local Resolution:** Traditional meteorological networks are sparse. IoT sensors enable the deployment of incredibly dense networks, providing unprecedented spatial resolution of atmospheric conditions, particularly in complex urban environments or remote areas.
- **Example:** A network of hundreds of low-cost temperature and humidity sensors across a city can map the Urban Heat Island effect with granular detail, revealing microclimates at a street or even building level.
- **Real-Time Data Streams:** Many IoT sensors are designed for continuous, real-time data transmission, offering up-to-the-minute information crucial for nowcasting and rapid decision-making.
- **Cost-Effectiveness:** Their relatively low cost makes it feasible to deploy far more sensors than traditional, expensive meteorological equipment.
- **Accessibility:** Data can often be accessed via web platforms or APIs, making it available to a wider range of users.
- **New Data Types:** Beyond standard meteorological variables, IoT sensors can collect data on specific parameters like particulate matter (PM2.5, PM10) or specific gases for air quality monitoring.

13.6.3 Applications in Meteorology

- **Urban Microclimate Monitoring:** Detailed mapping of urban heat islands, wind flow in street canyons, and localized air pollution hotspots.
- **Precision Agriculture:** Monitoring soil moisture, localized temperature, and humidity in specific fields to optimize irrigation and crop management.
- **Disaster Monitoring:** Early warning systems for localized flooding (smart rain gauges), landslides (soil moisture sensors), or wildfires (temperature/smoke sensors).
- **Smart Infrastructure:** Integrating weather data with smart buildings for optimized energy consumption (e.g., smart HVAC systems responding to real-time outdoor conditions).

13.6.1 Lmitations

- **Accuracy and Calibration:** Low-cost sensors can have lower accuracy and require frequent calibration compared to research-grade instruments.
- **Data Quality Control:** The sheer volume and potential variability in quality necessitate robust data quality control and validation methods.
- **Maintenance:** Managing and maintaining large-scale sensor networks can be challenging.
- **Drones (Unmanned Aerial Systems - UAS)**

What are Drones in Meteorology? Drones, or Unmanned Aerial Systems (UAS), are remotely piloted aircraft that can carry various meteorological sensors (e.g., temperature, humidity, pressure, wind, particulate matter, specialized atmospheric chemistry sensors) into the atmosphere.

13.6.2 How they are Changing Meteorology

- **Vertical Profiling:** Drones can provide high-resolution vertical profiles of the lower atmosphere (the atmospheric boundary layer) in areas and at times where traditional weather balloons are impractical or too expensive. This is crucial for understanding stability, mixing heights, and wind shear.
- **Example:** A drone can fly through a developing thunderstorm to collect in-situ data on updrafts and downdrafts, or map the vertical structure of an urban heat island.
- **Targeted Observations:** Drones can be deployed rapidly to specific locations of interest (e.g., near a severe storm, over a wildfire, within a pollution plume, over complex terrain) to gather targeted observations.

- **Hazardous Environments:** They can collect data in hazardous or inaccessible environments, such as inside volcanic plumes, near wildfires, or over contaminated areas.
- **Spatial Mapping:** Drones equipped with cameras or lidar can create high-resolution 3D maps of terrain, vegetation, and urban structures, which are valuable inputs for microscale atmospheric models.

13.6.3 Applications in Meteorology

- **Boundary Layer Research:** Detailed studies of atmospheric turbulence, energy fluxes, and pollutant dispersion.
- **Severe Weather Research:** Probing the near-storm environment of thunderstorms, potentially improving tornado warnings.
- **Air Quality Monitoring:** Mapping pollutant concentrations at different altitudes in urban areas or around industrial sites.
- **Wildfire Monitoring:** Providing real-time meteorological conditions within and around active wildfires to aid firefighting efforts.
- **Agricultural Meteorology:** Assessing crop health, water stress, and localized microclimates for precision agriculture.

13.6.4 Limitations

- **Flight Regulations:** Strict regulations regarding airspace, altitude limits, and line-of-sight operation.
- **Endurance and Payload:** Limited battery life and payload capacity restrict flight duration and sensor weight.
- **Weather Limitations:** Cannot operate in very strong winds, heavy precipitation, or icing conditions.
- **Cost:** While cheaper than manned aircraft, specialized meteorological drones and sensors can still be expensive.

13.6.5 Citizen Science Data

What is Citizen Science Data in Meteorology? Citizen science involves members of the public actively participating in scientific research. In meteorology, this often means volunteers collecting and sharing weather-related observations using personal weather stations (PWS), smartphone apps, or simple visual observations.

13.6.6 How it is Changing Meteorology

- **Massive Data Volume:** The sheer number of citizen scientists can generate an unprecedented volume of data, particularly from areas not covered by official networks.

- **Increased Spatial Density:** PWS networks (e.g., those affiliated with Weather Underground or CoCoRaHS) provide observations at resolutions far denser than national meteorological networks, filling in crucial data gaps.
- **Real-Time Observations:** Many PWS provide real-time data that can be assimilated into NWP models or used for nowcasting.
- **Community Engagement:** Fosters scientific literacy and community engagement in environmental issues.

13.6.7 Applications in Meteorology

- **Rainfall Mapping:** CoCoRaHS (Community Collaborative Rain, Hail & Snow Network) provides daily precipitation measurements that are vital for drought monitoring, flood forecasting, and hydrological modeling at high resolution.
- **Temperature and Humidity Mapping:** PWS data contributes to denser temperature and humidity maps, especially in urban or complex terrain areas.
- **Severe Weather Spotting:** Trained storm spotters provide real-time visual observations of severe weather (e.g., tornadoes, large hail, damaging winds) that are crucial for issuing warnings.
- **Phenology:** Tracking the timing of natural events (e.g., leaf-out, flowering) influenced by climate.

13.6.8 Limitations

- **Data Quality Control:** The primary challenge is ensuring the quality and accuracy of data from non-professional sensors and observers. Robust quality control algorithms and outlier detection are essential.
- **Representativeness:** Sensor siting may not always be ideal (e.g., PWS on rooftops can be affected by buildings).
- **Missing Data:** Gaps in data due to sensor malfunction or observer inconsistency.
- **Bias:** Potential biases from non-standardized equipment or procedures.

13.6.9 Explainable AI (XAI) in Meteorology

What is Explainable AI (XAI)? As Machine Learning (ML) and particularly Deep Learning (DL) models become increasingly complex and are deployed in critical applications like weather forecasting, they often operate as "black boxes." XAI is a field dedicated to developing methods and techniques that make ML models more understandable and transparent. It aims to answer questions like: "Why did the model make that prediction?""Which input features were most important for this specific forecast?""When is the model likely to be wrong?"

13.6.10 Why XAI is Crucial in Meteorology

- **Trust and Accountability:** Forecasters and decision-makers need to trust the models they use. If an AI model issues a severe weather warning, meteorologists need to understand the reasoning to confidently communicate it.
- **Scientific Discovery:** XAI can help scientists uncover new physical relationships or atmospheric patterns that the AI has learned but were previously unknown.
- **Model Improvement:** Understanding *why* a model makes errors helps pinpoint weaknesses in its training data, architecture, or underlying assumptions, guiding improvements.
- **Operational Implementation:** For models to be adopted in operational forecasting centers, their behavior must be predictable and understandable, especially in high-stakes situations.
- **Addressing "Black Box" Limitations:** ML excels at correlation, but meteorology requires understanding causation and physical processes. XAI helps bridge this gap.

13.6.11 How XAI is Applied in Meteorology

- **Feature Importance:** Techniques that identify which meteorological variables or input data regions (e.g., specific areas on a radar image) are most influential for a given forecast.
  - **Example:** For an AI model predicting severe thunderstorms, XAI might show that radar reflectivity and storm-relative helicity were the most important features in a specific case.
- **Sensitivity Analysis:** Perturbing input features to see how the model's output changes, revealing its sensitivity to different atmospheric conditions.
- **Activation Mapping (for CNNs):** Visualizing which parts of an input image (e.g., a satellite image of a hurricane) "activated" certain neurons in a CNN, revealing what features the network focused on for its prediction.
- **Rule Extraction:** For some simpler ML models, XAI can extract human-readable rules that explain the model's decision-making process.

- **Counterfactual Explanations:** Generating scenarios that show what minimal changes to the input would have led to a different prediction (e.g., "If the humidity were slightly lower, the model would not have predicted rain").

13.6.12 Challenges for XAI in Meteorology

- **Complexity of Models:** Explaining highly complex deep learning models trained on massive, multi-dimensional datasets is inherently difficult.
- **High-Stakes Decisions:** The consequences of a wrong weather forecast can be severe, raising the bar for required interpretability.
- **Physical Consistency:** Explanations must not only be understandable but also physically plausible to meteorologists.
- **Computational Cost:** Running XAI techniques on large meteorological models can be computationally intensive.

Emerging technologies like IoT sensors, drones, and citizen science data are rapidly expanding our observational capabilities, providing unprecedented resolution and coverage of the atmosphere. Simultaneously, advancements in Explainable AI are crucial for unlocking the full potential of complex machine learning models in meteorology, ensuring their trustworthiness, guiding their improvement, and fostering scientific discovery. The convergence of these innovations is propelling atmospheric science into an exciting future, enabling more precise predictions, deeper understanding, and ultimately, a more resilient society in the face of our dynamic environment.

13.7 Conclusion

Concluding our comprehensive journey through specialized applications and future trends in atmospheric science, it is abundantly clear that the field is undergoing a profound transformation. We've seen how the foundational knowledge and advanced analytical techniques developed throughout this book are being meticulously applied to critical real-world challenges, from understanding and predicting **air quality** and **pollutant dispersion** to managing vital **hydrometeorological data** for water resources. The crucial role of atmospheric science in the successful **integration of renewable energy** and the burgeoning field of **urban climatology** for building smart, resilient cities underscores the direct societal relevance of meteorological expertise.

Looking forward, the future of atmospheric science is characterized by an accelerating pace of innovation. The proliferation of **IoT sensors** and **drones** promises unprecedented observational density and flexibility, while the integration of **citizen science data** broadens our observational reach. Crucially, the rise of **Explainable AI (XAI)** is addressing the "black box" challenge of advanced models, ensuring that as our predictive capabilities grow, so too does our understanding and trust in the underlying science. This convergence of cutting-edge technology, interdisciplinary collaboration, and a continuous drive for greater resolution and accuracy is not just enhancing our ability to forecast weather and climate; it is actively shaping a more sustainable, resilient, and informed future for humanity.

14

Challenges, Ethics, and the Future of Meteorological Data Science

14.1 Introduction

As we reach the culmination of our exploration into meteorological data science, from foundational principles to specialized applications and emerging trends, it is imperative to pause and reflect on the broader context of this rapidly evolving field. The chapter, **"Challenges, Ethics, and the Future of Meteorological Data Science,"** shifts our focus from the "how" to the "why" and "what next," addressing the significant hurdles, crucial ethical considerations, and transformative opportunities that lie ahead.

The journey through meteorological data science has revealed a landscape of immense progress: unprecedented data volumes, sophisticated modeling capabilities, and increasingly accurate predictions. However, this progress is not without its complexities. We grapple with the inherent **chaotic nature of the atmosphere**, the persistent **uncertainties in our models**, and the sheer **computational demands** of simulating Earth's intricate systems. Beyond these scientific and technical challenges, the increasing societal reliance on meteorological data raises profound **ethical questions** concerning data privacy, equitable access to information, the responsible use of AI, and the potential for bias in algorithms.

This chapter will delve into these multifaceted aspects. We will first examine the enduring **scientific and technical challenges** that continue to push the boundaries of research, from improving sub-seasonal to seasonal predictability to seamlessly integrating diverse data sources. Following this, we will confront the critical **ethical considerations** that arise when dealing with vast amounts of environmental data and its direct impact on human lives and livelihoods. This includes discussions on data ownership, algorithmic transparency, and the potential for exacerbating existing societal inequalities if not handled with care. Finally, we will cast our gaze forward, envisioning the **future of meteorological data science**. This forward-looking perspective will explore the potential for truly integrated Earth system models, the transformative impact of artificial intelligence, and the growing imperative for interdisciplinary collaboration to address the grand challenges of climate change and environmental sustainability.

Ultimately, this chapter serves as a critical reflection, reminding us that while the technological advancements in meteorological data science are awe-inspiring, their true value lies in their responsible and ethical application for the benefit of all. It is a call to continuous innovation, thoughtful leadership, and a commitment to leveraging the power of data for a more resilient and informed future.

14.2 Data Privacy and Security in Meteorological Data Science

In an era defined by vast data collection and advanced analytics, the principles of **data privacy and security** have become paramount across all scientific disciplines, and meteorological data science is no exception. While much meteorological data (e.g., satellite imagery, global model outputs) is inherently public and non-sensitive, the increasing granularity of observations, the integration of diverse datasets, and the development of specialized applications introduce critical considerations, particularly concerning **sensitive data**.

14.2.1 Data Privacy: Protecting Individual and Proprietary Information

- **What is Data Privacy?** Data privacy refers to the rights of individuals and organizations to control how their personal or proprietary information is collected, stored, used, and shared. It's about ensuring that data is handled in a way that respects individual autonomy and organizational confidentiality.
- **Why is Data Privacy a Concern in Meteorology?**

  While traditional meteorological data (temperature, pressure, wind) is generally not considered "personal" in the same way as medical records, several scenarios raise privacy concerns:
- **Hyper-Local Observations (IoT, Citizen Science):**
  - **Location Privacy:** Dense networks of IoT sensors or personal weather stations (PWS) can collect data at very specific locations, potentially revealing the precise movements or presence of individuals at their homes or workplaces. If this data is associated with user accounts or combined with other datasets, it could lead to privacy breaches (Fig. 14.1).

Fig. 14.1: Citizen science and meteorology

- **Behavioral Inference:** Fine-grained environmental data, when combined with other information, could potentially be used to infer patterns of human behavior (e.g., energy consumption patterns linked to specific weather conditions).
    - **Example:** A smart home thermostat that shares localized temperature and humidity data, inadvertently revealing occupancy patterns.
- Integration with Socio-Economic Data
    - **Vulnerability Mapping:** When meteorological hazard data (e.g., flood risk maps, heatwave exposure) is overlaid with demographic data (e.g., income levels, age, health conditions) to assess vulnerability, this aggregated information, if not properly anonymized, could raise privacy concerns, especially if individual-level data is involved.
    - **Example:** A project mapping heat-related illness in vulnerable populations, where individual health records are linked to specific addresses and weather conditions.
- **Proprietary Commercial Data**
    - **Commercial Weather Services:** Private companies collect and use vast amounts of weather data, often for commercial purposes (e.g., optimizing logistics, energy trading). This data, or the algorithms derived from it, can be highly proprietary and require protection from unauthorized access or use.
    - **Example:** A private company's high-resolution wind forecast data used for optimizing wind farm operations.

- **Drone and High-Resolution Imagery**
  - **Visual Privacy:** Drones collecting high-resolution imagery (visible or thermal) in urban areas could inadvertently capture identifiable individuals or private property details, raising concerns about surveillance.
  - **Example:** A drone mapping urban heat islands that also captures images of people in their backyards.
- **Key Privacy Considerations**
  - **Anonymization/Pseudonymization:** Techniques to remove or obscure personally identifiable information (PII) from datasets.
  - **Data Minimization:** Collecting only the data that is absolutely necessary for the intended purpose.
  - **Consent:** Obtaining explicit consent from individuals when collecting data that could be linked to them.
  - **Data Governance Policies:** Clear rules and procedures for how data is handled, who has access, and for what purposes.
  - **Compliance with Regulations:** Adhering to data protection regulations like GDPR (General Data Protection Regulation) or CCPA (California Consumer Privacy Act) where applicable.

14.2.2 Data Security: Protecting Data from Unauthorized Access or Damage

- **What is Data Security?** Data security involves implementing measures to protect data from unauthorized access, use, disclosure, disruption, modification, or destruction. It's about safeguarding the integrity, confidentiality, and availability of data.
- **Why is Data Security Critical in Meteorology?**

  The integrity and availability of meteorological data are paramount, as disruptions can have severe consequences:
- **Operational Forecasting Integrity**
  - **Tampering:** Malicious actors attempting to alter real-time observations or model outputs could lead to incorrect forecasts, potentially causing widespread public harm (e.g., missed severe weather warnings, incorrect aviation forecasts).
  - **Denial of Service (DoS):** Attacks that prevent access to critical data streams or computing resources could cripple forecasting operations.

  - **Example:** A cyberattack on a national weather service's data assimilation system, leading to corrupted initial conditions for forecasts.
- **Critical Infrastructure Protection**
  - **Energy Grids:** The energy sector relies heavily on meteorological data for managing renewable energy (wind, solar) and optimizing grid operations. Compromised data could lead to blackouts or grid instability.
  - **Transportation:** Aviation, shipping, and road transport depend on accurate weather information. Security breaches could compromise safety.
  - **Example:** Hacking into a wind farm's meteorological data feed to disrupt its power output.
- **Research Data Integrity**
  - **Scientific Reproducibility:** Compromised or altered historical climate data could undermine scientific research, lead to erroneous conclusions about climate change, and erode public trust in science.
  - **Intellectual Property:** Protecting proprietary algorithms, model code, and specialized datasets developed by research institutions or private companies.
  - **Example:** A research climate model's historical forcing data being maliciously altered, leading to incorrect climate projections.
- **National Security**
  - **Military Operations:** Weather information is critical for military planning and operations.
  - **Intelligence:** Meteorological data can be part of broader intelligence gathering.
- **Key Security Considerations**
  - **Access Control:** Implementing strong authentication (e.g., multi-factor authentication) and authorization mechanisms to ensure only authorized personnel can access sensitive data and systems.
  - **Encryption:** Encrypting data both "at rest" (when stored) and "in transit" (when being transmitted) to protect it from interception.
  - **Network Security:** Using firewalls, intrusion detection systems, and secure network architectures to protect against cyber threats.

- **Regular Audits and Vulnerability Assessments:** Periodically checking systems for weaknesses and compliance with security policies.
- **Backup and Recovery:** Implementing robust data backup and disaster recovery plans to ensure data availability in case of system failures or attacks.
- **Supply Chain Security:** Ensuring that third-party data providers or software vendors also adhere to high security standards.
- **Physical Security:** Protecting physical infrastructure (e.g., data centers, supercomputers) from unauthorized access.

14.2.3 Considerations for Sensitive Data in Meteorology

When meteorological data becomes "sensitive" due to its granularity, integration with other information, or proprietary nature, specific considerations become paramount:

- **Data Minimization:** Collect only what is essential. If aggregated data is sufficient, avoid collecting individual-level details.
- **Purpose Limitation:** Use data only for the specific purpose for which it was collected and consented to.
- **Data Anonymization/Pseudonymization:** For public release or sharing, rigorously anonymize or pseudonymize data that could potentially identify individuals or reveal proprietary information. This is an ongoing challenge, as re-identification can sometimes occur even with anonymized data.
- **Access Control and Granular Permissions:** Implement strict access controls based on the "least privilege" principle, ensuring users only have access to the data they absolutely need.
- **Data Sharing Agreements:** For collaborations involving sensitive data, establish clear legal agreements outlining data usage, security protocols, and retention policies.
- **Ethical Review Boards:** For research involving human-related data (e.g., health impacts, vulnerability), seek approval from institutional review boards (IRBs) or ethics committees.
- **Transparency:** Be transparent with data subjects (individuals, organizations) about what data is being collected, why, and how it will be used and protected.
- **Regular Training:** Ensure all personnel handling data are trained on data privacy and security best practices.

As meteorological data science continues its rapid expansion, the issues of data privacy and security will only grow in prominence. While much atmospheric data is openly accessible, the increasing resolution of observations, the integration with socio-economic information, and the development of specialized commercial applications necessitate a proactive and robust approach to safeguarding sensitive information. By embedding strong privacy-by-design principles and implementing comprehensive security measures, the meteorological community can ensure that the immense benefits of data science are realized responsibly, maintaining public trust and protecting critical infrastructure in an increasingly data-driven world.

14.3 Ethical Considerations for AI/ML Applications in Meteorology: Bias and Transparency

The integration of Artificial Intelligence (AI) and Machine Learning (ML) into meteorological data science promises unprecedented advancements in forecasting, climate modeling, and specialized applications. However, as these powerful tools move from research labs to operational systems that directly impact human lives and livelihoods, critical **ethical considerations** come to the forefront. Among the most significant are the challenges of **bias** within AI/ML systems and the imperative for **transparency** in their decision-making processes.

14.3.1 The Ethical Imperative in AI/ML Meteorology

The decisions made based on meteorological forecasts and climate projections have profound societal consequences:

- **Public Safety:** Severe weather warnings, flood advisories, and air quality alerts directly affect human safety and health.
- **Economic Impact:** Forecasts influence decisions in agriculture, energy, transportation, and insurance, impacting livelihoods and economies.
- **Resource Allocation:** Climate projections guide long-term planning for water resources, infrastructure, and disaster preparedness.
- **Environmental Justice:** Unequal distribution of environmental burdens (like air pollution) can be exacerbated or mitigated by how data and models are used.

Given these high stakes, the ethical implications of AI/ML in meteorology cannot be overlooked.

14.3.2 Bias in AI/ML Applications

What is Bias in AI/ML? In the context of AI/ML, bias refers to systematic errors or unfair preferences within a model's output that lead to consistently inaccurate, discriminatory, or disadvantageous results for certain groups or conditions. AI/ML models learn from the data they are trained on, and if that data is biased, the model will learn and perpetuate those biases.

- **Sources of Bias in Meteorological AI/ML**
- **Observational/Data Bias**
    - **Uneven Data Distribution:** Meteorological observations are not uniformly distributed globally. There are often fewer high-quality observations in developing countries, remote regions, or over oceans. If an AI model is trained predominantly on data from well-monitored regions (e.g., North America, Europe), its performance might be poorer or biased when applied to data-sparse areas.
    - *Example:* An AI model trained to predict severe thunderstorms using radar data primarily from the U.S. might perform poorly or exhibit bias when applied to regions with different storm characteristics or sparser radar coverage (e.g., parts of Africa or South America).
    - **Historical Biases in Data Collection:** Past data collection practices might have inherent biases. For instance, early weather stations might have been preferentially located in certain areas (e.g., near airports or urban centers), leading to an unrepresentative historical record.
    - **Sensor Biases:** If certain types of low-cost IoT sensors consistently under-report or over-report values compared to reference instruments, and these are not properly corrected, an AI model trained on this data will inherit that bias.
- **Algorithmic/Model Bias**
    - **Algorithm Design:** The choice of algorithm or its specific configuration can introduce bias. For instance, some algorithms might implicitly favor majority classes in classification tasks.
    - **Parameterization Bias:** If AI is used to learn or improve parameterizations within NWP models, and the training data for these parameterizations is biased, the resulting model will inherit that bias.
    - **Example:** An AI model designed to downscale global climate model output to local scales might struggle with extreme events in regions where such events are rare in the training data, leading to under-prediction of risk in those areas.
- **Application/Societal Bias:**
    - **Unequal Impacts:** Even if a model is technically unbiased, its application can lead to biased societal outcomes if not considered

carefully. For example, if air quality forecasts are only provided at high resolution for affluent areas, but not for lower-income neighborhoods disproportionately affected by pollution, it exacerbates environmental injustice.

- **Resource Allocation:** If AI-driven risk assessments (e.g., for flood or heatwave vulnerability) are based on biased data or models, they could lead to misallocation of resources, leaving vulnerable communities unprotected.

- **Mitigating Bias**
  - **Diverse and Representative Data:** Actively seek out and incorporate diverse, high-quality data from all relevant geographical regions and atmospheric conditions. Address data gaps.
  - **Fairness Metrics:** Develop and use specific fairness metrics to evaluate model performance across different subgroups or conditions.
  - **Bias Detection Tools:** Employ techniques to detect and quantify bias in both training data and model outputs.
  - **Bias Correction Techniques:** Apply statistical or machine learning methods to correct for detected biases.
  - **Human-in-the-Loop:** Ensure human meteorologists and domain experts are involved in reviewing and validating AI model outputs, especially for critical decisions.
  - **Ethical AI Frameworks:** Develop and adhere to ethical guidelines and principles for AI development and deployment.

14.3.3 Transparency in AI/ML Applications

- **What is Transparency (Explainability)?** Transparency, often referred to as **explainability** (as discussed in Chapter 13), refers to the degree to which a human can understand the internal workings of an AI/ML model and comprehend *why* it made a particular prediction or decision. Transparent models allow users to trace the reasoning behind an output, rather than treating the model as a "black box."
- **Why is Transparency Crucial in Meteorology?**
  - **Trust and Confidence:** Meteorologists, emergency managers, and the public need to trust forecasts, especially for high-stakes events. If an AI model issues a severe weather warning, forecasters need to understand the meteorological rationale to confidently issue and communicate that warning.

- **Scientific Discovery and Improvement:** Understanding *how* an AI model arrives at its conclusions can reveal new physical insights into atmospheric processes or identify previously unrecognized patterns. This understanding is vital for improving both the AI model itself and the underlying scientific understanding.
- **Accountability and Debugging:** If a model makes a significant error, transparency allows scientists to diagnose the cause of the error (e.g., flawed data, incorrect feature weighting, model limitations) and fix it. Without transparency, debugging becomes a trial-and-error process.
- **Operational Adoption:** For AI/ML models to be widely adopted in operational forecasting centers, they must be interpretable by human forecasters, who are ultimately responsible for the forecasts issued.
- **Legal and Regulatory Compliance:** In some sectors, there may be future requirements for explainability, especially for automated decisions impacting individuals.

- **Challenges to Transparency in Meteorology:**
  - **Model Complexity:** Deep learning models, with millions of parameters and non-linear interactions, are inherently complex and difficult for humans to fully comprehend.
  - **High-Dimensional Data:** Meteorological data is often high-dimensional (spatial, temporal, vertical, multiple variables), making it challenging to visualize and interpret the model's processing of this data.
  - **Physical Consistency:** Explanations must not only be statistically sound but also physically plausible and consistent with meteorological principles.
  - **Promoting Transparency (Explainable AI - XAI):** As discussed in Chapter 13, XAI techniques are being developed to address this:
  - **Feature Importance Maps:** Visualizing which input features or regions of a map were most influential for a prediction.
  - **Partial Dependence Plots:** Showing how the model's output changes as a single input feature varies, while others are held constant.
  - **Local Interpretable Model-agnostic Explanations (LIME):** Explaining individual predictions by creating a simpler, interpretable model around that specific prediction.

- **SHAP (SHapley Additive exPlanations) Values:** Quantifying the contribution of each feature to a prediction.
- **Counterfactual Explanations:** Identifying the smallest changes to input data that would flip a prediction (e.g., "If wind speed had been 5 m/s lower, the model would not have predicted a severe storm").
- **Visualization Tools:** Developing specialized visualization tools to help meteorologists interact with and interpret AI model outputs.
- **Hybrid Approaches:** Combining the predictive power of AI with the physical consistency of traditional NWP models, or using AI to improve components of physical models, where the overall system remains more interpretable.

The ethical considerations of bias and transparency are not mere afterthoughts but fundamental pillars for the responsible and effective deployment of AI/ML applications in meteorological data science. Addressing bias requires proactive efforts in data collection, model design, and equitable application, ensuring that the benefits of advanced forecasting reach all communities fairly. Simultaneously, fostering transparency through Explainable AI techniques is crucial for building trust, facilitating scientific discovery, and enabling human meteorologists to confidently leverage these powerful tools. As we stride into the future of meteorological data science, a commitment to these ethical principles will be paramount in harnessing AI's potential for the greater good of society and the planet.

14.4 Interoperability and Standardization: The Need for Common Data Standards in Meteorology

In the increasingly data-driven world of meteorology and climate science, the ability for different systems, software, and datasets to seamlessly work together is not just a convenience—it's a fundamental necessity. This concept is known as **interoperability**. Central to achieving interoperability is the establishment and widespread adoption of **common data standards**. Without these standards, the vast and diverse landscape of meteorological data becomes a fragmented collection of isolated silos, hindering scientific progress, operational efficiency, and effective communication.

14.4.1 Interoperability: The Seamless Exchange of Information

- **What is Interoperability?** Interoperability refers to the ability of diverse information systems, applications, or components to communicate, exchange data, and use the information that has been exchanged, without requiring special effort from the end user. In essence, it means that data produced by one system can be easily understood and processed by another system.

- **Why is Interoperability Crucial in Meteorology?**

  The meteorological community relies on a highly complex and distributed ecosystem of data sources, models, and users:

- **Global Observation Networks:** Data is collected from thousands of ground stations, weather balloons, buoys, ships, aircraft, and a constellation of international satellites, operated by numerous countries and organizations.
- **Diverse Model Outputs:** Numerical Weather Prediction (NWP) and Climate Models are developed and run by various national meteorological services, research centers, and universities worldwide, each potentially using different internal data structures.
- **Varied User Applications:** Forecasts and climate data are used by a wide array of stakeholders, from government agencies and academic researchers to private companies (e.g., aviation, energy, agriculture) and the general public, each with different software and analytical needs.
- **Real-time Data Exchange:** For operational forecasting, real-time exchange of observations and model outputs across international borders is critical.
- **Data Assimilation:** Data assimilation systems need to ingest observations from highly diverse sources and formats to create a coherent initial state for models.
- **Climate Research:** Long-term climate studies often involve combining historical data from multiple archives and comparing outputs from different climate models.

Without interoperability, every time data moves between these different components or users, it would require time-consuming and error-prone manual conversion, reformatting, or custom programming. This creates significant bottlenecks, increases costs, and limits the pace of scientific discovery and operational efficiency.

14.4.2 Standardization: The Foundation for Interoperability

- **What are Common Data Standards?** Common data standards are agreed-upon rules, specifications, or protocols that define how data should be structured, formatted, described (metadata), and exchanged. They provide a universal language that allows different systems to "speak" to each other.

- **The Need for Common Data Standards in Meteorology:**

  The sheer volume, variety, and velocity of meteorological data necessitate robust standards to ensure consistency and usability. Without them, the "big data" challenge becomes insurmountable.

- **Ensuring Data Consistency and Quality**
  - **Uniformity:** Standards ensure that data from different sources measuring the same variable (e.g., temperature) are represented in the same units, format, and with consistent quality flags.
  - **Reduced Errors:** Standardized formats minimize errors that arise during data conversion or interpretation.
- **Facilitating Data Exchange and Sharing:**
  - **Seamless Flow:** Common standards enable automated, real-time exchange of observations and model outputs between national and international meteorological centers.
  - **Global Collaboration:** They are essential for international research collaborations (e.g., CMIP for climate models) where data from many different models needs to be combined and compared.
  - **Open Data Initiatives:** Standards underpin efforts to make meteorological and climate data openly accessible and usable by a wider community.
  - **Improving Data Discovery and Access:**
  - **Metadata Standards:** Standardized metadata (data about data) ensures that users can easily find, understand, and assess the suitability of a dataset for their needs (e.g., knowing its resolution, variables, time period, and quality).
  - **Searchability:** Consistent metadata makes data discoverable through automated search tools and data portals.
- **Supporting Model Development and Verification:**
  - **Common Input/Output:** Model developers can work with standardized input and output formats, reducing the effort needed to prepare data for models or process their results.
  - **Consistent Verification:** Standardized data facilitates objective and comparable verification of forecast skill across different models and regions.

- **Enabling New Technologies and Applications:**
- **AI/ML Readiness:** Machine learning algorithms thrive on clean, consistent data. Standards make it easier to prepare large, heterogeneous meteorological datasets for AI/ML training.
- **GIS Integration:** Standardized geospatial data formats allow for seamless integration of meteorological data into Geographic Information Systems (GIS) for advanced spatial analysis and visualization.

14.4.3 Key Meteorological Data Standards and Formats

The meteorological community has developed and adopted several crucial standards and formats:

- **GRIB (GRIdded Binary)**
  - **Description:** A highly efficient, self-describing binary format for gridded meteorological data, widely used for NWP model output and reanalysis products.
  - **Advantages:** Compact, efficient for storage and transmission, and self-describing (contains metadata about the data).
  - **Use:** The primary format for exchanging operational NWP forecasts globally (e.g., from ECMWF, NOAA, NCEP).
- **NetCDF (Network Common Data Form)**
  - **Description:** A widely adopted, self-describing, machine-independent data format for array-oriented scientific data. It's particularly popular in climate science.
  - **Advantages:** Flexible, supports metadata, widely compatible with many programming languages (Python, R, MATLAB), and can handle large, multi-dimensional datasets.
  - **Use:** Common for climate model outputs (e.g., CMIP project data), climate reanalysis products, and general scientific data archiving and exchange.
- **BUFR (Binary Universal Form for the Representation of meteorological data)**
  - **Description:** A compact and flexible binary format for observational data, designed to handle a wide variety of meteorological observations (surface, upper-air, satellite, radar).
  - **Advantages:** Highly efficient for transmitting observational data, self-describing, and adaptable to new observation types.

  - **Use:** The primary format for exchanging real-time meteorological observations globally via the WMO Global Telecommunication System (GTS).
- **ASCII/CSV**
  - **Description:** Simple text-based formats.
  - **Advantages:** Human-readable, easy to implement.
  - **Limitations:** Less efficient for large datasets, lacks self-description (metadata needs to be external), prone to errors.
  - **Use:** Still used for smaller datasets, simple station data, or for initial data exploration.
- **WMO (World Meteorological Organization) Standards:**
  - The WMO plays a central role in developing and promoting international standards for meteorological observations, data exchange, and processing (e.g., WMO Information System - WIS).

14.4.4 Challenges in Achieving Full Interoperability

Despite the progress, challenges remain:

- **Legacy Systems:** Many older systems and archives still use proprietary or non-standard formats.
- **Rapid Evolution of Data Types:** New observational technologies (e.g., IoT, drones) generate novel data types that require new or adapted standards.
- **Semantic Interoperability:** Beyond just format, ensuring that different systems interpret the *meaning* of data (e.g., what "cloud cover" precisely entails) consistently.
- **Computational Overhead:** Converting between formats can sometimes add computational overhead, especially for very large datasets.
- **Human Factor:** Resistance to change or lack of training can hinder adoption of new standards.

Interoperability and standardization are not merely technical jargon; they are the invisible infrastructure that enables the seamless flow of meteorological information across a complex global ecosystem. By embracing common data standards, the meteorological community ensures data consistency, facilitates efficient exchange, enhances discoverability, and accelerates scientific progress. This commitment to a shared language for data is fundamental for maximizing the utility of our ever-growing environmental datasets, fostering global collaboration, and ultimately strengthening our collective ability to understand, predict, and respond to the challenges of our dynamic systems.

14.5 Future Directions in Meteorological Data Science

The field of meteorological data science is on the cusp of a transformative era, driven by exponential growth in data, unprecedented computational power, and groundbreaking advancements in artificial intelligence. The future will see a deepening of interdisciplinary collaboration, a relentless pursuit of higher resolution and longer lead times, and a fundamental shift towards more open and integrated scientific practices.

14.5.1 Data Fusion: Towards a Unified View of the Earth System

What is Data Fusion? Data fusion is the process of combining data from multiple, disparate sources to produce a more consistent, accurate, and comprehensive understanding of a phenomenon than could be achieved by using any single data source alone. In meteorology, this means intelligently merging observations from satellites, ground stations, radar, weather balloons, aircraft, IoT sensors, and even citizen science (Fig. 14.2).

- **Why it's a Future Direction:**
  - **Overcoming Limitations:** Each observational platform has strengths and weaknesses (e.g., satellites offer broad coverage but indirect measurements; ground stations are precise but sparse). Data fusion aims to leverage the strengths of all.
  - **Filling Gaps:** Combining data helps fill spatial and temporal gaps in observations, particularly in data-sparse regions (e.g., oceans, remote areas).
  - **Enhanced Accuracy and Consistency:** By integrating diverse data, data fusion techniques can reduce uncertainties, correct biases, and create a more physically consistent representation of the atmosphere.
  - **Improved Initial Conditions:** Better initial conditions for NWP models lead to more accurate forecasts.
  - **New Insights:** Fusing different types of data can reveal relationships and phenomena that are not apparent from individual datasets.
- **Future Trends in Data Fusion:**
  - **Advanced Data Assimilation:** Next-generation data assimilation systems will be even more sophisticated in handling heterogeneous data types, non-linear relationships, and quantifying observation uncertainties.
  - **Machine Learning for Fusion:** AI/ML algorithms are increasingly being used to learn optimal ways to combine diverse datasets, identify subtle signals, and even infer missing information.

- **Multi-Sensor Integration:** Seamless integration of data from new and existing sensors (e.g., combining drone profiles with satellite imagery and ground-based radar).

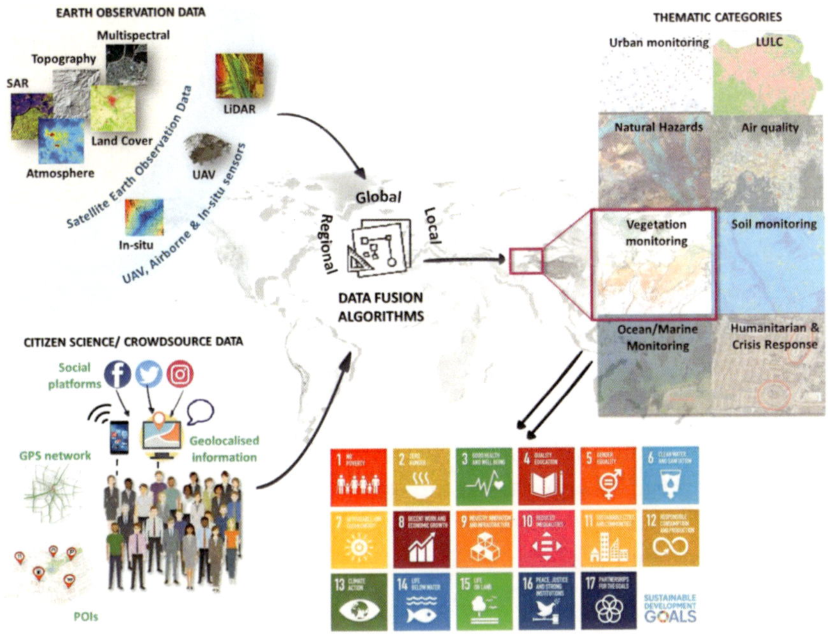

Fig. 14.2: Data fusion in earth observation system and citizen science

14.5.2 Advanced AI and Machine Learning: Beyond Prediction

What is Advanced AI/ML in Meteorology? Beyond simply improving forecast accuracy, advanced AI/ML is poised to revolutionize meteorological data science in more fundamental ways, moving beyond traditional statistical methods to tackle complex, non-linear problems.

- **Future Trends in Advanced AI/ML:**
  - **AI-Native Models:** Developing entirely new atmospheric models based on AI/ML principles, potentially learning the equations of motion directly from data, or emulating computationally expensive physical processes. This could lead to significantly faster and more efficient models.
  - **Hybrid Modeling:** The most likely near-term future involves "hybrid" models, where AI/ML components are seamlessly integrated into traditional physics-based NWP models to improve specific parameterizations (e.g., clouds, convection, radiation) or to correct biases.

- **Extreme Event Prediction and Attribution:** AI/ML will excel at identifying subtle precursors to extreme events (e.g., heatwaves, flash floods, severe thunderstorms) from vast datasets, and in attributing specific events to climate change.
- **Uncertainty Quantification:** Advanced ML techniques will provide more robust and nuanced quantification of forecast uncertainty, crucial for probabilistic forecasting.
- **Personalized and Actionable Forecasts:** AI will enable the generation of highly localized, personalized forecasts tailored to specific user needs (e.g., "What's the wind forecast for my specific rooftop solar panel?", "What's the pollen forecast for my exact walking route?").
- **Automated Discovery:** AI could help discover new climate patterns, teleconnections, or physical relationships within complex Earth system data that are difficult for humans to identify.
- **Explainable AI (XAI) Maturity:** As AI models become more integrated, XAI will mature to provide clearer, more physically consistent explanations for AI's decisions, fostering trust and enabling scientific insight.

14.5.3 Coupled Earth System Models: A Holistic View

- **What are Coupled Earth System Models (ESMs)?** As discussed in Chapter 12, ESMs are the most comprehensive climate models, integrating not just the atmosphere, ocean, land, and sea ice, but also biogeochemical cycles (e.g., carbon, nitrogen), aerosols, and atmospheric chemistry.
- **Future Trends in Coupled ESMs:**
    - **Higher Resolution:** ESMs will continue to increase their spatial resolution, allowing them to explicitly resolve more processes and provide more detailed regional climate projections, blurring the lines with NWP models.
    - **More Comprehensive Components:** Further integration of previously separate components, such as dynamic ice sheets, interactive vegetation, and human systems (e.g., energy, agriculture), to provide a more complete picture of Earth system feedbacks.
    - **Seamless Prediction:** Developing ESMs that can operate across a wide range of timescales, from weather forecasts to decadal predictions and long-term climate projections, using a single, unified framework.

- **Improved Feedbacks:** Enhanced representation of critical feedbacks within the Earth system (e.g., cloud feedbacks, carbon cycle feedbacks, permafrost thaw) to reduce uncertainty in climate projections.
- **Coupled Human-Earth System Models:** Integrating socio-economic models (e.g., population growth, energy demand, land use decisions) directly into ESMs to explore the two-way interactions between human society and the Earth system.

14.5.4 Open Science Initiatives: Collaboration and Accessibility

- **What are Open Science Initiatives?** Open science promotes transparency, collaboration, and accessibility throughout the scientific research lifecycle. In meteorological data science, this translates to open access to data, open-source software and models, and open publication of research.
- **Future Trends in Open Science:**
  - **Increased Data Accessibility:** More meteorological and climate data will be made freely and openly available, often in cloud-optimized formats, facilitating access for researchers, developers, and the public worldwide.
  - **Example:** Cloud-based data archives (e.g., Google Earth Engine, AWS Open Data) for large climate datasets.
  - **Open-Source Software and Models:** A growing trend towards developing and sharing open-source NWP models (e.g., WRF, ICON), climate models, and data analysis tools. This fosters community development, transparency, and reproducibility.
  - **Reproducible Research:** Emphasis on making research fully reproducible by sharing code, data, and detailed methodologies, often through platforms like Jupyter notebooks or online repositories.
  - **Citizen Science Integration:** Further development of platforms and methodologies to effectively integrate and utilize citizen science data into operational and research workflows.
  - **International Collaboration:** Strengthening international efforts to standardize data, share computing resources, and collaborate on major modeling and observational projects.
  - **Cloud Computing and High-Performance Computing (HPC) as a Service:** Increasing use of cloud platforms for running models and analyzing large datasets, democratizing access to powerful computing resources.

The future of meteorological data science is characterized by a relentless pursuit of deeper understanding, higher precision, and broader impact. Driven by the convergence of massive data streams, advanced AI, and increasingly comprehensive Earth System Models, the field is poised to deliver unprecedented insights into our dynamic planet. Crucially, this scientific advancement will be underpinned by a growing commitment to open science, fostering global collaboration and ensuring that the immense power of meteorological data is harnessed ethically and effectively for the benefit of all humanity in the face of complex environmental challenges.

14.6 Conclusion: The Era of Weather Intelligence Driven by Data Science and Analytics

As we bring this comprehensive exploration of meteorological data science to a close, it is abundantly clear that we stand at the precipice of a new era: the era of **Weather Intelligence**. This is a paradigm where raw atmospheric data is no longer merely observed or forecasted, but actively transformed into **actionable insights and strategic foresight**. At the very heart of this transformation lies the indispensable power of **data science and analytics**.

Throughout these chapters, we have journeyed from the foundational principles of meticulous data collection and rigorous quality control, through the sophisticated analytical techniques of machine learning and statistical modeling, to the intricate world of numerical weather prediction and long-term climate analysis. We've seen how data science breathes life into abstract numbers through compelling visualizations, translating complex atmospheric dynamics into understandable patterns. Beyond traditional forecasting, the specialized applications we've explored—from precise air quality analysis and vital hydrometeorological insights for water management, to optimizing renewable energy integration and shaping resilient urban climatology—all underscore how data science and analytics are the engines generating this crucial weather intelligence.

This weather intelligence is not confined to scientific institutions; it permeates every facet of modern life. The ability to predict a storm's path with greater precision, assess flood risk at a hyper-local scale, project future temperatures for agricultural planning, or understand the dispersion of pollutants in real-time, directly translates into saved lives, protected livelihoods, and more resilient communities. It empowers daily decisions, guides long-term strategic planning, and fundamentally shapes our collective response to the profound challenges of a changing climate.

Yet, the pursuit of superior weather intelligence is an ongoing endeavor, fraught with scientific and technical challenges—from the inherent chaotic nature of the atmosphere and persistent model uncertainties to the ethical complexities surrounding data privacy, algorithmic bias, and equitable access to information. These challenges, however, are not deterrents but rather catalysts for further innovation, pushing the boundaries of what data science and analytics can achieve.

Looking ahead, the trajectory of meteorological data science is one of relentless advancement and deepening integration. The convergence of hyper-resolution modeling, cutting-edge AI, and comprehensive Earth System Models, coupled with the proliferation of diverse observational platforms and a growing commitment to open science, promises a future where weather intelligence is even more precise, comprehensive, and widely accessible. This journey is fundamentally about leveraging the power of data to foster a more nuanced understanding of our planet, empowering truly informed decision-making, and ultimately, ensuring a sustainable and resilient future for all. The enduring strength and societal impact of meteorological data science lie in its unwavering commitment to transforming raw data into the invaluable intelligence that guides us forward.

Bibliography

Books

- Albert, L. 1976. Weather Wisdom: Facts and Folklore of Weather Forecasting. Doubleday Publisher. ISBN: 9780385110167. Pp 180.
- Duchon C. and Robert H. 2012. Time Series Analysis in Meteorology and Climatology: An Introduction. John Wiley & Sons Publisher. ISBN: 978-1119960997. Pp 262.
- Gibbons G. 2010. Weather Forecasting. Good Wings, USA. ISBN: 978-0785705475. Pp 32.
- James R. H. and Hakim G.J. 2012. An Introduction to Dynamic Meteorology. Academic Press Publisher. ISBN: 978-0123848666. Pp 538.
- Kingsbury, S.A.; Kingsbury, M.E. and Mieder, W. 1996. Weather Wisdom - Proverbs, Superstitions, and Signs. Peter Lang, USA. ISBN: 9780820433264. Pp 478.
- Mavi, HS. 1986. Introduction to Agrometeorology. Oxford & IBH Publishing Co Pvt. Ltd. ISBN: 978-8120409101. Pp 296.
- Ramana Rao, B.V. and V. Uma Maheswara Rao. 2024. Agricultural Meteorology. Jain Brothers. ISBN: 978-9390576975. Pp 154.
- Ramana Rao B.V., Surender Singh, V. Uma Maheswara Rao, Anu Bulusu. 2025. Farming Weather: Harnessing Citizen Science for Smart Agriculture. Publisher: Professional Prints. New York USA. ISBN: 978-1966695561. Pp 154.
- Ramana Rao B.V., Surender Singh, V. Uma Maheswara Rao. 2025. Conceptual Meteorology. Publisher: New India Publishing Agency, New Delhi. ISBN 978-9358874501. Pp 152.
- Ramana Rao B.V., Surender Singh, V. Uma Maheswara Rao. 2025. Forecasting the Future: Effective Climate Guidance on Monsoons for Farm Prosperity. Publisher: Professional Prints. New York USA. ISBN: 978-1966695912. Pp 246.
- Ramana Rao B.V., Surender Singh, V. Uma Maheswara Rao. 2025. Preparatory Microclimatology. Publisher: Professional Prints. New York USA. ISBN: 9781966695189. Pp 167.
- Ramana Rao B.V., Surender Singh, V. Uma Maheswara Rao. 2025. Primer on Agrohydrometeorology. Publisher: Professional Prints. New York USA. ISBN: 978-1966695301. Pp 144.
- Ramana Rao, B.V., Surender Singh, V. Uma Maheswara Rao. 2025. System Simulation and Agromet Advisory Services. Publisher: New India Publishing Agency, New Delhi. Pages 487.
- Seemann, J.; Chirkov, Y.I.; Lomas, J. and Primault, B. 1979. Agrometeorology. Springer Berlin, Heidelberg. ISBN: 978-3642672903. Pp 326.

- Sievert T. 2005. Weather Forecasting. Bridgestone Books. ISBN: 978-0736837392. Pp 24.
- Surender S., V. Uma Maheswara Rao, B.V. Ramana Rao. 2025. Conceptual Meteorology. Publisher: New India Publishing Agency, New Delhi. ISBN: 978-9358871487. Pp 392.
- Surender S., V. Uma Maheswara Rao, B.V. Ramana Rao. 2025. Weather Forecasting. Publisher: New India Publishing Agency, New Delhi. ISBN: 978-9358875492. Pp 392.
- Surender S. 2025. MCQs on Agricultural Meteorology. Publisher: New India Publishing Agency, New Delhi. ISBN: 978-9372196450. Pp 338.

Journals

- **Agricultural and Forest Meteorology**

  https://www.sciencedirect.com/journal/agricultural-and-forest-meteorology
- **Chinese Journal of Agrometeorology**

  https://zgnyqx.ieda.org.cn/EN/1000-6362/home.shtml
- **Italian Journal of Agrometeorology**

  https://riviste.fupress.net/index.php/IJAm/issue/view/15
- **Journal of Agril Meteorology**

  https://journals.indexcopernicus.com/search/details?id=46231
- **Journal of Agrometeorology**

  https://journal.agrimetassociation.org/index.php/jam/issue/view/73
- **Journal of Climatology & Weather Forecasting**

  https://www.iomcworld.com/climatology-weather-forecasting.html
- **MAUSAM**

  https://mausamjournal.imd.gov.in/index.php/MAUSAM
- **Weather and Forecasting**

  https://journals.ametsoc.org/view/journals/wefo/wefo-overview.xml
- **Journal of Applied Meteorology and Climatology** https://journals.ametsoc.org/view/journals/apme/apme-overview.xml
- **International Journal of Climatology**

  https://rmets.onlinelibrary.wiley.com/journal/10970088

Web Resources

- https://climateanalytics.org/
- https://www.ncei.noaa.gov/cdo-web/
- https://community.wmo.int/en/climate-data-and-monitoring
- https://www.india.gov.in/website-climate-data-service-portal-cdsp
- https://mausam.imd.gov.in/
- https://imdagrimet.gov.in/
- https://www.tropmet.res.in/
- https://nwp.ncmrwf.gov.in/

- https://www.skymetweather.com/
- https://niasm.icar.gov.in/
- https://www.sac.gov.in/Vyom/
- https://www.isro.gov.in/
- https://www.icrisat.org/
- https://www.ecmwf.int/
- https://wmo.int/
- https://www.fao.org/climate-smart-agriculture/en/
- https://www.agrimetassociation.org/
- https://safoam.org.in/
- https://www.insam.org.in/
- https://medium.com/weather-wisdom
- https://www.theatlantic.com/magazine/archive/1895/10/weather-and-weather-wisdom/635504/
- https://open.spotify.com/show/7jVDLu4zYivr2xhVKeNgkm
- https://www.intechopen.com/chapters/88975

Subject Index